Y0-DXK-435

SOIL AMENDMENTS

Impacts on Biotic Systems

SOIL AMENDMENTS

Impacts on Biotic Systems

Edited by
Jack E. Rechcigl

WITHDRAWN

LEWIS PUBLISHERS
Boca Raton Ann Arbor London Tokyo

Library of Congress Cataloging-in-Publication Data

Soil Amendments: Impacts on Biotic Systems / editor, Jack E. Rechcigl.
p. cm. -- (Agriculture and environment series)
Includes bibliographical references and index.
ISBN 0-87371-860-7
1. Soil amendments--Environmental aspects. 2. Agricultural ecology. I. Rechcigl, Jack E. II. Series.
IN PROCESS
574.5′222--dc20

94-17628
CIP

This book contains information obtained from authentic and highly regarded sources. Reprinted material is quoted with permission, and sources are indicated. A wide variety of references are listed. Reasonable efforts have been made to publish reliable data and information, but the author and the publisher cannot assume responsibility for the validity of all materials or for the consequences of their use.

Neither this book nor any part may be reproduced or transmitted in any form or by any means, electronic or mechanical, including photocopying, microfilming, and recording, or by any information storage or retrieval system, without prior permission in writing from the publisher.

All rights reserved. Authorization to photocopy items for internal or personal use, or the personal or internal use of specific clients, may be granted by CRC Press, Inc., provided that $.50 per page photocopied is paid directly to Copyright Clearance Center, 27 Congress Street, Salem, MA 01970 USA. The fee code for users of the Transactional Reporting Service is ISBN 0-87371-360-7/95/$0.00+$.50. The fee is subject to change without notice. For organizations that have been granted a photocopy license by the CCC, a separate system of payment has been arranged.

CRC Press, Inc.'s consent does not extend to copying for general distribution, for promotion, for creating new works, or for resale. Specific permission must be obtained in writing from CRC Press for such copying.

Direct all inquiries to CRC Press, Inc., 2000 Corporate Blvd., N.W., Boca Raton, Florida 33431.

© 1995 by CRC Press, Inc.
Lewis Publishers is an imprint of CRC Press

No claim to original U.S. Government works
International Standard Book Number 0-87371-860-7
Library of Congress Card Number 94-17628
Printed in the United States of America 1 2 3 4 5 6 7 8 9 0
Printed on acid-free paper

353123

H
45
63
65
?95

AGRICULTURE AND ENVIRONMENT SERIES

Agriculture is an essential part of our economy on which we all depend for food, feed, and fiber. With the increased agricultural productivity in this country as well as abroad, the general public has taken agriculture for granted while voicing their concern and dismay over possible adverse effects of agriculture on the environment. The public debate that has ensued on the subject has been brought about, in part, by the indiscriminate use of agricultural chemicals and, in part, by disinformation, based largely on anecdotal evidence.

At the national level, recommendations have been made for increased research in this area by such bodies as the Office of Technology Assessment, the National Academy of Sciences, and the Carnegie Commission on Science, Technology, and Government. Specific issues identified for attention include: contamination of surface and ground water by natural and chemical fertilizers, pesticides, and sediment, the continued abuse of fragile and nutrient-poor soils and suitable disposal of industrial and agricultural waste.

Although a number of publications have appeared recently on specific environmental effects of some agricultural practice, no attempt has been made to approach the subject systematically and in a comprehensive manner. The aim of this series is to fill the gap by providing the synthesis and critical analysis of the state of the art in different areas of agriculture bearing on environment and vice versa. Efforts will also be made to review research in progress and comment on perspectives for the future. From time to time methodological treatises as well as compendia of important data in handbook form will also be included. The emphasis throughout the series will be on comprehensiveness, comparative aspects, alternative approaches, innovation, and worldwide orientation.

Specific topics will be selected by the Editor-in-Chief with the council of an international advisory board. Imaginative and timely suggestions for the inclusion in the series from individual scientists will be given serious consideration.

Jack E. Rechcigl
Editor-in-Chief

Jack E. Rechcigl is an Associate Professor of Soil and Environmental Sciences at the University of Florida and is located at the Research and Education Center in Ona, Florida. He received his B.S. degree (1982) in Plant Science from the University of Delaware and his M.S. (1983) and Ph.D. (1986) degrees in Soil Science from Virginia Polytechnic Institute and State University. He joined the faculty of the University of Florida in 1986 as Assistant Professor and in 1991 was promoted to Associate Professor.

Dr. Rechcigl has authored over 100 publications, including contributions to books, monographs, and articles in periodicals in the fields of soil fertility, environmental quality, and water pollution. His research has been supported by research grants totaling over $1.5 million from both private sources and government agencies. Dr. Rechcigl has been a frequent speaker at national and international workshops and conferences and has consulted in various countries including Brazil, Nicaragua, Venezuela, Australia, Taiwan, Philippines, and Czechoslovakia.

He is currently an Associate Editor for the *Journal of Environmental Quality* and is Editor-in-Chief of the "Agriculture and Environment Book Series". He is also serving as an invitational reviewer for manuscripts and grant proposals for scientific journals and granting agencies.

Dr. Rechcigl is a member of the Soil Science Society of America, American Society of Agronomy, International Soil Science Society, Czechoslovak Society of Arts and Sciences, various trade organizations, and the Honorary Societies of Sigma Xi, Gamma Sigma Delta, Phi Sigma, and Gamma Beta Phi.

Contributors

Amadu Ayebo
Center for International and Rural Environmental Health
University of Iowa
Iowa City, Iowa

Henry Campa, III
Department of Fisheries and Wildlife
Michigan State University
East Lansing, Michigan

Dale W. Cole
College of Forest Resources
University of Washington
Seattle, Washington

David J. Glass
Office of Technology Affairs
Massachusetts General Hospital
Charlestown, Massachusetts

Larry G. Hansen
Department of Veterinary Biosciences
University of Illinois at Urbana-Champaign
Urbana, Illinois

Jonathan B. Haufler
Department of Fisheries and Wildlife
Michigan State University
East Lansing, Michigan

Robert B. Harrison
College of Forest Resources
University of Washington
Seattle, Washington

K. E. Havens
Department of Research
South Florida Water Management District
West Palm Beach, Florida

Charles L. Henry
College of Forest Resources
University of Washington
Seattle, Washington

J. Kent Johnson
Center for International and Rural Environmental Health
University of Iowa
Iowa City, Iowa

Burton C. Kross
Center for International and Rural Environmental Health
University of Iowa
Iowa City, Iowa

C. C. Mitchell
Department of Agronomy and Soils
Auburn University
Auburn, Alabama

G. L. Mullins
Department of Agronomy and Soils
Auburn Univesity
Auburn, Alabama

Michael L. Olson
Center for International and Rural Environmental Health
University of Iowa
Iowa City, Iowa

W. D. Pitman
Rosepine Research Station
Louisiana State University Agricultural Center
Rosepine, Louisiana

David J. Schaeffer
Department of Veterinary Biosciences
University of Illinois at Urbana-Champaign
Urbana, Illinois

A. D. Steinman
Department of Research
South Florida Water Management District
West Palm Beach, Florida

Table of Contents

Preface

The quality of our environment has become the predominant issue of today. The continuous loss of the precious natural resources workwide coupled with the uncontrolled pollution of land, water, and air are a frequent subject of public discussions and concern.

The newspaper articles which frequently target agriculture for most of the environmental ills are usually based on anecdotal evidence rather than on solid scientific data.

The purpose of this book is to present a comprehensive and balanced synthesis of our knowledge pertaining to the environmental effects of soil amendments on various biotic systems, including crops, livestock, wildlife, forestry, aquatic systems, and human beings. Separate chapters focus on the remedial effects of alternative farming systems and biotechnology with reference to specific biotic systems.

This publication should be of interest and practical use to students and professionals in agriculture and environmental sciences, as well as to the policy makers and general public concerned with environmental issues.

The contributors have undertaken their assignments with pioneering spirit and they thus deserve our sincere appreciation and gratitude. The Lewis Publishers and CRC Press, Inc. and their editorial staff, particularly Skip DeWall and Paul Gottehrer were always very helpful and cooperative from the beginning to the end.

Jack Rechcigl

CHAPTER 1

Crops

G. L. Mullins and C. C. Mitchell

TABLE OF CONTENTS

0-87371-860-7/95/$0.00+$.50
© 1994 by CRC Press, Inc.

I. INTRODUCTION

Soils represent our most important natural filtration system. They can effectively decompose organic compounds, recycle nutrients, and protect the environment by removing substances from percolating water.

Prior to the industrial age, animal manures including human waste were the primary amendments to agricultural soils for the purpose of improving crop yields. With increasing industrial activity and increasing human populations the use of sewage sludge and industrial wastes has become more popular. With increasing costs associated with waste disposal, land application of wastes will continue to be a desirable outlet. Land application of wastes is economically desirable and it prevents the wasteful loss of essential plant nutrients and soil conditioning agents.

Waste materials are desirable due to their plant nutrient content and their ability to serve as soil conditioners which also improve crop performance. However, some wastes have undesirable characteristics which could lead to detrimental effects on crops. In this chapter we will address both the beneficial and adverse effects on crops that are associated with the land application of various organic waste materials.

II. BENEFICIAL EFFECTS

Except for land farming systems and dedicated disposal sites, the only reason to apply an amendment to the soil is the beneficial effects it may have on the crops/plants grown on the land. Records of the application of soil amendments to improve crop performance are as old as civilization. In the *Odyssey*, Homer mentions the manuring of vineyards during the 9th century B.C. Truck gardens and olive groves around ancient Athens were fertilized by city sewage. In fact, Theophrastus (372 to 287 B.C.) listed human manure above that of swine, goats, sheep, cows, oxen, and horses for its value as a soil amendment. The Bible mentions the value of wood ashes, and the early Romans advocated its use on grain.[1]

The reasons for beneficial effects of organic soil amendments on crops are often complex. Most of the benefits are related to chemical, biological, and physical changes in the soil. These result in improved root growth, better soil water relations, deeper rooting depth, increased N fixation (legumes), reductions of toxic minerals (e.g., Al, Zn, Mn, etc.), higher nutrient availability, decreases in soil-borne pathogens (e.g., parasitic nematodes), and many others. By altering the chemical, biological, and physical characteristics of the soil, the intention is to improve the growth, yield, or quality of the crops being produced. Amendments may also have long-term effects on the soil-crop systems which will improve the long-term sustainability of cropping systems.

A. CHEMICAL EFFECTS

1. Soil pH and Liming Materials

The pH of soil can have a dramatic effect on crop performance by influencing the solubility and availability of nutrient and non-nutrient elements present. Liming of acid soils is intended to primarily reduce the toxicity of aluminum and/or manganese by increasing soil pH, an easily measured, chemical property. Crushed calcitic or dolomitic limestone, which is rich in Ca and Mg, is the liming material of choice but many other soil amendments have been shown to work equally well when used at a comparable neutralizing value (usually reported as the $CaCO_3$ equivalent). As with most by-products used as soil amendments, the neutralizing value can be quite variable. Most regulations control the manufacture and sale of crushed limestone as a soil liming agent. However, the factors appropriate for determining the effective soil neutralizing value of crushed limestone (% $CaCO_3$ and fineness) may be inappropriate for some by-products used as alternative soil liming materials. Fly ash and bottom ash from the burning of wood and precipitated lime by-products are normally high in metal oxides and hydroxides. Therefore, using a neutralizing value based primarily on the total alkalinity or $CaCO_3$ equivalency is usually all that is needed.

Some materials which have shown promise or are being used primarily as soil liming materials are:

- Calcium silicate slags (basic slag)[2]
- Fly ash (from fossil coal combustion)[3]
- Boiler wood ash and/or mixed ash[4–9]
- Cement flue dust[10]
- Lime-stabilized sludges (Table 1)
- Sludges treated with ash[11]
- By-product lime (paper mill lime, calcium carbide lime)
- Paper mill sludge[12]

2. Plant Nutrients

Most organic by-products have been evaluated as a source of plant nutrients. When amendments are used primarily as a soil liming material or source of soil organic matter, plant nutrients are applied. Again, the chemical effect of these added nutrients on crops will depend on the soil physical and chemical properties and to a lesser extent, the physiology of the crop being grown. Results of research on plant nutrient availability from organic soil amendments are as varied as the by-products evaluated and the crops and soils available for bioassays (Tables 1 to 3).

Most research reported in Tables 1 to 3 is positive regarding the increase of certain plant nutrients in plant tissue. Yield increases may be attributed solely to the increased availability of N, P, K or (rarely) secondary and micronutrients. In some cases, yield increases or increases in plant growth parameters were measured with no

Table 1 A Summary of Research Relating to Agronomic Crop Response to Soil Amendments

Amendment	Location	Rate applied/crop	Results
		Corn (*Zea mays* L.)	
Sewage sludge	TN	11.2 dry Mg ha^{-1}	Increased grain yield (N avail. and soil water); N mineralization 50% 1st year and 30% 2nd year.[13]
Aer. dig. ss	NJ	22 and 45 dry Mg ha^{-1} $year^{-1}$ over 4 years	All ss rates greatly increased yields; heavy metal increases modest and within normal range; Zn increases consistently above control.[14]
Aer. dig. ss	VA	0–84 dry Mg ha^{-1}	1.8, 304, 17.2, and 248 kg ha^{-1} Cd, Cu, Ni, and Zn applied; Cd<0.01 mg kg^{-1} in grain; Cd and Cu unaffected in grain; Ni and Zn increased; no Cu phytotoxicity.[15]
Sewage sludge	AL, KY, MS, TN, NC	Variable depending upon site and other crops	ss effective source of plant nutrients where soil nutrients were inadequate; no harmful effects of metals.[16]
Composted ss	MD	0–1500 kg P ha^{-1} as ss and TSP fertilizer in field and greenhouse	No difference in P sources at rates<100 kg P ha^{-1}; TSP 4–7X more effective than compost in increasing earleaf P.[17]
Paper sludge (C/N = 150)	—	448–1344 Mg ha^{-1} + N for C/N of 10–100	Yields reduced 1st year; increased 2nd.[18]
Composted municipal	TN	90–2240 Mg ha^{-1} over 5 years	55 and 153% total yield increase over 14 years with 90 and 2240 Mg ha^{-1} rates, respectively.[19]
Nutrasweet sludge	GA	0–150 Mg N kg^{-1} as a sludge, am. sulfate and urea	Sludge dry matter yields equal to or higher than a.s. or urea; sludge should be managed as ammoniacal N fertilizer.[20]
Poultry manure	DE	0–270 kg PAN ha^{-1} (predicted available N)	Earleaf N lower for manure than am. nitrate; no differences in yield at 1 site.[21]
Dairy manure slurry	IN	112–336 Mg ha^{-1} (304–914 kg N ha^{-1})	Max. grain and silage with 224 Mg ha^{-1}; plant composition not affected except for N 2nd year.[22]
Dairy manure slurry	NC	Based on N	Manure yield comparable to fertilizer N; spring application best; nitrification inhibitor had no effect; 224 kg N ha^{-1} max. rate.[23]
Cu-enriched swine manure and $CuSO_4$	VA (3 soils)	0–67.2 Mg ha^{-1} $year^{-1}$ for 3 years	Application of manure and or $CuSO_4$ did not decrease corn grain yields on either soil. Minimal effects of treatments on Cu in corn ear leaves and grain.[24]

Cu-rich swine manure and $CuSO_4$	VA (3 soils)	1100 Mg ha^{-1} over 11 years (235 kg Cu ha^{-1})	Normal Cu in leaves and grain; no yield decrease due to Cu.[25]
Sewage sludge	MN	0-45 Mg ha^{-1}	Cd and Zn uptake was increased by sludge application. For corn silage, *in vitro* digestible dry matter, cell wall constituents, acid detergent fiber and lignin, and silica accumulation were not affected.[26]
Phosphogypsum	AL	0–112 Mg ha^{-1} (925 Bq kg^{-1}, ^{226}Ra)	Cadmium and ^{226}Ra uptake by corn, wheat and soybean was not affected by phosphogypsum when applied at a rate that was 200 times above the normal rate for fertilizer Ca and S.[27]
		Small Grains	
An. Dig. Liq. ss	AZ	112–224 kg N ha^{-1} as ss on wheat (*Triticum aestivum* L.)	No effect on grain yield; increased days to heading; no effect on quality of grain or metal concentration.[28]
An. Dig. Liq. ss	AZ (greenhouse)	0–5x recommended plant available N rate on barley (*Hordeum vulgare* L.)	Increased vegetative growth; no effect on grain yield; decreased stand when >3x N rate applied (salts); delayed maturity at high rates; no effect on plant Cd, Cu, Pb, Ni or Zn.[29]
An. Dig. Liq. ss	AZ	78–224 kg N ha^{-1} as fertilizer and ss on wheat (*Triticum aestivum* L.)	Delayed maturity; no effect on grain yield or feeding quality; high rate produced most hay; metals low in hay and grain.[30]
ss high in Zn, Cu, and Ni vs. uncontaminated ss	England (greenhouse)	5–10 g kg^{-1} on barley (*Hordeum vulgare* L.)	Yields higher from all sources (N); Ni and Zn in tissue declined with time; plants control Cu in tissues more than Zn and Ni.[31]
Composted ss	CA	0–180 Mg ha^{-1} $year^{-1}$ for 8–10 year on barley (*Hordeum vulgare* L.)	No measurable uptake of Se in leaf and grain.[32]
Composted ss vs. An. dig. ss	CO (greenhouse)	0–140 Mg ha^{-1} on oats (*Avena sativa* L.)	Higher yields with composted ss; salinity in ss decreased grain yields; higher Cd and Zn in crop with ss due to lower soil pH; Cu in tissue not affected.[33]
Co-composted ss (4 types)	DE	0–44 Mg ha^{-1} on wheat (*Triticum aestivum* L.)	Net N immobilization in soil and N deficiency in wheat; reduced dry matter yields; metals not factor.[34]

Table 1 (Continued) A Summary of Research Relating to Agronomic Crop Response to Soil Amendments

Amendment	Location	Rate applied/crop	Results
Aer. dig. ss	VA	0–84 dry Mg ha^{-1} on barley (*Hordeum vulgare* L.)	1.8, 304, 17.2, and 248 kg ha^{-1} Cd, Cu, Ni, and Zn applied; Cd<0.01 mg kg^{-1} in grain; Cd and Cu unaffected in grain; Ni and Zn increased; no Cu phytotoxicity.[15]
Primary and secondary paper mill sludges	WI	2.5–10% by weight on oats (*Avena sativa* L.)	Increase in yield at low rate of secondary sludge but decrease at all rates for high C/N sludge (150).[35]
Nutrasweet sludge	GA	0–150 Mg N kg^{-1} as sludge, am. sulfate and urea on wheat (*Triticum aestivum* L.)	Sludge dry matter yields equal to or higher than a.s. or urea; sludge should be managed as ammoniacal N fertilizer.[20]
Hog and cow manure	WV (greenhouse)	44.8 wet Mg ha^{-1} on barley (*Hordeum vulgare* L.)	Rooting depth in acid subsoil increased due to soluble organics complexing toxic Al.[36]
Gamma-irradiated ss	India	10% of soil on rice (*Oryza sativa* L.)	ss had positive effect on grain yield, dry weight, protein, soluble sugars, starch, and chlorophyll.[37]
Paper and power ash sludges	MS	0–224 Mg ha^{-1} on wheat (*Triticum aestivum* L.)	Negative effects on yield of rates >67 Mg ha^{-1}; negative effects decreased with time; metals in normal range in leaves.[11]
Sewage sludge	IN	0–446 Mg ha^{-1} on wheat (*Triticum aestivum* L.)	Sludge rates in excess of federal regulations increased the protein content of wheat grain and flour. Test weight, thousand-kernel weight and particle size index was not affected. Sludge applications decreased baking quality due to the increased protein content. Sludge also increased the concentration of Cd, Zn, Mn, and Ni in the grain.[38]
Pig slurry and poultry manure compost	Spain	20 Mg ha^{-1} on wheat (*Triticum aestivum* L.)	Compost increased yields and leaf P, K, and Na.[39]
Cattle dung and wool waste compost	India	10 Mg ha^{-1} on wheat (*Triticum aestivum* L.)	Growth increased with reduction in C:N ratio.[40]
Phosphogypsum	AL	0–90 kg S ha^{-1} as phosphogypsum on wheat (*Triticum aestivum* L.)	Phosphogypsum increased wheat forage yields. Mined and phosphogypsum produced the same yields. Content of ^{226}Ra and ^{210}Po in forage tissue was not affected by phosphogypsum.[188–189]

		Forages	
An. dig. liq. ss	AZ	78–224 kg N ha^{-1} on wheat (*Triticum aestivum* L.) hay	ss increased yields over N fertilizer.[28,30]
An. dig. liq. ss	OR	110–880 kg N ha^{-1} as ss vs. NH_4NO_3 on tall fescue (*Festuca arundinacea* Schreb.)	ss 27–44% as effective as fertilizer for N source; metals increased in forage but below toxic levels.[41]
ss	Nova Scotia	45 wet Mg ha^{-1} on mixed grasses and legumes	ss increased Cu in forage but concentrations in normal range.[42]
ss on mine spoil	PA	—	ss resulted in increase in plant height and dry matter production.[43]
Composted ss	MD	0–135 Mg ha^{-1} on tall fescue (*Festuca arundinacea* Schreb.)	Fescue used 8% of compost N and 80% of fertilizer N during 167 d; authors provide regression of N availability.[44]
Aer. dig. ss and pig manure	Quebec	75–150 wet Mg manure ha^{-1}; 150–300 wet Mg ss ha^{-1} on timothy (*Phleum pratense* L.)	Similar yields with fertilizer and higher rates; no metal contamination.[45]
Municipal waste water labeled ^{15}N	Alberta	11-year study with alfalfa and reed canarygrass (*Phalaris arundinacea* L.)	Excess waste water did not increase alfalfa yields; no phytotoxicity; improved forage quality.[46]
An. dig. dairy slurry	IN	112–336 Mg ha^{-1} (304–914 kg N ha^{-1}) on alfalfa *(Medicago)* — orchardgrass (*Dactylis glomerata* L.)	Cropping reduced soil components except P and K to initial levels after 1 year; 112 Mg ha^{-1} rate supported maximum yields.[22]
Poultry litter	AL	5.6–22.4 Mg ha^{-1} (150–600 kg N ha^{-1}) on bermudagrass (*Cynodon dactylon* (L.) Pers.)	High rate produced same yields and quality as 336 kg fertilizer N ha^{-1}.[47]
Pine bark and straw	NM	18 Mg bark ha^{-1}; 5 mg barley straw ha^{-1} on wheatgrass establishment	Amendments produced outstanding vegetative cover of mine spoil on nonsodic soil.[48]
Phosphogypsum	FL	0–4.0 Mg ha^{-1} to bahiagrass (*Paspalum notatum* Flugge) pasture	Phosphogypsum increased forage yields by as much as 28%. The content of ^{226}Ra, ^{210}Po, and ^{210}Pb in the forage were not increased beyond the control of phosphogypsum.[49]

Table 1 (Continued) A Summary of Research Relating to Agronomic Crop Response to Soil Amendments

Amendment	Location	Rate applied/crop	Results
		Legumes	
ss	MD	0–312 kg available N ha^{-1} on soybeans (*Glycine max* (L.) Merr.)	No adverse effects of ss; highest rate did not affect nodulation; ss increases foliar N 1st year; ss acceptable practice on soybeans.[50]
Gamma-irradiated ss	India (greenhouse)	10% of soil used for chickpea	Decreased growth with non-irradiated ss (ammonia toxicity/phenolics); positive growth with irradiated ss over non-amended soil.[51]
ss high in Zn, Cu, and Ni vs. uncontaminated ss	England (greenhouse)	5–10 g kg^{-1} on white clover (*Trifolium pratense* L.)	Yields higher from all sources (N); Ni and Zn in tissue declined with time; plants control Cu in tissues more than Zn and Ni.[31]
Lime stabilized ss	MO (greenhouse)	0–7.5 g kg^{-1} on alfalfa (*Medicago sativa* L.)/red clover (*Trifolium pratense* L.)	High rates decreased germination; yield increased up to 5.0 g kg^{-1}; ss decreased exchangeable Al and increased soil pH.[52]
Cattle dung and wool waste compost	India	10 Mg ha^{-1} on chickpea (*Cicer arietinum* L.)	Growth increased with reduction of C:N ratio.[40]
Boiler wood ashes vs. ag. lime	AL	10.3 Mg ash ha^{-1}	Ash slightly more effective than ag. lime on soil pH and yields slightly higher than ag. lime; no effect on plant nutrient uptake.[53]
Paper mill sludge	AL	0–134 Mg ha^{-1} $year^{-1}$, 3 years, soybeans (*Glycine max* (L.) Merr.)	Soybean yields were not affected by the sludge treatments.[12]
Paper sludge (C/N=150)	—	448–1344 Mg ha^{-1} + N for C/N of 10–100 on *Phaseolus vulgaris*	Yields reduced 1st year; increased 2nd.[18]
Paper and power ash sludges	MS	0–224 Mg ha^{-1} on soybeans (*Glycine max* (L.) Merr.)	Positive growth and yield with <224 Mg sludge ha^{-1}; metals in normal range in leaves.[11]
		Cotton	
An. dig. dried ss	AZ	20–80 Mg ha^{-1} $year^{-1}$ for 3 years	Yields with highest rate comparable to fertilized fields; no effect on leaf or seed Cd, Ni, Zn, and Cu.[54]

Non-composted, organic waste (newsprint, wood chips, yard waste, cotton gin trash, poultry manure)	AL	44.8 Mg C ha^{-1} adjusted	All sources reduced yield; yield reduction less with poultry manure; newsprint reduced weeds; wastes had no effect on herbicide efficacy; 4–6 weeds needed for biological stabilization.[55]
Newsprint and broiler litter	AL	Various ratios with soil and broiler litter on surface and in trenches	Metal concentration in cotton leaves within normal range; Cr and Pb higher with newsprint.[56]
Newsprint and broiler litter	AL	Various ratios with soil and broiler litter on surface and in trenches	50:40:10 ratio of soil:newsprint:litter increased lint yields 340–450 kg ha^{-1}.[198]
Broiler litter	AL	0–9 Mg ha^{-1} $year^{-1}$ (0–270 kg N ha^{-1})	In 3 of 5 site years, N in litter as effective as NH_4NO_3 on yields; generally, litter enhances yields; growth regulator not necessary to control vegetative growth.[57]
		Sorghum (*Sorghum bicolor* (L.) Moench)	
An. dig., gamma-irradiated, dried, ss and dairy manure	NM	34–67 Mg ha^{-1}	ss improved yields and corrected Fe chlorosis.[58]
Beef cattle manure compost	KS	0–16.1 Mg ha^{-1} $year^{-1}$ for 4 years	Fertilizer efficiency 13% for compost vs. 36% for N fertilizer.[59]
Urban compost	Spain	14 Mg ha^{-1}	Yields not affected; increase in available Cu and Zn.[60]
Municipal wastewater	AZ	Wastewater mixed with irrigation water	Taller, more head, higher yield with mix; higher forage yields.[61]
Slag from a brass recycling plant	AL	Applied at rates to supply 0–6 mg Zn/kg to a Hybrid sorghum (*Sorghum bicolor* (L.) Moench)	On an acid soil with low cation exchange capacity, the brass recycling plant slag was as effective as water soluble chemicals in supplying B and Zn to the Sorghum.[62]
		Other Crops	
Residual vegetable and animal waste compost	Italy	0–30 Mg ha^{-1} applied 6 years previous on sunflower (*Helianthus annuus* L.)	After 6 years cropping, improvements in yield and crop quality still apparent in high compost mixtures.[63]

Table 2 Response of Vegetable Crops (Yield and Heavy Metal Uptake) to Soil Amendments

Crop	Amendment	Location	Rate applied	Results
Broccoli (*Brassica oleracea*) Lettuce (*Lactuca sativa* L.) Eggplant (*Solanum melongena* L.) Tomato (*Lycopersicon esculentum*) Potato (*Solanum tuberosum* L.) Cabbage (*Brassica oleracea*) Carrot (*Daucus carota*) Cantaloupe (*Cucumis melo* L.) Pepper (*Capsicum* spp.)	Anaerobically digested sewage sludge	Alabama	0,224 mt/ha	Increasing soil temperature increased Cd and Zn concentration in broccoli and potatoes. Liming reduced the concentration of Cd and Zn all crops. Lettuce was determined to be an accumulator of Cd. In general, crop yields were highest when sludge was applied.[66]
Lettuce Radish (*Raphanus sativus*) Carrots Tomatoes	Anaerobically digested sewage sludge	Rhode Island	0,20,60 mt/ha	Lettuce had the highest accumulation of Cd, Cu, and Zn. Carrot roots had the highest uptake of Ni. Sludge application increased Cd in lettuce and Cu in tomatoes during the second year of the test. Zinc and Ni were increased in most of vegetables by the application of sludge. No phytotoxic symptoms were observed.[67]
Potatoes Lettuce Carrots Peas (*Piscum sativum* L.) Radish Sweet corn (*Zea mays* L.) Tomatoes	Anaerobically digested sewage sludge	Minnesota	0,112,225,450 mt/ha	No adverse effects of sewage sludge on yields, and no physiological imbalances were observed. Concluded that lettuce is a metal accumulator, but potatoes and carrots are nonaccumulators and should be acceptable crops for sludge treated soils. In edible fruits, the uptake of Fe, Mn, Cu, B, and Pb was not affected by sludge. Pea vines accumulated three times as much Zn as the fruit.[68]
Potatoes	Anaerobically digested sewage	New York	0, 100–120 mt/ha/year for 5 years	Total N and protein N in the cortex was increased by the addition of sludge, but the reverse was observed for the pith tissues. Ascorbic acid was slightly higher due to sludge application in one of 2 years. Phenolic content and enzymatic activity was not affected by the addition of sludge.[69]
Potatoes	Anaerobically digested sewage	New York	0, 100–120 mt/ha/year for 5 years	PCBs in potato leaves and tuber pulp were below detection limits. Potato peels contained 0.04 mg/kg. Concluded that potatoes grown in PCB contaminated soils pose no significant health risk especially if they are peeled.[70]

Carrots	Municipal and industrial sludges	Wisconsin	0,90 mt/ha	Test included three hybrids and three bed types. Heavy metal uptake was hybrid selection dependent. Highest yields and elemental concentrations were obtained in the municipal sludge treatments. In the second year, the industrial sludge resulted the highest level of Cd in carrot roots. Concluded that hybrid selection was the important factor affecting root nutrient levels.[71]
Cucumber *(Cucumis sativus)*	Municipal and industrial sludges	Wisconsin	0,90 mt/ha	Test included three hybrids and three bed types. Highest yields were obtained with flat beds that were treated with municipal sludge. Sludge treatments showed no deficiency or toxicity in cucumber plants. There were strong genotypic differences for elemental accumulation in the fruit and peel.[72]
Cucumber	Digested and composted sewage sludge	Maryland	0, 25, 50% of volume of growth media as sludge Elemental S: final pH range of 3.4–7.2	Fruit yields in the greenhouse were not affected by any treatments. Cadmium in the leaves and fruit were not affected by treatments. Zinc in the fruit was affected by the treatments. Authors concluded that cucumbers can be grown in media amended with up to 50% composted sewage sludge that has a low heavy metal content.[73]
Green beans *(Phaseolus vulgaris)* Radish Carrots Cabbage	1 aerobically, 2 anaerobically, 1 composted sewage sludges	W. Va.	0,90, & 180 Mg/ha	Concentrations of Cd, Cr, and Pb in edible parts were increased by no more than 1.0 mg/kg above the control. Cu and Zn was increased in some of the sludge treatments, not enough to be of concern. The authors concluded that the edible portions of the vegetables showed no appreciable accumulation of heavy metals on a soil treated with four sewage sludges.[74]
Lettuce	Municipal and industrial sludges	Wisconsin	0, 90 mt/ha/year for 4 years	Sixty cultivars and plant introductions were compared. Highly significant genotype and sludge effects on Cd in leaves. Cadmium in lettuce leaves ranged from 3.1 to 11.9 mg/kg, 1.4 to 5.5 mg/kg, and from 1.2 to 2.5 mg/kg in industrial sludge, municipal sludge and control treatments, respectively.[75]

Table 2 (Continued) Response of Vegetable Crops (Yield and Heavy Metal Uptake) to Soil Amendments

Crop	Amendment	Location	Rate applied	Results
Tomatoes	Anaerobically digested municipal sewage sludge	New York	0,224 mt/ha	Two tomato varieties were compared. Sludge grown tomatoes were similar in weight, volume, height, and diameter. Sludge produced tomatoes had lower tissue firmness, and titratable acidity. The overall quality of the sludge produced tomatoes was rated lower as compared to the control.[76]
Sweet corn	Feedlot manure and sewage sludge, compost	Maryland	0,40,90,160 Mg/ha	Feedlot manure and a sewage sludge compost improved sweet corn yields on a spoil area that had been mined for sand and gravel. Higher yields were attributed to a more favorable pH, higher water content, and increased nutrient availability.[77]
Lettuce Broccoli	Chrome tannery waste	Oregon	0 to 192 Mg/ha (0 to 1750 kg Cr/ha)	Except for N, the application of the tannery waste had no consistent effect on the concentration of any element or heavy metals (including Cr) in the plant tissues. Authors concluded that the tannery waste, with proper safeguards, may be an effective and safe crop fertilizer. Also concluded that tannery waste application should be based on its N content.[78]
Cabbage Radish Tomato Cucumber Spinach (*Spinacia oleracea* L.) Onion (*Allium cepa* L.) Snapbean (*Phaseolus vulgaris* L.) Mustard (*Brassica juncea* L.)	Spent mushroom compost	Washington	0,2,10,20 kg/m^2	In the first year, cucumber and snapbean yields increased with increasing compost rates, but onion yields decreased. Tomato and cabbage yields were not affected by treatments. During the second year, yields of these crops peaked at a rate of 10 kg/m^2. Yield reductions of salt sensitive crops (snap bean, onion, radish) were attributed to the high K content of the compost.[79]
Watermelons *(Citrullus vulgaris)* Tomatoes Potatoes	Phosphogypsum	Florida	0,1.68,2.24 Mg/ha	Phosphogypsum increased tomato, potato, and watermelon yields by 6, 19, and 49%, respectively.[80]

Sugar Beets (*Beta vulgaris* L.)	Beef feedlot waste	Texas	0 to 1608 Mg/ha	All treatments that had received the beef feedlot waste produced significantly higher sugar beet root and sucrose yields as compared to the fertilized control.[81]
Lettuce	Composted vs. an. dig. ss	Colorado (greenhouse)	—	Higher yields with compost (higher pH); ss decreased yields (salinity/Zn/low pH); leaf Cd/Zn increased at low pH; Cu availability unchanged.[33]
Red Beet	ss high in Zn, Cu and Ni	England (greenhouse)	5–10 g/kg	Higher yield from all ss (N); Ni low.[31]
Swiss chard	Composted ss	California	—	No measurable uptake of Se; chard lower in Se than radish.[32]
Radish	Residue compost vs. vermicompost	California		Increased rates increased yields and N accumulation but decreased N use efficiency; low rates of compost fertilizer best.[64]
Broccoli	Compost and pig manure on high Pb soils	China		Organic amendments reduced Pb in plants.[65]

Table 3 Response of Vegetable Transplants to Soil Amendments

Crop	Amendment	Rate applied	Results
Broccoli Cabbage Lettuce Eggplant Pepper Tomato	Composted sewage sludge	Volume basis, mixed with equal parts of perlite and peat	Transplants were grown for 8 weeks. Cadmium and Zn in the transplants did not reach excessive levels for plant nutrition or human consumption. Concluded that composted sewage sludge could supply trace elements, P, Ca, and Mg to transplants for a minimum of 5 weeks. Additional N and K would be needed.[82]
Broccoli Cabbage Lettuce Eggplant Pepper Tomato	Composted sewage sludge	Mixed with equal volumes of perlite and peat. Treated with rates of N and K.	Lettuce, broccoli, and cabbage transplants reached marketable size with the addition of N alone to the mixtures. Eggplant, tomato and pepper transplants required by N and K. Cadmium concentrations in the transplants were within acceptable levels.[83]
Lettuce Tomato	Composted sewage sludge	Mixed with equal volumes of peat and perlite or vermiculite	Concluded that composted sewage sludge in mixtures containing vermiculite may eliminate the need for complete fertilizers. This mixture needed only N to produce marketable size transplants. The seedlings did not contain toxic levels of Zn, Cd, Pb, Ni, Mn, Fe, or Cu.[84]
Tomato Cabbage	Sewage sludge compost: low-metal, residential; high-metal, industrial	Mixed with equal volumes of peat and vermiculite	Transplants grown in the mixture containing the low-metal sludge had similar quality as compared to plants in peat-vermiculite mixtures without sludge. Mixtures containing the high metal sludge compost produced plants that were small, less developed, and with visual symptoms of heavy metal toxicity. Analysis of the transplants showed that the tomatoes had toxic levels of Zn and Cu while cabbage transplants had toxic levels of Zn.[85]

Cabbage Tomato Muskmelon (*Cucumis melo* L.)	Sewage sludge compost: low-metal, residential; high-metal, industrial	Mixed with equal volumes of peat and vermiculite	Transplant quality was lower for cabbage and tomatoes in mixtures with high metal compost. Muskmelon transplant quality was not affected by sludge treatments. After planting the field, marketable yields of all crops were the same regardless of the transplant mixture. Transplant media had little effect on heavy metals in foliage or edible plant parts of field grown plants. Cadmium concentrations were low regardless of treatment. Concluded that sewage sludge compost of low heavy metal content can be used safely for the production of vegetable transplants.[86]
Tomato	Composted sewage sludge	Mixed with peat: 0, 10, 20, 30, and 40% of volume. Nitrogen: poultry litter or osmocote	Concluded that composted sewage sludge is a satisfactory growth medium for tomato transplant production. Composted sewage should not exceed 30% of the total volume when used to amend sphagnum peat. The use of poultry litter may require leaching of the mixture to reduce soluble salts.[87]
Cabbage Pansy (*Viola tricolor*) Snapdragon (*Antirrhinum majus*)	Composted sewage sludge	Mixed with organic and inorganic components to produce 19 growth media.	When compared to five commercial potting mixes, plant yields were generally higher in media containing composted sewage sludge. Concluded that sewage sludge compost may be substituted for more expensive components of potting mixes.[88]

corresponding increase in selected plant nutrients. This may have been due to a dilution effect, since only plant nutrient concentration and not total nutrient uptake was measured in plant tissue. In general, the amendment was evaluated because it had the potential of supplying a particular plant nutrient which was yield limiting under certain cropping conditions. Predicting nutrient availability from organic amendments has been very difficult, and no universally applicable recommendations on organic amendments have been set.

Nitrogen availability and utilization by crops has received most attention in the literature. Of all soil-derived essential nutrients, nitrogen is needed in the largest quantity by most crops. It usually limits yields of most crops if not applied as a fertilizer and is the most expensive primary plant nutrient which growers must purchase. Nitrogen can also become a source of surface or groundwater contamination. In addition, the complexity of the N cycle in nature makes it the most difficult nutrient to manage in cropping systems.

The amount of an organic amendment to be applied to a crop is generally based on the N needs of the crop and estimates of N availability from the amendment.[89] The N needs have been established based upon extensive field research using inorganic fertilizers containing nitrate, ammonium, or urea forms of N. Estimating N availability from organic amendments is more elusive.

Attempts to develop a N availability index based upon soil analyses have been difficult but some are being used. All depend upon calibration to local soil-crop-climate conditions.[90] The presidedress nitrate test for corn has shown promise in humid regions of the U.S.[91–92] Other biological and chemical indexes of soil N availability include aerobic incubation of soils, total Kjeldahl N, autoclave-extractable N, KCl extractable NH_4-N, and $KMnO_4$- oxidizable N.[90]

Municipal sewage sludges have been extensively evaluated and managed primarily as a source of crop N. Estimates of first year N availability range from 10% for composted sludges to over 40% for lime stabilized, raw, or waste activated sludges. Second-year N availability ranges from 5 to 20%. By the fourth and fifth year after application, most N availability is less than 5% of the total applied.[93] Over a 3-year period, sewage sludge in Oregon was found to be between 27 and 44% as effective as NH_4NO_3 as a source of N for tall fescue (*Festuca arundinacea* Shreb. 'Alta').[41] A study in Tennessee on corn found organic N mineralization rates near 50% the first year and 30% in the second year.[13] However, beneficial residual effects of N can be seen after more than 5 years of cropping.[63]

Manures are usually higher in urea and ammonium forms of N. Therefore, first-year N availability from manures can be very high. Bitzer and Sims[21] propose using a predicted available N (PAN) index for poultry manure where:

$$\text{PAN in manure} = 0.80\ (NH_4\text{-N} + NO_3\text{-N}) + 0.60\ (\text{total N} - \text{inorganic N})$$

In their studies, PAN resulted in the attainment of comparable corn yields with three poultry manures and ammonium nitrate. Residual soil available N from manures was determined to be very low. The presidedress nitrate-N test on corn in

Pennsylvania has proved particularly effective on sites with a history of organic soil amendments.[91]

Less attention has been devoted to P, K, Ca, Mg, and S availability to crops from organic soil amendments because these are generally applied in large enough quantities that they are not likely to be yield-limiting factors. Most research, however, does include some or all of these measurements in its assay of crop nutrient uptake or soil chemical properties.

Micronutrients applied in sludges are often mentioned along with heavy metal analyses. High levels of Zn, Cu, B, or Mo in the soil amendment could be potentially toxic to sensitive crops. However, in all the references listed in Table 1, none reported reduced crop yield or growth parameters which were linked to toxic effects of these micronutrients. In most studies, soil levels of Zn and/or Cu were increased while plant tissue concentrations or plant uptake were unaffected or only slightly increased. A long-term study in Virginia where over 235 kg Cu ha^{-1} has been applied as Cu-rich swine manure to the soil over 11 years has shown normal Cu in corn leaves and grain.[25] Increases in Zn and Ni concentrations in plants from high sludge application is more common than increased Cu concentrations. This suggests that plants control their internal copper concentration much more closely than their Zn and Ni concentrations.[31]

B. BIOLOGICAL EFFECTS

1. Pathogenic Bacteria and Fungi Control

Certain soil-borne diseases of crops may be partially controlled with organic soil amendments.[94] Chicken manures, alfalfa meal, and certain other nitrogenous organic amendments are known to reduce the activity and/or survival of *Phytophthora cinnamomi* and *P. parasitica* in soil.[95–99] Other pathogens which have been controlled using organic amendments include Fusarium,[100] *Sclerotium rolfsii*,[100–103] *S. sclerotiorum* and *Alternaria solani*,[104] and *Rhizoctonia solani*.[105–108] The exact mechanisms involved in the biological control of these crop pathogens have been the subject of extensive speculations and intensive investigations. Huber and Watson[94] summarized the possible mechanisms involved by concluding that organic soil amendments influence the severity of soil-borne crop diseases by (1) increasing the biological buffering capacity of the soil, (2) reducing pathogen numbers during anaerobic decomposition of organic matter, and (3) affecting nitrification which influences the dominant form of N in the soil. In addition, the form of N available can also influence microbial composition and host physiology.[95]

2. Parasitic Nematode Control

Organic soil amendments can be used to control phytoparasitic nematodes in soil, thus improving growth and crop yield potential where these soil pathogens are present. Prior to the 1980's most of the research with organic amendments to control plant parasitic nematodes was principally outside of the U.S.[109–110] However, since the early 1980's, several organic wastes have been shown to have nematocidal activity. These included spent coffee grinds and pecan shells,[111] dry olive *(Olea europaea)* pomace,[112] oil cakes and chicken litter,[113] chitinous wastes such as crab and shrimp shells,[113–117] and hemicellulosic (paper) waste.[118–119] Rodriguez-Kabana and Morgan-Jones[120] acknowledge that the most effective amendments are those with narrow C:N ratios and high protein or amine-type N compounds. However, at levels of ammonia, urea, and low C:N ratio materials required to be nematocidal, ammonia toxicity to crops is likely to occur. Ratios of C:N in the range of 11:1 to 23:1 were found to be nematocidal and non-phytotoxic.

Chitin (crab and shrimp shells) seem to stimulate a select microflora in a soil capable of degrading chitin polymers. The resulting reduction in *Meloidogyne sp.* (root knot nematodes) and *Heterodera glycines* (soybean cyst nematodes) may be explained in part by the chitinous composition of nematode egg masses.[120] Several fungal species isolated from soils treated with chitin are able to decompose the polymer and colonize nematode eggs.[114–115]

C. PHYSICAL EFFECTS

The physical effects of organic soil amendments on soil properties is well documented and frequently mentioned as having positive effects on crop growth and yield. These include reducing the bulk density of the topsoil, increased water holding capacity of coarse-textured soils, increased water infiltration and drainage in fine-textures soils, and improved soil aggregation (structure). The physical improvements in soil properties relate directly to the physical benefits to the crop itself in developing a larger, deeper, and more extensive root system which results in a healthier plant with higher yield potential. The benefits of improved soil physical conditions cannot easily be separated from improvements in plant nutrition, reduction of toxic substances in the soil, or beneficial biological effects. Cotton yield increases in Alabama from the use of poultry litter as a source of N were greater than that explained by N alone. In other cases, inorganic (fly ash) and organic soil amendments (sludges and manures) are used solely to improve soil physical conditions for vegetating damaged land.[3,48,121]

III. FACTORS RESPONSIBLE FOR ADVERSE EFFECTS ON BIOTIC SYSTEMS

As demonstrated in the previous section, waste materials contain organic matter and essential plant nutrients that are beneficial to plants when land applied. Land application represents an economically desirable outlet for the producers of a waste and a potential cheap source of organic matter and fertilizer elements for the land owner. In addition to the potentially beneficial components, some waste materials may also contain non-essential elements, persistent organic compounds and microorganisms that may be harmful to crops. In this section we will briefly address some

Table 4 Metal Concentrations (mg/kg) in Sewage Sludges from the North Central United States

Metal	Range	Median	Mean
Pb	13–19,700	500	1360
Zn	101–27,800	1740	2790
Cu	84–10,400	850	1210
Ni	2–3,520	82	320
Cd	3–3,410	16	110
Cr	10–99,000	890	2,620
Ba	<0.01–0.9	0.02	0.06
Fe	<0.1–15.3	1.1	1.3
Al	0.1–13.5	0.4	1.2
Mn	18–7,100	260	380
B	4–760	33	77
As	6–230	10	43
Co	1–18	4.0	4.3
Mo	5–39	30	28
Hg	0.5–10,600	5	733

From Sommers, L.E., Nelson, D.W., and Yost, K.J., *J. Environ. Qual.*, 5, 303, 1976. With permission.

of the potential adverse effects associated with the land application of waste materials.

A. CHEMICAL FACTORS

1. Heavy Metals

In addition to essential elements that will be beneficial to plant growth, many, if not all, wastes contain heavy metals that have the potential of being toxic to plants and animal life. The waste material that has received the most attention has been sewage sludge. The composition of sewage sludge is highly dependent on the type of treatment, the composition of the sewage influent, and the type and amount of industrial activity in an area.[122] Many heavy metals have multiple industrial uses[123] and can be found in high concentrations in some industrial wastes. Sommers et al.[124] demonstrated that the composition of sewage sludge will vary considerably from one city to another and also within a given treatment plant. The variable nature of sewage is illustrated in Table 4 which summarizes heavy metal concentrations in sewage sludge samples collected in the North Central U.S. Sewage sludge (Table 4) typically contains higher levels of heavy metals as compared to many other wastes as demonstrated in Table 5. Of interest is the high level of Cu reported in some of the poultry litter samples.[125] In a review article by Edwards and Daniel,[126] Cu in poultry litter was reported to range from 25 to 127 mg/kg with a mean of 55.8 mg/kg. The high levels of Cu in the study of Stephenson et al.[125] reflects the addition of $CuSO_4$ to the poultry feed. Manure from pigs has also been reported to have elevated levels of Cu when the pigs are given dietary supplements of Cu as $CuSO_4$. In Virginia, manure from pigs receiving feed treated with $CuSO_4$ had Cu concentrations that ranged from 899 to 1398 mg/kg.[24]

Table 5 Heavy Metal Content (mg/kg) of Various Wastes for the Southern United States

	Textile waste[127]			Fermentation waste[127]			Wood processing waste[127]			Poultry litter[125]	
Metal	Range	Median	Mean	Range	Median	Mean	Range	Median	Mean	Range	Mean
Pb	9–250	135	129	<1–95	6	29	<1–90	36	42	—	—
Zn	40–1,800	940	864	5–975	40	255	22–337	73	122	106–669	473
Cu	149–760	416	390	3–210	13	81	<1–91	58	53	25–1003	473
Ni	31–155	40	63	<1–34	18	18	6–492	60	119	—	—
Cd	<1–9	4	4	<1–3	<1	2	<1–4	<1	2	—	—
Cr	41–5,560	1830	2490	<1–540	10	117	<1–362	30	81	—	—

Table 6 EPA Limits for Metal Concentrations and Loading Rates for the Land Application of Sewage Sludge[130]

	Agricultural land, forests, public contract sites, land reclamation		Lawns and home gardens	
Metal	Maximum Concentration (mg/kg)	Maximum Loading (kg/ha)	Maximum Concentration (mg/kg)	Annual Loading (kg/ha)
Arsenic	75	41	41	2.0
Cadmium	85	39	39	1.9
Chromium	3000	3000	1200	150
Copper	4300	1500	1500	75
Lead	840	300	300	15
Mercury	57	17	17	0.85
Molybdenum	75	18	18	0.90
Nickel	420	420	420	21
Selenium	100	100	36	5.0
Zinc	7000	2800	2800	140

When considering land application of waste materials, Cd, Cu, Ni, Pb, and Zn are the heavy metals that are considered to pose the greatest threat to the environment.[128] Cadmium is the metal of most concern since it has been associated with renal failure and pulmonary emphysema.[129] Cadmium can be taken up by food crops, enter the food chain, and accumulate in human tissue causing health care problems. Manganese, Fe, Al, Cr, As, Hg, Sb, Se, Co, and V[122,128] are metals considered to pose little hazard when sludge is land applied since they have low availability in soils and since sludge has low levels of these elements. Molybdenum can pose a threat in some situations since plants can accumulate enough Mo to cause molybdenosis in ruminant animals without any visible toxic effects on the plants.[128] Boron can also be phytotoxic if applied at high rates. In 1993, the Environmental Protection Agency (EPA) established ceiling concentrations and cumulative loading rates for sludge-borne heavy metals.[130] Limits were established for As, Cd, Cr, Cu, Pb, Hg, Mo, Ni, Se, and Zn and are summarized in Table 6. The EPA guidelines limit both maximum concentrations of these metals as well as loading rates, but they also recognize exceptional quality sludge which can be applied with no restrictions on use, the rate of application, or application record keeping.

Another potentially significant source of Cd is phosphate fertilizers. Phosphate rock can contain appreciable amounts of Cd with most of the Cd being associated with the phosphate fertilizers. In the U.S., Florida phosphate rock contains 3 to 15 mg/kg Cd while Western phosphate rock contains up to 130 mg Cd/kg.[131] The Cd content of P fertilizers is closely related to the P content of the fertilizers.[132]

In Australia, phosphate fertilizers contain 18 to 91 mg Cd/kg.[133] Research has shown that the use of Australian superphosphate will increase Cd uptake by pasture species, cereals, and edible portions of vegetables. Williams and David[133] concluded, however, that wheat and cereal foods produced in Australia on soils with long term P fertilizer usage have low contents of Cd (0.012 to 0.036 mg/kg).

In the U.S., Mulla et al.[134] evaluated Cd uptake in a soil that received 175 kg P/ha/year as treble superphosphate for 36 years. This rate of P is considered to be a high rate of fertilizer P. The concentration of Cd in the grain and leaves of barley *(Hordenum vulgare)* grown in the P treated soil was the same as the control. Swiss chard *(Beta vulgaris)* grown in a greenhouse had a higher level of Cd (1.6 mg/kg) when grown in the P-treated soil as compared to the control. Yields of swiss chard were not affected. Mortvedt[135] evaluated Cd uptake in nine long-term (> 50 years) soil fertility experiments in the U.S. In these studies annual Cd rates were estimated to range from 0.3 to 1.2 g/ha. Analysis of corn (*Zea mays* L.), soybean (*Glycine max* L. Merr.), wheat (*Triticum aestivum* L.), and timothy (*Phleum pratens* L.) tissue showed that the increase in plant uptake of Cd resulting from the use of P fertilizers containing < 10 mg Cd/kg is negligible. Triple superphosphate and diammonium phosphates manufactured from Florida phosphate rock contained from 3 to 10 mg Cd/kg.[135] In most of the studies cited, Cd uptake decreased with liming.[132–133,135]

Several reviews[122,127–128,136–137] have demonstrated that land application of wastes containing heavy metals will result in increased plant uptake of heavy metals. Availability of the applied metals to plants will depend on the soil characteristics, the composition of the waste, and the waste application rate.[122] Heavy metal uptake by plants varies considerably among species and also among genotypes within a species. For example, leafy vegetables such as lettuce (*Lactuca sativa* L.), swiss chard *(Beta vulgaris)*, tobacco (*Nicotina tabacum* L.), spinach (*Spinacia oleracea* L.), and curlycress (*Arabido psisthaliana* L.) take up more Cd than crops like corn (*Zea mays* L.), soybean (*Glycine max* L. Merr.), wheat (*Triticum aestivum* L.), and forages.[122,127] Most of the studies evaluating varietal effects on metal uptake have dealt with Cd. These studies have shown that when treated with the same rates of Cd, uptake and Cd distribution in the plant can vary widely among genotypes. Six varieties of head lettuce (*Lactuca sativa* L.), two varieties of leaf lettuce, and one romaine lettuce varied both in total Cd uptake and in how the Cd was distributed among the roots and shoots.[138] Yuran and Harrison[75] evaluated Cd uptake by 60 lettuce cultivars and also observed a highly significant varietal effect on Cd uptake. When a Cd-accumulating corn hybrid was grown on a sludge-treated soil, the Cd concentration in the grain was 13 to 18 times higher that the concentration of Cd in the grain of a low Cd-accumulating corn hybrid.[139] In an earlier study by the same research group,[140] grain from 20 corn inbreds that were grown in sludge-treated soil had Cd concentrations that ranged from 0.08 to 3.87 mg/kg. Zinc concentration in the grain varied from 33.8 to 70 mg/kg among the 20 inbreds. Soybean also differ in their ability to accumulate Cd.[141]

Heavy metals may also affect soil microorganisms that indirectly affect crop performance. Reddy et al.[142] observed a significant reduction in *Rhizobium japonicum* in soils treated with sewage sludge. The authors speculated that the reduction in rhizobia may have been due to the heavy metals in the sludge. Madaringa and Angle[143] concluded, however, that in soils receiving high sludge rates, sludge-borne soluble salts and not heavy metals were responsible for short-term reductions in *Bradyrhizobium japonicum*. Brookes et al.[144] measured N_2 fixation (C_2H_2 reduction) by blue green algae in soils that had been treated with high-metal sewage sludge or a low-metal farm yard manure. Nitrogen fixation on the high-metal soil measured over a period of 118 days was only 1/3 of that measured on the low-metal soil. The total biomass carbon in the high metal soil, as measured by $CHCl_3$-fumigation, was only 40% of that measured in the low-metal soil.[145]

If present at high levels, many heavy metals contained in wastes can be toxic to plants. Concern relative to the toxicity of these metals have been addressed in previous reviews.[122,128,136] Plants respond differently to each heavy metal. Marschner has provided an adequate discussion of essential and beneficial heavy metals, their tolerance levels within a given plant species and the respective unique toxicity symptoms.[146]

As previously noted, the reaction of plants to heavy metals will depend on several factors including the sensitivity of the individual plant and the form of the applied metal.[147] Review articles demonstrate that heavy metals can have pronounced effects on plant root systems and water and nutrient uptake.[147–150] Toxic levels of heavy metals can result in a reduction in root biomass production, enhanced lateral root formation and changes in root growth rates.[147–150] Heavy metals can inhibit root extension and elongation by interfering with cell division and/or cell wall extension.[147–150] Excess heavy metals can result in a thickening of plant root tips, decreased root hair development, a stimulation of root lignification, structural changes in the root, and damaged cell membranes. Heavy metals can also reduce water[149] and nutrient uptake.[150]

Much of the evidence regarding the phytotoxicity of heavy metals has been generated using short-term studies and high rates of metals added to soil or nutrient solution. For sewage sludges, Baker stated that phytotoxicities and yield depressions could be caused by prolonged use of sludges with high levels of Cu, Zn, and Ni.[151] In practice, however, phytotoxicities due to sludge-borne metals are rarely observed. Juste and Mench conducted an exhaustive literature survey of long-term (> 10 years) field experiments involving land application of sewage sludges.[137] Results of 23 studies were surveyed. The authors concluded that Cd, Ni, and Zn are the most bioavailable, but Cr and Pb uptake was insignificant. Plant uptake of heavy metals was related to the total amount of metals that was soil applied. Phytotoxicities resulting from sludge-borne heavy metals were rare for grain crops. In one instance detrimental effects to legumes was attributed to a reduction in nitrogen fixation. Reddy et al. reported a decline in *Rhizobium japonicum* in soils treated with sewage sludge.[142] In 65% of the cases surveyed by Juste and Mench, sewage sludge applications resulted in a positive effect on plant growth.[137]

2. Toxic Organics

Sludges and other waste materials may contain toxic organic compounds. Primary concern regarding land application of wastes containing toxic compounds is with the incorporation of the compound into the food chain and the associated risks to human health. Polyhalogenated biphenyl compounds (PCBs), for example, are known to be present in the food chain.[152] Health risks associated with toxic organics in land-applied sewage sludge have been reviewed.[153–155]

Sewage sludge has been known to contain organochlorine insecticides, polyhalogenated biphenyl compounds (PCBs), polynuclear aromatic hydrocarbons, and many other chlorinated compounds.[153] Dioxins are known to form during the bleaching process of paper production[156–157] and have been speculated by Dacre[155] to be present in some sewage sludge.

Some plants have been shown to accumulate waste-borne toxic organics[158] such as PCBs.[159–160] Carrots *(Daucus carota)* grown in soil treated with 100 mg/kg PCBs were reported to contain 7 to 16 mg/kg PCBs.[161] Moza et al.[160] evaluated plant uptake of ^{14}C-labeled 2,4′,5-trichlorobiphenyl and ^{14}C-labeled 2,2′,4,4′,6-pentachlorobiphenyl from soil receiving a one time application of each compound. Carrots grown during the year of application accumulated 3.1% of the radioactivity from the applied trichlorobiphenyl. Sugarbeet (*Beta vulgaris* L.) grown the year following treatment accumulated only 0.2% of the applied radioactivity. For pentachlorobiphenyl the total recovery during the first year was only 0.2%. When grown on soil treated with sewage sludge, potato (*Solanum tuberosum* L.) leaves and potato tuber pulp were below detectable levels of PCBs.[70] The potato peels contained 0.04 mg/kg PCBs. In a greenhouse test O'Connor et al. evaluated sludge-borne PCB uptake by three food-chain crops and a grass species.[162] The sludge applied was heavily contaminated with PCB (52 mg/kg). Carrots were the only plants that were contaminated with PCB and the contamination was restricted to the peels. Fries and Marrow treated soil with ^{14}C-labeled Tri-, tetra, and pentachlorobiphenyls.[163] No ^{14}C was detected in the aerial plant parts of soybean when a vapor barrier was provided between the soil and aerial plant parts. The authors concluded that aerial plant parts could be contaminated with PCBs by volatilization and redeposition. Beall and Nash using the same technique reported similar results for DDT.[164] Pal et al. in a review article concluded that PCBs can be taken up by plants but there is no evidence of biomagnification in aerobic soil plant systems for crops such as soybean, fescue (*Festuca*), tomato (*Lycopersicon esculentum*), radish (*Raphanus sativus*), beet (*Beta vulgaris* L.), and carrot.[159]

Weber and Mrozek treated a sand with 0, 1, 10, 100 and 1000 mg/kg of analytical grade PCB (Aroclor 1254).[165] Soybean plant height and plant weights were decreased by the addition of PCB. After a 26-day growth period the 100-mg/kg treatment reduced plant height, plant weight, and water uptake by 23, 27, and 45%, respectively as compared to the control. For fescue (*Fescue arundinacea* Schrib), PCB inhibited plant weight only at the highest rate of PCB. The 1000-mg/kg treatment decreased the fresh plant weight of fescue by 16%.

There is no evidence of plant uptake of sludge-borne Di-(2-Ethylhexyl) Phthalate (DEHP) by lettuce, carrot, chile pepper (*Capsicum annuum* L.), and tall fescue when grown in soil treated with ^{14}C-DEHP-contaminated sludge.[166] In a greenhouse test using four soils treated with the flame retardant polybrominated biphenyls (PBB's) there was little if any uptake by orchard grass (*Dactylis glomerata* L.) and carrot.[167]

The effect of pesticides applied to soil plant systems are mainly transient on nontarget soil microorganisms when applied at normal field rates.[168] The effects of pesticides applied as a component of waste materials on soil biology is uncertain.

B. PATHOGENS

Municipal and animal wastes may contain organisms that can be pathogenic to humans and animals.[169–170] Treatment of sewage sludge (i.e., anaerobic digestion) results in a significant reduction in pathogens which usually makes the waste safe for soil application.[171] The preferred practice for killing microorganisms in sewage sludge is by lime stabilization, where lime is added to increase sludge pH above 12 for a period of time. Increased instances of plant diseases associated with land application of wastes was not addressed in any of the literature that was reviewed for this chapter.

C. SALINITY

All wastes contain soluble salts and vary considerably in their content, and excessive salt in soils (applied and natural) can have detrimental effects on plant growth.[172–173] Excess salts can reduce plant growth and germination in crops and high Na levels can result in a degradation of soil physical properties. The Na hazard of wastes and irrigation waters is evaluated by the Sodium Adsorption Ratio (SAR). The SAR is expressed as: $SAR = Na/[(Ca + Mg)/2]^{1/2}$, where concentrations are expressed in meq/L. Waste waters having a SAR > 15 should be avoided.[174]

Increased soil salinity has been reported to decrease the emergence rate and emergence of sugar beets and onions (*Allium cepa* L.).[175] El-Sharkawi and Springuel reported an interaction between temperature and salinity tolerance of wheat, barley, and sorghum (*Sorghum bicolor* (L.)).[176] Plants vary widely in their tolerance of salinity[172,177] and their tolerance can be dependent on the growth stage of the plants.[177] Salinity will be tolerated by crops up to a threshold level after which yields will decrease linearly with increasing salt content.[177]

Concerns regarding the salt content of wastes have been addressed in review articles.[178–179] Minimizing salt problems associated with land application of wastes can be accomplished by analyzing the salt content of the waste and monitoring the change in the salt content of the soil with time. Salt buildup will be dependent on the salt content of the waste, the rate of application, and the amount of leaching.

D. OTHERS

A potential environmental concern for the phosphate fertilizer industry in the U.S. is the production and disposal of phosphogypsum. The annual production of phosphogypsum in the U.S. is 45 million Mg[180] with Florida accounting for 74% of the total production.[181] Phosphate rock naturally contains low levels of radioactive elements, especially radium and uranium.[182–184] During the production of wet process phosphoric acid most of the uranium is associated with the phosphoric acid and most of the ^{226}Ra associates with the phosphogypsum.[182] Phosphate fertilizers are considered to present little if any appreciable radiation hazard.[184]

Phosphogypsum from Florida phosphate rock contains an average of 814 Bq ^{226}Ra/kg which is above the EPA limit of 185 Bq ^{226}Ra/kg for hazardous waste.[181,185] Current EPA regulations prohibit the agricultural use of phosphogypsum that contains more than 370 Bq ^{226}Ra/kg.[186]

Phosphogypsum contains 85 to 93% pure gypsum and has been shown to be an excellent source of fertilizer Ca and S, and a suitable amendment for sodic soils.[187] The benefits resulting from agricultural use of phosphogypsum were recently reviewed by Alcordo and Rechcigl.[49]

Published data suggests that the radioactive elements in phosphogypsum will have minimal impact on plant growth. Mays and Mortvedt added phosphogypsum at rates of 0, 22, and 112 Mg/ha to a silt loam soil in Alabama.[27] A rate of 112 Mg/ha was considered to be about 200 times the normal rate for applying fertilizer Ca and S. The phosphogypsum contained 0.23 mg Cd/kg and 925 Bq/kg ^{226}Ra. Cadmium and ^{226}Ra in the grain of corn, soybean, and wheat were not affected by the phosphogypsum. Mullins and Mitchell applied phosphogypsum (777 Bq/kg ^{226}Ra) to supply 0, 22, and 45 kg S/ha to wheat.[188–189] Wheat forage produced on the phosphogypsum treatments had the same content of ^{226}Ra and ^{210}Po (polonium) as the control. In Florida, phosphogypsum (671.5, 900.2, and 1139 Bq/kg of ^{226}Ra, ^{210}Po, and ^{210}Pb, respectively) was applied at rates of 0, 0.4, 2.0, and 4.0 Mg/ha to a bahiagrass (*Paspalum notatum* flugge) pasture.[49] The contents of ^{226}Ra, ^{210}Po, and ^{210}Pb in bahiagrass forage from the phosphogypsum treatments were not significantly different from the control. Thus, the use of phosphogypsum in agriculture should have minimal effects on radionuclide concentrations in plants or on plant growth. Even at a solid radioactive waste disposal area in Idaho, accumulation of radionuclides by vegetation was considered not to be a major environmental hazard.[190]

IV. EFFECTS ON CROPS

A. AGRONOMIC CROPS

Research on the effects of organic and inorganic by-products and wastes on crops is extensive. Selected studies were summarized in Table 1. Most of these studies were designed to either demonstrate the effectiveness of a particular product or by-product as a soil amendment for the crop(s) studied and/or to determine if certain properties of the material (e.g., metals) would result in phytotoxicity, high plant concentrations, or damaging environmental effects (water quality). In general, most of the studies successfully demonstrated that use of the amendment had positive effects on plant growth parameters, yields, or plant nutrient uptake.

B. VEGETABLE CROPS

Numerous studies have been conducted to evaluate the response of vegetables and vegetable transplants to soil amendments. Most of this work dealt with vegetables that are commonly grown in home gardens or sold commercially. The primary focus of this work was on heavy metal uptake by both the edible and nonconsumable plant parts, plant growth and yield, and the effects of plant genotype on metal uptake. Results from selected articles are summarized in Tables 2 and 3.

A review of the work summarized in Table 2 leads one to conclude that various waste materials having low metal content can be safely applied to soils that are used for vegetable production. In the late 1970's maximum metal concentrations for low metal sludges were suggested by Chaney and Giordano to be < 2500 mg/kg Zn, 1000 mg/kg Cu, 200 mg/kg Ni, 25 mg/kg Cd, and 1000 mg/kg Pb.[191] In 1993 the EPA set maximum concentration limits for sludge-borne heavy metals when sludge is applied to lawns or home gardens. These limits are: 41 mg/kg As, 39 mg/kg Cd, 1200 mg/kg Cr, 1500 mg/kg Cu, 300 mg/kg Pb, 17 mg/kg Hg, 18 mg/kg Mo, 420 mg/kg Ni, 36 mg/kg Se, and 2800 mg/kg Zn (Table 6). The EPA has also set maximum cumulative and annual loading rates for these metals when applied to lawns and gardens (Table 6). The results also demonstrate that lettuce is an accumulator of heavy metals as compared to other commonly grown vegetable crops. Others have shown that leafy vegetables (i.e., lettuce, spinach (*Spinacia oleracea* L.), curlycress (*Arabidopsis thalina* L.), swiss chard) tend to accumulate greater amounts of Cd as compared to many other crops.[122] The results show that heavy metal uptake is highly dependent on the plant genotype, and that vegetable transplants can be safely grown in potting mixtures that contain up to 30% by volume of low metal composted sewage sludge. Effects of toxic organics and waste-borne pathogenic organisms on plant growth were not addressed in any of these studies.

C. FRUIT CROPS

Results of selected studies evaluating the response of fruit crops to various soil amendments are summarized in Table 7. Most of the trials demonstrated that the respective waste material(s) studied were satisfactory soil amendments that increased fruit yields without an accumulation of heavy metals. In one instance apple quality was decreased by long term irrigation with municipal waste water; however, the agents causing a reduction in quality were not identified.

V. WAYS TO MINIMIZE ADVERSE EFFECTS AND MAXIMIZE BENEFICIAL EFFECTS OF AMENDMENTS

Analytical technology is easily accessible to all who desire to use an organic or unconventional inorganic amendment on cropland. Very simple tests may not be able to accurately predict nutrient availability, but they can easily be used to screen samples for potential problems. The U.S. EPA recognized this in the 1993, 503 regulations for application of municipal sludges to agricultural lands, when maximum concentration for certain metals were established for clean sludges (Table 6). This allows use of sludges as a beneficial soil amendment without burdensome attention to metal loading rates.

Table 7 Effects of Soil Amendments on Fruit Trees

Crop	Amendment	Location	Rate applied	Results
Apple (*Malus Domestica* Borkh.)	Secondary municipal effluent or well water	British Columbia	Growing season, daily. Applied for 5 years.	Municipal effluent increased N, P, and K concentrations in the leaves. Effluent resulted in increased trunk diameter, and higher fruit yield and numbers during the first 2 years of yield measurement.[192]
Apple	Municipal waste water or well water	British Columbia	Growing season, applied daily for 7 years.	Irrigation with municipal waste water decreased flesh firmness, and increased core flush. The authors concluded that irrigation with municipal waste water had detrimental effects on apple quality. Casual agents were not identified.[193]
Apple	Fluidized-bed combustion material	Maryland	0, 9.2 and 36 kg/m^2	Cumulative apple yields over a six year period were increased for 3 of 4 tree types by the application of fluidized bed material. A 15% reduction in yield obtained with the fourth tree type. No nutritional problems were observed for any tree type. Authors concluded that high rates (112 Mg/ha) of fluidized bed material could be applied to apples.[194]
Pear (*Pyrus communis*)	Composted sewage sludge	Maryland	0, 18 mg/m^2	Site included six pear cultivars. Addition of the lime-stabilized sewage sludge compost increased leaf Ca, but trace metals were not affected. Compost addition did not increase the level of Cd in the leaves, fruit flesh or peels. Authors concluded that the addition of low-metal sludge is suitable to renovate pear orchards.[195]
Citrus (*Citrus sinesis*)	Phosphogypsum	Florida	0–1.12 Mg/ha	At one location, phosphogypsum application increased juice brix, brix:acid ratio, and the content of Ca. At a second location, phosphogypsum also resulted in a reduction in titratable acidity.[196]
Pineapple (*Ananas comosus*)	Phosphogypsum	Brazil	0–75 g/plant (up to 3.1 Mg/ha) plus KCl compared to K_2SO_4	Similar yields, but K_2SO_4 resulted in better juice quality as compared to the phosphogypsum treatments.[197]

For use on crops, the following analytical information is desirable on most amendments:

- *Moisture (% of wet weight) or % solids* — The physical condition of the amendment determines how it is handled and the amount of water that has to be transported is a primary consideration in its handling and spreading.
- *Ash or loss on ignition (%)* — This is an indication of mineral matter (soil) in the material which can decrease its value as a source of plant nutrients or organic matter. Some animal manures contain over 50% ash.
- *Neutralizing value as % $CaCO_3$* — This is the most important analysis on wood ashes, fly ash, and some sludges treated with lime or ash. These amendments may be most appropriately applied as an alternative soil liming material. Neutralizing values can range from over 70% for some hardwood ashes and lime derived by-products (e.g., basic slag and cement flue dust) to less than 10% for boiler ashes derived from burning mostly coal. Acidic by-products will, of course, have negative values. Over-application of sludges with a high neutralizing value could result in micronutrient deficiencies of Fe, Zn, or Mn on some soils and crops.
- *Soluble salts or specific conductance (mmhos cm^{-1} or siemens m^{-1})* — This is determined by measuring electrical conductivity (EC) on liquid samples or (usually) mixing dry samples in a 1:2 ratio with water. This gives a relative measurement of the collective effects of the soluble salts in the material. Excess salts can damage salt-sensitive crops and affect soil structure. This is a particularly serious consideration in arid regions.
- *Total nutrients present* — Although total nutrients present are related to nutrient availability to crops, there are so many factors that affect availability that these analyses must be used with caution. Nevertheless, total nutrient analyses enable the user to calculate total loading rates in order to reduce environmental risks from gross over applications of nutrients such as N and P. U.S. EPA guidelines (Table 6) specify maximum concentrations of heavy metals that are allowed in sewage sludges. Total nutrients that should be analyzed on all samples are: N, P, K, Ca, Mg, S, Cu, Zn, Mn, B, and (rarely) Mo and Cl.
- *Other metals and potential contaminants* — These may be required by U.S. EPA regulations on certain waste materials because of their potential phytotoxicity or potential environmental risk. These include: Cd, Cr, Pb, As, Se, Ni, and Hg. Sodium (Na) should be determined if soluble salts are high or a source of Na is suspected in the material because of the problems encountered in some soils from excess Na application.

VI. SUMMARY

Recycling of non-hazardous, agricultural, industrial, and municipal wastes/by-products for agricultural crop production is becoming a common practice. Widespread public concern about their potential negative effects on crops is largely unsupported by research. These concerns, particularly regarding heavy metal applications, may be based upon past government regulations and limited research. Only 5 of 74 summaries in Tables 1 to 3 indicate negative effects of soil amendments on crop yield, quality, or metal concentrations. The negative effects were predictable based upon known properties of the amendment and intentionally gross over applications. High metal concentrations in crops can be predicted based upon an analysis of the amendment and rate applied to the soil. However, metal uptake is tempered by crop genotype, plant part sampled, soil, and climate.

Almost half of the summaries in Tables 1 to 3 reported only positive crop effects from the amendments studied, and most of the other summaries reported that use of a nonconventional amendment was no better than commercial fertilizers or lime used for comparisons. Like fertilizers, lime, or conventional organic soil amendments, the soil amendments reviewed in this chapter influenced crops by changing the soil's chemical and physical properties, adding or immobilizing plant nutrients, or adding or controlling soil-borne pathogens.

Organic wastes/by-products and other nonconventional amendments may be safely and effectively used to enhance crop performance if applied at rates based upon analyses of the product.

REFERENCES

1. Tisdale, S.L., Nelson, W.L., and Beaton, J.D., *Soil Fertility and Fertilizers,* 4th ed., Macmillan Publishing, New York, 1985.
2. Wolt, J.D. and Adams, F., Whatever happened to basic slag?, *Highlights Agric. Res.,* 25(1), 15, 1978.
3. Ferrialo, G., Zilli, M., and Converti, A., Fly ash disposal and utilization, *J. Chem. Tech. Biotechnol.,* 281, 1990.
4. Etiegni, L., Mahler, R.L., Campbell, A.G., and Shafii, B., Evaluation of wood ash disposal on agricultural land. I. Potential as a soil additive and liming agent, *Commun. Soil Sci. Plant Anal.*, 22, 243, 1991.
4. Etiegni, L., Mahler, R.L., Campbell, A.G., and Shafii, B., Evaluation of wood ash disposal on agricultural land. II. Potential toxic effect on plant growth, *Commun. Soil Sci. Plant Anal.,* 22, 257, 1991.
5. Ohno, T. and Erich, M.S., Effect of wood ash application on soil pH and soil test nutrient levels, *Agric. Ecosyst. Environ.,* 32, 223, 1990.
6. Naylor, L.M. and Schmidt, E., Paper mill wood ash as a fertilizer and liming material: field trials, *Tech. Assoc. Pulp Paper Ind. (TAPPI) J.,* 72, 199, 1989.
7. Lerner, B.R. and Utzinger, J.D., Wood ash as soil liming material, *HortScience*, 21, 76, 1986.
8. Huang, H., Campbell, A.G., Folk, R., and Mahler, R.L., Wood ash as a soil additive and liming agent for wheat: field studies, *Commun. Soil Sci. Plant Anal.,* 23, 25, 1992.
9. Mitchell, C.C. and Muse, J.K., Boiler wood ash as an alternative soil liming material, Ala. Coop. Ext. Serv. New Tech. Demo. Rep. no. S-05–93, Auburn University, AL, 1993.
10. Mitchell, C.C., Evans, C.E., and Kee, D.D., Cement flue dust as an alternative lime source, Ala. Coop. Ext. Ser. Timely Information Agron. Ser. no. S-07–91, Auburn University, AL, 1991.

11. Lytton, D.L., Effects of land disposal of combined power ash and paper waste sludges on crops and soils, Ph.D. diss., Mississippi State Univ., Mississippi State, MS, 1991.
12. Burrows, D.C, Mullins, G.L., and Odom, J.W., Soybean response to paper mill sludge applications, in *Agron. Abstr.,* 1988, p. 36.
13. Cripps, R.W., Winfree, S.K., and Reagan, J.L., Effects of sewage sludge application method on corn production, *Commun. Soil Sci. Plant Anal.,* 23, 1705, 1992.
14. Higgins, A.J., Land application of sewage sludge with regard to cropping systems and pollution potential, *J. Environ. Qual.,* 13, 441, 1984.
15. Rappaport, B.D., Martens, D.C., Reneau, Jr., R.B., and Simpson, T.W., Metal accumulation in corn and barley grown on a sludge-amended Typic Ochraqualf, *J. Environ. Qual.,* 16, 29, 1987.
16. Mays, D.A. and Giordano, P.M., Benefits from land application of municipal sewage sludge, Cir. Z-238, TVA-Nat. Fert. Development Ctr., Muscle Shoals, AL, 1988.
17. McCoy, J.L., Sikora, E.J., and Weil, R.R., Plant availability of phosphorus in sewage sludge compost, *J. Environ. Qual.,* 15, 404, 1986.
18. Aspitarte, T.R., Rosenfeld, A.S., Samle, B.C., and Amberg, H.R., Pulp and paper mill sludge disposal and crop production, *Tech. Assoc. Pulp Paper Ind. (TAPPI),* 56, 140, 1973.
19. Mays, D.A. and Giordano, P.M., Landspreading municipal waste compost, *Biocycle,* Mar., 37, 1989.
20. Shahandeh, H. Cabrera, M.L., and Sumner, M.E., Evaluation of nutrasweet sludge as a nitrogen fertilizer for corn and wheat, *Commun. Soil Sci. Plant Anal.,* 23, 1911, 1992.
21. Bitzer, C.C. and Sims, J.T., Estimating the availability of nitrogen in poultry manure through laboratory and field studies, *J. Environ. Qual.*, 17, 47, 1988.
22. Sutton, A.L., Nelson, D.W., Moelleer, J.J., and Hill, D.L., Applying liquid dairy waste to silt loam soils cropped to corn and alfalfa-orchardgrass, *J. Environ. Qual.,* 8, 515, 1979.
23. Safley, Jr., L.M., Westerman, P.W., Barker, J.C., King, L.D., and Bowman, D.T., Slurry dairy manure as a corn nutrient source, *Agric. Wastes,* 18, 123, 1986.
24. Mullins, G.L., Martens, D.C., Miller, W.P., Kornegay, E.T., and Hallock, D.L., Copper availability, form, and mobility in soils from three annual copper-enriched hog manure applications, *J. Environ. Qual.*, 11, 316, 1982.
25. Anderson, M.A., McKenna, J.R., Martens, D.C., Donohue, S.J., Kornegay, E.T., and Lindemann, M.D., 1991. Long-term effects of copper rich swine manure application on continuous corn production, *Commun. Soil Sci. Plant Anal.*, 22, 993, 1991.
26. Dowdy, R.H, Bray, B.J., Goodrich, R.D., Marten, G.C., Pamp, D.E., and Larson, W.E., Performance of goats and lambs fed corn silage produced on sludge-amended soil, *J. Environ. Qual.*, 12, 467, 1983.
27. Mays, D.A., and Mortvedt, J.J., Crop response to soil applications of phosphogypsum, *J. Environ. Qual.*, 15, 78, 1986.
28. Day, A.D., Thompson, R.K., and Swingle, R.S., Wheat hay grown with sludge, *Biocycle,* Aug., 40, 1987.
29. Day, A.D., Solomon, M.A., Ottman, M.J., and Taylor, B.B., Crop response to sludge loading rates, *Biocycle,* Aug., 72, 1989.
30. Day, A.D., Ottman, M.J., Taylor, B.B., Pepper, I.L., and Swingle, R.S., Wheat responds to sewage sludge as fertilizer in an arid environment, *J. Arid Environ.,* 18, 239, 1990.

31. Sanders, J.R., McGrath, S.P., and Adams, T.M., Zinc, copper and nickel concentrations in soil extracts and crops grown on four soils treated with metal-loaded sewage sludges, *Environ. Pollution,* 44, 193, 1987.
32. Logan, T.J., Chang, A.C., Page, A.L., and Gange, T.J., Accumulation of selenium in crops grown on sludge-treated soil, *J. Environ. Qual.,* 16, 349, 1987.
33. Simeoni, L.A., Barbarick, K.A., and Sabey, B.R., Effect of small-scale composting of sewage sludge on heavy metal availability to plants, *J. Environ. Qual.,* 13, 264, 1984.
34. Sims, J.T., Nitrogen mineralization and elemental availability in soils amended with cocomposted sewage sludge, *J. Environ. Qual.,* 19, 669, 1990.
35. Dolar, S.G., Boyle, J.R., and Keeney, D.R., Paper mill sludge disposal on soils: effects on the yield and mineral composition of oats (*Avena sativa L.), J. Environ. Qual.*, 1, 405, 1972.
36. Wright, R.J., Hern, J.L., Baligar, V.C., and Bennett, O.L., The effect of surface applied soil amendments on barley root growth in an acid subsoil, *Commun. Soil Sci. Plant Anal.,* 16, 179, 1985.
37. Pandya, G.A., Prakash, L., Devasia, P., and Modi., V.V., Effect of gamma irradiated sludge on the growth and yield of rice (*Oryza sativa L*., var. GR-3), *Environ. Pollution,* 51, 63, 1988.
38. Kirleis, A.W., Sommers, L.E., and Nelson, D.W., Yield, heavy metal content, and milling and baking properties of soft red winter wheat grown on soils amended with sewage sludge, *Cereal Chem.,* 61(6), 518, 1984.
39. Gonzalez, J.L., Benitez, I.C., Perez, M.I., and Medina, M., Pig-slurry composts as wheat fertilizers, *Bioresource Technol.,* 40, 125, 1992.
40. Tiwari, V.N., Pathak, A.N., and Lehri, L.K., Response to differently amended wool-waste composts on yield and uptake of nutrients by crops, *Biological Wastes,* 28, 313, 1989.
41. Kiemnec, G.L., Jackson, T.L., Hemphill, Jr., D.D., and Volk, V.V., Relative effectiveness of sewage sludge as a nitrogen fertilizer for tall fescue, *J. Environ. Qual.,* 16, 353, 1987.
42. MacLean, K.S., Robinson, A.R., and MacConnell, H.M., *Commun. Soil Sci. Plant Anal.,* 18, 1303, 1987.
43. Sopper, W.E. and Kerr, S.N., Revegetating strip-mined land with municipal sewage sludge, Proj. summary, EPA-600/52–81–182, US EPA, Cincinnati, OH, 1981, 7 pg.
44. Tester, C.F., Sikora, E.J., Taylor, J.M., and Parr, J.F., Nitrogen utilization by tall fescue from sewage sludge compost amended soils, *Agron. J.,* 74, 1013, 1982.
45. Warman, P.R., Effects of fertilizer, pig manure, and sewage sludge on timothy and soils, *J. Environ. Qual.,* 15, 95, 1986.
46. Bole, J.B., Gould, W.D., and Carson, J.A., Yields of forages irrigated with wastewater and the fate of added nitrogen-15-labeled fertilizer nitrogen, *Agron. J.*, 77, 715, 1985.
47. Wood, C.W., Torbert, H.A., and Delaney, D.P., Poultry litter as a fertilizer for bermudagrass: effects on yield and quality, *J. Sustainable Agric.,* 3, 21, 1993.
48. Scholl, D.G. and Pase, C.P., Wheatgrass response to organic amendments and contour furrowing on coal mine spoil, *J. Environ. Qual.,* 13, 479, 1984.
49. Alcordo, I.S. and Rechcigl, J.E., Phosphogypsum in agriculture: a review, *Adv. Agron.,* 49, 55, 1993.
50. Angle, J.S., Madariaga, G.M., and Heger, E.A., Sewage sludge effects on growth and nitrogen fixation of soybean, *Agric. Ecosystems and Environ.*, 41, 231, 1992.

51. Pandya, G.A., Sachidanand, S., and Modi, V.V., Potential of recycling gamma-irradiated sewage sludge for use as a fertilizer: a study on chickpea (*Cicer arietinum*), *Environ. Pollution,* 56, 101, 1989.
52. Vivekanandan, M., Brown, J.R., Williams, J., Clevenger, T., Belyea, R., and Tumbleson, M.E., Tolerance of forage legumes to lime-stabilized sludge, *Commun. Soil Sci. Plant Anal.,* 22, 449, 1991.
53. Mitchell, C.C. and Carroll, S.D., Effect of papermill boiler ash as an alternative liming material in Dallas County, Ala. Coop. Ext. Serv. New Tech. Demo. Rep. no. S-2–89, Auburn University, AL, 1989.
54. Watson, J.E., Pepper, I.L., Unger, M., and Fuller, W.H., Yields and leaf elemental composition of cotton grown on sludge-amended soil, *J. Environ. Qual.,* 14, 174, 1985.
55. Edwards, J.H., Walker, R.H., Mitchell, C.C., and Bannon, J.S., Effects of soil-applied noncomposted organic wastes on upland cotton, *Proc. Beltwide Cotton Conferences,* National Cotton Council, Memphis, TN, Vol. 3, 1993, p. 1354.
56. Edwards, J.H., Burt, E.C., Raper, R.L., and Hill, D.T., Recycling newsprint on agricultural land with the aid of poultry litter, *Compost Sci. Util.,* 1, 79, 1993.
57. Mitchell, C.C., Burmester, C.H., and Wood, C.W., Broiler litter as a source of N for cotton, *Proc. Beltwide Cotton Conferences,* National Cotton Council, Memphis, TN, Vol. 3, 1357, 1993.
58. McCaslin, B.D., Davis, J.G., Cihacek, L., and Schluter, L.A., Sorghum yield and soil analysis from sludge-amended calcareous iron-deficient soil, *Agron. J.*, 79, 204, 1987.
59. Schlegel, A.J., Effect of composted manure on soil chemical properties and nitrogen use by grain sorghum, *J. Prod. Agric.,* 5, 153, 1992.
60. Cabrera, F., Diaz, E., and Madrid, L., Effect of using urban compost as manure on soil contents of some nutrients and heavy metals, *J. Sci. Food Agric.,* 47, 159, 1989.
61. Day, A.D. and Cluff, C.B., Municipal wastewater increases crop yields, *Biocycle,* Jan./Feb., 48, 1985.
62. Odom, J.W. and Mullins, G.L., Evaluation of a brass recycling plant slag as a micronutrient fertilizer, *Agron. Abstr.,* p. 44, 1988.
63. Allievi, L., Marchesini, A., Salardi, C., Piano, V., and Ferrari, A., Plant quality and soil residual fertility six years after a compost treatment, *Bioresource Tech.*, 43, 85, 1993.
64. Buchanan, M. and Gliessman, S.R., How compost fertilization affects soil nitrogen and crop yield, *Biocycle,* Dec., 72, 1991.
65. Wong, M.H. and Lau, W.M., The effects of applications of phosphate, lime, EDTA, refuse compost, and pig manure on the Pb contents of crops, *Agric. Wastes,* 12, 61, 1985.
66. Giordano, P.M., Mays, D.A., and Behel, A.D. Jr., Soil temperature effects on uptake of cadmium and zinc by vegetables grown on sludge-amended soil, *J. Environ. Qual.*, 8, 233, 1979.
67. Schauer, P.S., Wright, W.R., and Pelchat, J., Sludge-borne heavy metal availability and uptake by vegetable crops under field conditions, *J. Environ. Qual.*, 9, 69, 1980.
68. Dowdy, R.H. and Larson, W.E., The availability of sludge-borne metals to various vegetable crops, *J. Environ. Qual.*, 4, 278, 1975.
69. Mondy, N.I., Naylor, L.M., and Phillips, J.C., Quality of potatoes grown in soils amended with sewage sludge, *J. Agric. Food Chem.*, 33, 229, 1985.
70. Gosselin, B., Naylor, L.M., and Mondy, N.I., Uptake of PCBs by potatoes grown on sludge-amended soils, *Am. Potato J.,* 63, 563, 1986.

71. Harrison, H.C., Carrot response to sludge application and bed type, *J. Am. Soc. Hort. Sci.,* 111(2), 211, 1986.
72. Harrison, H.C. and Staub, J.E., Effects of sludge, bed, and genotype on cucumber growth and elemental concentrations in fruit and peel, *J. Am. Soc. Hort. Sci.*, 111(2), 205, 1986.
73. Falahi-Ardakani, A., Corey, K.A., and Gouin, F.R., Influence of pH on cadmium and zinc concentrations of cucumber grown in sewage sludge, *HortScience*, 23(6), 1015, 1988.
74. Keefer, R.F., Singh, R.N., and Horvath, D.J., Chemical composition of vegetables grown on an agricultural soil amended with sewage sludges, *J. Environ. Qual.*, 15, 146, 1986.
75. Yuran, G.T. and Harrison, H.C., Effects of genotype and sewage sludge on cadmium concentration in lettuce leaf tissue, *J. Am. Soc. Hort. Sci.,* 111(4), 491, 1986.
76. Vecchio, F.A., Armbruster, G., and Lisk, D.J., Quality characteristics of New Yorker and Heinz 150 tomatoes grown in soil amended with a municipal sewage sludge, *J. Agric. Food Chem.*, 32, 364, 1984.
77. Hornick, S.B., Use of organic amendments to increase the productivity of sand and gravel spoils: Effect on yield and composition of sweet corn, *Am. J. Alt. Agric.,* 3(4), 156, 1988.
78. Hemphill, D.D., Jr., Volk, V.V., Sheets, P.J., and Wickliff, C., Lettuce and broccoli response and soil properties resulting from tannery waste applications, *J. Environ. Qual.*, 14, 159, 1985.
79. Wang, S.H., Lohr, V.I., and Coffey, D.L., Spent mushroom compost as a soil amendment for vegetables, *J. Am. Soc. Hort. Sci.,* 109(5), 598, 1984.
80. Hunter, A.H., Use of phosphogypsum fortified with other selected essential elements as a soil amendment on low cation exchange soils, Publ. 01–034–081, Fla. Inst. of Phosphate Research, Bartow, 1989.
81. Eck, H.V., Winter, S.R., and Smith, S.J., Sugarbeet yield and quality in relation to residual beef feedlot waste, *Agron. J.*, 82, 250, 1990.
82. Falahi-Ardakani, A., Bouwkamp, J.C., Gouin, F.R., and Chaney, R.L., Growth response and mineral uptake of vegetable transplants grown in a composted sewage sludge amended medium. I. Nutrient supplying power of the medium, *J. Environ. Hort.*, 5(3), 107, 1987.
83. Falahi-Ardakani, Gouin, F.R., Bouwkamp, J.C., and Chaney, R.L., Growth response and mineral uptake of vegetable transplants grown in a composted sewage sludge amended medium. II. Influenced by time of application of N and K, *J. Environ. Hort.*, 5(3), 112, 1987.
84. Falahi-Ardakani, A., Bouwkamp, J.C., Gouin, F.R., and Chaney, R.L., Growth response and mineral uptake of lettuce and tomato transplants grown in media amended with composted sewage sludge, *J. Environ. Hort.*, 6(4), 130, 1988.
85. Sterrett, S.B., Chaney, R.L., Reynolds, C.W., Schales, F.D., and Douglass, L.W., Transplant quality and metal concentrations in vegetable transplants grown in media containing sewage sludge compost, *HortScience*, 17(6), 920, 1982.
86. Sterrett, S.B., Reynolds, C.W., Schales, F.D., Chaney, R.L., and Douglass, L.W., Transplant quality, yield, and heavy-metal accumulation of tomato, muskmelon, and cabbage grown in media containing sewage sludge compost, *J. Am. Soc. Hort. Sci.,* 108(1), 36, 1983.

87. Sims, J.T. and Pill, W.G., Composted sewage sludge and poultry manure as growth media amendments for tomato transplant production, *Appl. Agric. Res.*, 2(3), 158, 1987.
88. Hemphill, D.D., Jr., Ticknor, R.L., and Flower, D.J., Growth response of annual transplants and physical and chemical properties of growing media as influenced by composted sewage sludge amended with organic and inorganic materials, *J. Environ. Hort.*, 2(4), 112, 1984.
89. King, L.D., Availability of nitrogen in municipal, industrial, and animal wastes, *J. Environ. Qual.*, 13, 609, 1984.
90. O'Keefe, B.E., Axley, J., and Meisinger, J.J., Evaluation of nitrogen availability indexes for a sludge compost amended soil, *J. Environ. Qual.*, 15, 121, 1986.
91. Fox, R.H., Roth, G.W., Iversen, K.V., and Piekielek, W.P., Soil and tissue nitrate tests compared for predicting soil nitrogen availability to corn, *Agron. J.*, 81, 971, 1989.
92. Magdoff, F.R., Ross, D., and Amadon, J., A soil test for nitrogen availability in corn, *Soil Sci. Soc. Am. J.*, 48, 1301, 1984.
93. US-EPA, Land application of municipal sludges, *U.S. Environmental Protection Agency Process Design Manual*, EPA, Washington, DC, 1983.
94. Huber, D.M. and Watson, R.D., Effect of organic amendment on soil-borne plant pathogens, *Phytopathology*, 60, 22, 1970.
95. Tsao, P.H. and Oster, J.J., Relation of ammonia and nitrous acid to suppression of Phytophthora in soils amended with nitrogenous organic substances, *Phytopathology*, 71, 53, 1981.
96. Tsao, P.H. and Zentmyer, G.A., Suppression of *Phytophthora cinnamomi* and *Ph. parasitica* in urea-amended soils, in *Soil-Borne Plant Pathogens*, Shippers, B. and Gams W., Eds., Academic Press, London, England, 1979, 191.
97. Tsao, P.H., Bhalla, H.S., and Zentmyer, G.A., Effect of certain organic amendments on survival and activity of *Phytophthora cinnamomi* in soil, *Proc. Am. Phytopathol. Soc.*, 2, 40, 1975.
98. Broadbent, P. and Baker, K.F., Soils suppressive to Phytophthora root rot in eastern Australia, in *Biology and Control of Soil-Borne Plant Pathogens*, Bruehl, G.W., Ed., Am. Phytopathol. Soc., St. Paul, MN, 1975, p. 152.
99. Gilpatrick, J.D., Effect of soil amendments upon inoculum survival and function in Phytophthora root rot of avocado, *Phytopathology*, 59, 979, 1969.
100. Henis, Y. and Chet, I., The effect of nitrogenous amendments on the germinability of sclerotia of *Sclerotium rolfsii* and on their accompanying microflora, *Phytopathology*, 58, 209, 1968.
101. Henis, Y. and Chet, I., Mode of action of ammonia on *Sclerotium rolfsii*, *Phytopathology*, 57, 425, 1967.
102. Canullo, G.H. and Rodriguez-Kabana, R., Evaluation of soil amendments for control of *Sclerotium rolfsii*, *Phytopathology*, 81, 696.
103. Canullo, G.H., Rodriguez-Kabana, R., and Kloepper, J.W., Changes in the populations of microorganisms associated with the application of soil amendments to control *Sclerotium rolfsii Sacc.*, *Plant Soil*, 144, 59, 1992.
104. Kokalis-Burelle, N., and Rodriguez-Kabana, R., Effects of pine bark extracts and pine bark powder on fungal pathogens, soil enzyme activity, and microbial populations, *Biological Control* (in review), 1994.

105. Hoitink, H.A.J., Inbar, Y., and Boehm, M.J., Status of compost-amended potting mixes naturally suppressive to soilborne diseases of floricultural crops, *Plant Dis.,* 75, 869, 1991.
106. Nelson, E.B., Kuter, G.A., and Hoitink, H.A.J., Effects of fungal antagonists and compost age on suppression of Rhizoctonia damping-off in container media amended with composted hardwood bark, *Phytopathology,* 73, 1457, 1983.
107. Nelson, E.B. and Hoitink, H.A.J., The role of microorganisms in the suppression of *Rhizoctonia solani* in container media amended with composted hardwood bark, *Phytopathology,* 73, 274, 1983.
108. Kuter, G.A., Nelson, E.B., Hoitink, H.A.J., and Madden, L.V., Fungal populations in container media amended with composted hardwood bark suppressive and conducive to Rizhoctonia damping-off, *Phytopathology,* 73, 1450, 1983.
109. Singh, R.S. and Sitaramaiah, K., Control of plant parasitic nematodes with organic soil amendments, PANS, 16, 287, 1970.
110. Sitaramaiah, K., Control of root-knot nematode with organic soil amendments, *Indian Farmer Digest,* 11(4), 19, 1978.
111. Mian, I.H. and Rodriguez-Kabana, R., Soil amendments with oil cakes and chicken litter for control of Meloidogyne arenaria, *Nematropica,* 12, 205, 1982.
112. Rodriguez-Kabana, R., Pinochet, J., and Calvert, C., Olive pomace for control of plant-parasitic nematodes, *Nematropica,* 22, 149, 1992.
113. Mian, I.H. and Rodgriguez-Kabana, R., Organic amendments with high tannin and phenolic contents for control of Meloidogyne arenaria in infested soil, *Nematropica,* 12, 221, 1982.
114. Godoy, G., Rodriguez-Kabana, R., Shelby, R.A., and Morgan-Jones, G., Chitin amendments for control of *Meloidogyne arenaria* in infested soil, II. Effects on microbial population, *Nematropica,* 12, 63, 1983.
115. Rodriguez-Kabana, R., Morgan-Jones, G., and Gintis, B.O., Effects of chitin amendments to soil on Heterodera glycines, microbial populations, and colonization of cysts by fungi, *Nematropica,* 14, 10, 1984.
116. Rodriguez-Kabana, R., Boube, D., and Young, R.W., Chitinous materials from blue crab for control of root-knot nematode: I. Effect of urea and enzymatic studies, *Nematropica,* 19, 53, 1989.
117. Rodriguez-Kabana, R., Boube, D., and Young, R.W., Chitinous materials from blue crab for control of root-knot nematode: II. Effect of soybean meal, *Nematropica,* 20, 153, 1989.
118. Culbreath, A.K., Rodriguez-Kabana, R., and Morgan-Jones, G., The use of hemicellulosic waste matter for reduction of phytotoxic effects of chitin and control of root-knot nematodes, *Nematropica,* 5, 49, 1985.
119. Huebner, R.A., Rodriguez-Kabana, R., and Patterson, R.M., Hemicellulosic waste and urea for control of plant parasitic nematodes: effect on soil enzyme activities, *Nematropica,* 13, 37, 1983.
120. Rodriguez-Kabana, R. and Morgan-Jones, G., Biological control of nematodes: soil amendments and microbial antagonists, *Plant Soil,* 100, 237, 1987.
121. Adriano, D.E., Page, A.L., Elseewi, A.A., Change, A.C., and Straughan, I., Utilization and disposal of fly ash and other coal residues in terrestrial ecosystems: a review, *J. Environ. Qual.,* 9, 333, 1980.
122. Sommers, L.E., Toxic metals in agricultural crops, in *Sludge-Health Risks of Land Application*, Bitton, G., Damron, B.L., Edds, G.T., and Davidson, J.M., Eds., Ann Arbor Science Publishers, Inc, Ann Arbor, 1980, chap. 5.

123. Moore, J.W. and Ramamoorthy, S., *Heavy Metals in Natural Waters*, Springer-Verlag, New York, 1984.
124. Sommers, L.E., Nelson, D.W., and Yost, K.J., Variable nature of chemical composition of sewage sludges, *J. Environ Qual.*, 5, 303, 1976.
125. Stephenson, A.H., McCaskey, T.A., and Ruffin, B.G., A survey of broiler litter composition and potential value as a nutrient resource, *Biological Wastes*, 34, 1, 1990.
126. Edwards, D.R. and Daniel, T.C., Environmental impacts of on-farm poultry waste disposal — a review, *Bioresource Technol.*, 41, 9, 1992.
127. King, L.D. and Giordano, P.M., Effect of sludges on heavy metals in soils and crops, in *Agricultural Use of Municipal and Industrial Sludges in the Southern United States*, King, L.D., Ed., Southern Cooperative Series Bulletin 314, North Carolina State Univ., Raleigh, 1986, chap. 3.
128. CAST, Application of sewage sludge to cropland: Appraisal of potential hazards of the heavy metals to plants and animals, U.S. EPA, MCD-33, 1976.
129. Ryan, J.A., Pahren, H.R., and Lucas, J.B., Controlling cadmium in the human food chain: a review and rationale based on health effects, *Environ. Res.*, 28, 251, 1982.
130. US-EPA, Standards for use or disposal of sewage sludge: final rule, *Fed. Reg.*, 58(32), 9248, 1993.
131. Mortvedt, J.J., Mays, D.A., Osborn, G., Uptake by wheat of cadmium and other heavy metal contaminants in phosphate fertilizers, *J. Environ. Qual.*, 10, 193, 1981.
132. Williams, C.H. and David, D.J. The accumulation in soil of cadmium residues from phosphate fertilizers and their effect on the cadmium content of plants, *Soil Sci.*, 121, 86, 1976.
133. Williams, C.H. and David, D.J., The effect of superphosphate on the cadmium content of soils and plants, *Aust. J. Soil Res.*, 11, 43, 1973.
134. Mulla, D.J., Page, A.L., and Ganje, T.J., Cadmium accumulation and bioavailability in soils from long-term phosphorus fertilization, *J. Environ. Qual.*, 9, 408, 1980.
135. Mortvedt, J.J., Cadmium levels in soils and plants from some long-term soil fertility experiments in the United States of America, *J. Environ. Qual.*, 16, 137, 1987.
136. Baker, D.E., et al., *Criteria and Recommendations for Land Application of Sludges in the Northeast*, Bull. 851, Pa. Agr. Expt. Sta., University Park, 1985.
137. Juste, C. and Mench, M., Long-term application of sewage sludge and its effects on metal uptake by crops, in *Biogeochemistry of Trace Metals*, Adriano, D.C., Ed., Lewis Publishers, Boca Raton, FL, 1992, chap. 6.
138. John, M.T. and van Laerhoven, C.J., Differential effects of cadmium on lettuce varieties, *Environ. Pollut.*, 10, 163, 1976.
139. Hinesly, T.D., Alexander, D.E., Redborg, K.E., and Ziegler, E.L., Differential accumulations of cadmium and zinc by corn hybrids grown on soil amended with sewage sludge, *Agron. J.*, 74, 469, 1982.
140. Hinesly, T.D., Alexander, D.E., Ziegler, E.L, and Barrett, G.L., Zinc and Cd accumulation by corn inbreds grown on sludge amended soil, *Agron. J.*, 70, 425, 1978.
141. Boggess, S.F., Willavize, S., and Koeppe, D.E., Differential response of soybean varieties to soil cadmium, *Agron. J.*, 70, 756, 1978.
142. Reddy, G.B., Cehng, C.N., and Dunn, S.J., Survival of *Rhizobium japonicum* in soil-sludge environment, *Soil Biol. Biochem.*, 13, 343, 1983.
143. Madaringa, G.M. and Angle, J.S., Sludge-borne salt effects on survival of *Bradyrhizobium japonicum*, *J. Environ. Qual.*, 21, 276, 1992.

144. Brookes, P.C., McGrath, S.P., and Helinen, C., Metal residues previously treated with sewage sludge and their effects on growth and nitrogen fixation by blue-green algae, *Soil Biol. Biochem.*, 18, 345, 1986.
145. Brookes, P.C., Heijnen, C.E., McGrath, S.P., and Vance, E.D., Soil microbial biomass estimates in soils contaminated with metals, *Soil Biol. Biochem.*, 18, 383, 1986.
146. Marschner, H., *Mineral Nutrition in Higher Plants*, Academic Press, London, 1986.
147. Punz, W.F. and Sieghardt, H., The response of roots of herbaceous plant species to heavy metals, *Environ. Exp. Botany,* 33(1), 85, 1993.
148. Breckle, S.W., Growth under stress — heavy metals, in *Plant Roots: The Hidden Half*, Waisel, Y., Eshel, A, and Kafkafi, U., Eds., Marcel Dekker, Inc., 1991, chap. 17.
149. Barcelo, J. and Poschenrieder, C.H., Plant water relations as affected by heavy metal stress: a review, *J. Plant Nutr.*, 13(1), 1, 1990.
150. Kahle, H., Response to roots of trees to heavy metals, *Environ. Exp. Botany,* 33(1), 99, 1993.
151. Baker, D.E., Trace metal interactions in relation to soil loading capacities, in *Criteria and Recommendations for Land application of Sludges in the Northeast*, Baker, D.E., et al., Bull. 851, Pa. Agr. Expt. Sta., University Park, 1985, p. 52–54.
152. Klein, W. and Weisgerber, I., PCBs and environmental contamination, *Environ. Qual. Safety*, 5, 237, 1976.
153. Dacre, J.C., Potential health hazards of toxic organic residues in sludge, in *Sludge-Health Risks of Land Application*, Bitton, G., Damron, B.L., Edds, G.T., and Davidson, J.M., Eds., Ann Arbor Science Publishers, Ann Arbor, 1980, chap 4.
154. Pahren, H.R., Lucas, J.B., Ryan, J.A., and Dotson, G.K., Health risks associated with land application of municipal sludge, *J. Water Poll. Control Fed.*, 51, 2588, 1979.
155. Babish, J.G., Health risks associated with the organic fraction of municipal sewage sludges, in *Criteria and Recommendations for Land Application of Sludges in the Northeast*, Baker, D.E. et al., Eds., Bull. 851, Pa. Agr. Exp. Sta., University Park, 1985, p. 54–61.
156. EPA, Integrated risk assessment for dioxins and furans from chlorine bleaching in pulp and paper mills. U.S. EPA Environmental Protection Agency, Office of Toxic Substances, Washington, D.C., EPA 560/5–90–011, 1990, 73 p.
157. Hrutfiord, B.F. and Negri, A.R., Dioxin sources and mechanisms during pulp bleaching, *Chemosphere*, 25(1–2), 53, 1992.
158. Nash, R.G., Plant uptake of insecticides, fungicides, and fumigants in soils, in *Pesticides in Soil and Water*, Guenzi, W.D., Ed., Soil Sci. Soc. Amer., Madison, 1974, chap. 11.
159. Pal, D., Weber, J.B., and Overcash, M.R., Fate of polychlorinated biphenyls (PCBs) in soil plant systems, *Residue Rev.*, 74, 45, 1980.
160. Moza, P., Scheunert, I., Klein, W., and Korte, F., Studies with 2,4′, 5-trichlorobiphenyl-^{14}C and 2,2′, 4,4′, 6-pentachlorobiphenyl-^{14} in carrots, sugar beets, and soil, *J. Agric. Food Chem.*, 27, 1120, 1979.
161. Iwata, Y, Genther, F.A., and Westlake, W.E., Uptake of a PCB (Aroclor 1254) from soil by carrots under field conditions, *Bull. Environ. Contam. Toxicol.,* 11, 523, 1974.
162. O'Connor, G.A., Eiceman, G.A., and Kiehl, D., Plant uptake of sludge-borne PCBs, *Agron. Abstr.,* p. 44, 1988.
163. Fries, G.F. and Marrow, G.S., Chlorobiphenyl movement from soil to soybean plants, *J. Agric. Food Chem.*, 29, 757, 1981.
164. Beall, M.L. and Nash, R.G., Organochlorine insecticide residues in soybean plant tops: root vs. vapor sorption, *Agron. J.*, 63, 460, 1971.

165. Weber, J.B. and Mrozek, E., Jr., Polychlorinated biphenyls: Phytotoxicity, absorption and translocation by plants and inactivation by activated carbon, *Bull. Environ. Contam. Toxicol.*, 23, 412, 1979.
166. Aranda, J.M., O'Connor, G.A., and Eiceman, G.A., Effects of sewage sludge on Di-(2-Ethylhexyl) Phthalate uptake by plants, *J. Environ. Qual.*, 18, 45, 1989.
167. Jacobs, L.W., Chou, S.F., and Tidje, J.M., Fate of polybriminated biphenyls (PBBs) in soils. Persistence and plant uptake, *J. Agric. Food Chem.*, 24(6), 1198, 1976.
168. Anderson, J.R., Pesticide effects on non-target soil microorganisms, in *Pesticide Microbiology*, Hill, I.R. and Wright, S.J.L., Eds., Academic Press, New York, 1978, chap 7.
169. Menzies, J.D., Pathogen considerations for land application of human and domestic animal wastes, in *Soils for Management and Utilization of Organic Wastes and Waste Waters*, Elliott, L.F. and Stevenson, F.J., Eds., Soil Sci. Soc. Am., Madison, 1977, chap. 22.
170. Sagik, B.P., Duboise, S.M., and Sorber, C.A., Health risks associated with microbial agents in municipal sludge, in *Sludge-Health Risks of Land Application*, Bitton, G., Damron, B.L., Edds, G.T., and Davidson, J.M., Eds., Ann Arbor Science Publishers, Ann Arbor, 1980, chap. 2.
171. Hunt, C.M., Pathogens in sewage sludges and sludge-amended soils, in *Criteria and Recommendations for Land Application of Sludges in the Northeast*, Baker, D.E. et al., Bull. 851, Pa. Agr. Exp. Sta., University Park, 1985, p. 62.
172. Bernstein, L., Salt tolerance of plants, U.S.DA Agric. Inform. Bull No. 283, 1964, 24 p.
173. U.S.D.A. Salinity Laboratory Staff, *Diagnosis and Improvement of Saline and Alkali Soils*, U.S.DA, Handbook 60, 1954.
174. Ellis, B.G., Analyses and their interpretation for wastewater application on agricultural land, in *Application of Sludges and Wastewaters on Agricultural Land*, Knezek, B.D. and Miller, R.H, Eds., Ohio Agric. Res. and Dev. Center Res. Bul. 1090, 1976, sec. 6.
175. Ayers, A.D., Seed germination as affected by soil moisture and salinity, *Agron. J.*, 44, 82, 1952.
176. El-Sharkawi, H.M. and Springuel, I.V., Germination of some crop plant seeds under salinity stress, *Seed Sci. Technol.*, 7, 27, 1979.
177. Maas, E.V. and Hoffman, G.J., Crop salt tolerance-current assessment, *J. Irrig. Drainage Div. ASCE,* 103, 115, 1977.
178. Stewart, B.A. and Meek, B.D., Soluble salt considerations with waste application, in *Soils for Management and Utilization of Organic Wastes and Waste Waters*, Elliott, L.F. and Stevenson, F.J., Eds., Soil Sci. Soc. Am., Madison, 1977, chap. 9.
179. Morrill, L.G. and Carlile, B.L., Considerations for the application of salt-containing wastes to agricultural lands, in *Agricultural Use of Municipal and Industrial Sludges in the Southern United States*, King, L.D., Ed., Southern Cooperative Series Bulletin 314, North Carolina State Univ., Raleigh, 1986, chap. 4.
180. Arman, A. and Seals, R.K, A preliminary assessment of utilization alternatives of phosphogypsum, in *Proc. Third Int. Symp. on Phosphogypsum,* Chang, W.F., Ed., Fla. Inst. Phosphate Research, Bartow, Publ. 01–060–083, 1990, p. 562–583.
181. May, A. and Sweeney, J.W., Assessment of environmental impacts associated with phosphogypsum in Florida, in *Proc. Int. Symp. on Phosphogypsum,* Fla. Inst. of Phosphate Research, Bartow, Publ. 01–001–017, 1980, p. 415.
182. Berish, C.W., Potential environmental hazards of phosphogypsum storage in central Florida, in *Proc. Third Int. Symp. on Phosphogypsum,* Chang, W.F., Ed., Fla. Inst. Phosphate Research, Bartow, Publ. 01–060–083, 1990, p. 1–29.

183. Lardinoye, M.H. and Weterings, K., Unexpected Ra-226 buildup in wet process phosphoric acid plants, *Health Phys.*, 42(4), 503, 1982.
184. Menzel, R.G., Uranium, radium, and thorium content in phosphate rocks and their possible radiation hazard, *J. Agric. Food Chem.*, 16, 231, 1968.
185. Federal Register, V. 43, No. 243 (Dec. 18, 1978), pp. 58957–58959, 1978.
186. Federal Register, V. 57, No. 107 (June. 3, 1992), pp. 23305–23320, 1992.
187. Appleyard, F.C., Gypsum industry in the United States: An overview including potential for use of chemical gypsum, in *Proc. Int. Symp. on Phosphogypsum,* Fla. Inst. of Phosphate Research, Bartow, Publ. 01–001–017, 1980, p. 57–85.
188. Mullins, G.L. and Mitchell, C.C., Jr., Use of phosphogypsum to increase yield and quality of annual forages, Fl. Inst. Phosphate Research, Bartow, Publ. 01–048–084, 1990.
189. Mullins, G.L. and Mitchell, C.C., Jr., Wheat forage response to tillage and sulfur applied as phosphogypsum, in *Proc. Third Int. Symp. on Phosphogypsum,* Chang, W.F., Ed., Fla. Inst. Phosphate Research, Bartow, Publ. 01–060–083, 1990, p. 362.
190. Arthur, W.J., III., Radionuclide concentrations in vegetation at a solid radioactive waste-disposal area in southeastern Idaho, *J. Environ. Qual.*, 11, 394, 1982.
191. Chaney, R.L. and Giordano, P.M., Microelements as related to plant deficiencies and toxicities, in *Soils for Management and Utilization of Organic Wastes and Waste Waters*, Elliott, L.F. and Stevenson, F.J., Eds., Am. Soc. Agron., Madison, 1977, p. 233–280.
192. Neilsen, G.H., Stevenson, D.S., Fitzpatrick, J.J., and Brownlee, C.H., Nutrition and yield of young apple trees irrigated with municipal waste water, *J. Am. Soc. Hort. Sci.,* 114(3), 377, 1989.
193. Meheriuk, M. and Neilsen, G.H., Fruit quality of McIntosh apples irrigated with well or municipal waste water, *Can. J. Plant Sci.*, 71, 1267, 1991.
194. Korcak, R.F., Fluidized bed material applied at disposal levels: effects on an apple orchard, *J. Environ. Qual.*, 17, 469, 1988.
195. Korcak, R.F., Renovation of a pear orchard site with sludge compost, *Comm. Soil Sci. Plant Anal.*, 17(11), 1159, 1986.
196. Myhre, D.L., Martin, H.W., and Nemec, S., Yield, ^{226}Ra concentration, and juice quality of oranges in groves treated with phosphogypsum and mined gypsum, in *Proc. Third Int. Symp. on Phosphogypsum,* Chang, W.F., Ed., Fla. Inst. Phosphate Research, Bartow, Publ. 01–060–083, 1990, p. 11.
197. Bianco, S., Ruggiero, C., Vitti, G.C., and Santos, P.R.R.S., Effects of phosphogypsum and potassium chloride on the nutritional status, production, and organoleptical quality of pineapple fruits, in *Proc. Third Int. Symp. on Phosphogypsum,* Chang, W.F., Ed., Fla. Inst. Phosphate Research, Bartow, Publ. 01–060–083, 1990, p. 348–361.
198. Edwards, J.H., Recycling newsprint in agriculture, *Biocycle*, Jul., 33, 1992.

CHAPTER 2

Livestock and Domestic Animals

Larry G. Hansen and David J. Schaeffer

TABLE OF CONTENTS

0-87371-860-7/95/$0.00+$.50
© 1995 by CRC Press, Inc.

I. INTRODUCTION

A. SOIL AND LEGAL LOADING RATES

Soil is produced by the interaction of climate and living organisms on rocks. These plus the mineralogical constituents of the parent materials determine primary soil composition. The constituents of soil are divided into several fractions.[1] Colloidal inorganic particles have a diameter less than 0.0002 cm and are separated by suspending them in water. These colloidal particles may constitute 10 to 80% of the dry weight of most soils and often give soils their color, texture, and ion-exchange properties. Organic matter in the soil (density about 0.5) usually is 1 to 40% of the dry weight and also contributes to soil color, texture, and ion exchange properties.

For agricultural soils, repeated tillage, cropping, and amendments exert considerable additional influence on the primary soil composition and characteristics. Consequently, areas with similar climate and geology have similar types of soils at a landscape (or "ecoregion") scale, although composition varies widely at sub-ecoregion scales and with depth; agricultural soils frequently differ markedly from adjacent undisturbed soils, especially with regard to organic matter. Table 1 presents ranges of selected elements present in typical soils, compares these elements to ranges found in typical sewage sludges and presents recent U.S. EPA ceiling concentrations for sewage sludges and cumulative loading rates for soil amendments with sewage sludges. Table 2 presents loading limits for selected organic chemicals.

B. AGRICULTURAL SOIL AMENDMENTS

A wide variety of materials are applied to agricultural soils to change physical and chemical characteristics. Many of these, including manures, sludges, compost,

Table 1 Elemental Composition of Soil,[1] Sewage Sludge,[2] and Application Limits for Soil Amendments[3]

	Dry soil mean (range) (mg/kg)	Mean levels in municipal sludge[a] (mg/kg)	Sludge pollutant concentration: Ceiling[b] (mg/kg)	Sludge pollutant concentration: Average[c] (mg/kg)	Cumulative pollutant loading rate (kg/ha)	Cumulative pollutant loading rate (ppm)[d]
As	6 (0.1–40)	10.1	75	41	41	20.5
Cd	0.06 (0.01–0.7)	16.5	85	39	39	19.5
Cr	100 (5–3000)	268	3000	1200	3000	1500
Cu	20 (2–100)	1028	4300	1500	1500	750
Pb	10 (2–200)	259	840	300	300	150
Hg	0.03 (0.01–0.3)	19.8	57	17	17	8.5
Mo	2 (0.2–5)	22.5	75	18	18	9
Ni	40 (10–1000)	148	420	420	420	210
Se	0.2 (0.01–2)	6.2	100	36	100	50
Zn	300 (60–2000)	1618	7500	2800	2800	1000

[a] The annual application rate for domestic septage applied to agricultural land, forest, or a reclamation shall not exceed N/0.0026, where N is the amount of nitrogen in pounds per acre per 365 day period needed by the crop or vegetation.

[b] All pollutants must be below the ceiling concentration.

[c] Monthly average concentrations cannot exceed the listed value.

[d] Concentration in top 15 cm of soil, assuming soil mass of 2×10^6 kg soil/ha.

Table 2 Concentration and Loading Limits for Industrial Organic Compounds in Sewage Sludge Used for Land Application

Compound	Agriculture U.S. EPA annual pollutant loading rate: kg/ha	Agriculture U.S. EPA annual pollutant loading rate: ppm[a]	Nonagriculture U.S. EPA annual pollutant loading rate: kg/ha[a]	Nonagriculture U.S. EPA annual pollutant loading rate: ppm
Aldrin/dieldrin (total)	0.016	0.72	0.0073	0.33
Benzo[a]pyrene	0.13	5.90	0.15	6.9
Chlordane	1.2	54.54	0.53	24
DDT/DDE/DDD (total)	0.0055	0.25	0.0024	0.11
Dimethyl nitrosamine	0.039	1.77	0.030	1.4
Heptachlor	0.073	3.18	0.69	1.5
Hexachlorobenzene	0.039	1.77	0.062	2.8
Hexachlorobutadiene	0.034	1.54	0.15	6.8
Lindane	4.6	209.09	2.02	92
Polychlorinated biphenyls	0.0056	0.25	0.0024	0.11
Toxaphene	0.048	2.18	0.021	0.97
Trichloroethylene	0.013	0.59	3.96	180

[a] Calculated based on 22 metric tons/dry ha application rate.

and ash mimic natural cycles involved in rejuvenating soils. The use of waste materials for maintaining and improving soil productivity presents a desirable and economically attractive alternative as well as assisting in the disposal of otherwise troublesome wastes.

Commercial fertilizers, insecticides, herbicides, and fungicides, however, are applied in specific situations under controlled conditions. These specialized mostly

Table 3 Representative Macronutrients and Other Major Elements Found in Digested Sewage Sludges from Various Municipalities in 1973–1974[4] with Some Values for the Early 1980s[7]

Element	Concentration (% dry weight)						
	Atlanta	Denver	Miami	Philadelphia		Seattle	
				1970s	1980s	1970s	1980s
Al	3.5	1.3	0.9	2.3	2.3	2.7	1.8
Cl	0.05	0.7	1.0	0.1	—	0.2	—
Fe	2.4	8.3	1.0	2.0	1.9	2.7	2.1
K	0.6	3.9	0.3	1.4	0.06	1.6	0.01
N	2.4	3.1	3.3	2.3	0.6	2.1	3.3
P	1.4	1.3	1.7	1.2	0.7	1.8	1.8
Ash	57.3	37.8	39.7	53.2	76.0	56.7	45.2

Table 4 Concentrations of Selected Trace Elements in Digested Sewage Sludges from Various Municipalities in 1973–1974

Element	Concentration (mg/kg dry weight)						
	Atlanta	Denver	Miami	Philadelphia		Seattle	
				1970s	1980s	1970s	1980s
As	4	14	10	16	7	30	11
Cd	104	46	150	192	12	64	73
Cr	1320	936	1430	2320	1290	1320	1511
Cu	1463	1370	1200	2680	907	1170	1261
Hg	7	4	16	5	3	8	10
Mn	267	224	32	95	529	350	241
Mo	6	18	37	8	10	2	10
Ni	169	562	453	432	305	153	265
Pb	1445	1011	1467	7627	475	2411	557
Sb	15	4	15	44	12	11	11
Se	2	4	3	3	4	3	3
Zn	2838	2860	1400	6890	863	1830	2344

synthetic additives have been well-reviewed and should present minimal health hazards if label recommendations are followed. They will be discussed only in relationship to the "organic" soil amendments.

The benefits of using waste materials as soil amendments justify the extensive investigation required to determine reasonable limits for the cumulative adventitious addition of potential toxicants and the close scrutiny required to assure that these limits are not exceeded. The more variable the amendment material, the more complex is the task of determining suitable application rates and maximum cumulative applications.

Municipal sewage sludges present perhaps the greatest challenge due to the great variability[3-6] in nutrient as well as potentially toxic elements (Tables 3 and 4). Wood ash is a prized and valuable soil amendment. Ashes from the incineration of mixed waste or coal have generally been determined to be potentially too toxic to justify their limited utility as soil amendments; however, the desirability of concentrating dispersed toxicants into a confined and abiotic medium is too great to ignore other non-agricultural uses for these materials. Ash from less intense and less controlled burning such as brush clearing and prairie restoration projects should be considered in the same context, depending on the nutrient history of the vegetation (e.g., grown in metal-enriched acid soils); however, heavy metal content is generally low relative to nutrients and occurrence is limited in most situations. Liming and acidification materials must be carefully monitored for potentially toxic or otherwise undesirable substances because of repeated heavy application rates; with liming materials, the desirability of co-application with sewage sludges makes it even more important to detect co-contaminants to avoid exceeding safe application rates. The greatest danger with composted materials is the perception that, since they are "natural" or "organic", they are ultrapure.

1. Municipal Sludges

Municipal sewage sludges present the most complex of the soil amendment scenarios and will be discussed in the greatest detail. Most municipalities are faced with the enormous problem of disposing of 47 lb (dry weight) of sewage sludge per year for every individual in the U.S. (5.4 million dry metric tons annually by various means).[3] Landfilling (34% co-disposal, about 10% other) and incineration (16%) are expensive and wasteful. Land application (agricultural 22%, other 11.4%) provides the opportunity for rational dispersion which makes use of the nutrient content and may permit ecological assimilation. The once common option for coastal cities of ocean dumping has been discontinued in most areas. Therefore, a massive movement to sell or donate municipal sewage sludges for large scale agricultural soil amendments and small scale home use began with little sound information regarding possible consequences. Some municipalities even went as far as to purchase large tracts of distressed lands wherein the benefits of sludge amendment should be readily apparent. Since each municipal sludge (Tables 3 and 4) is individualized by virtue of local industry, life styles, treatment particulars, and even water supply characteristics,[4] there were very mixed reviews regarding the safety/hazard of soil amendments.[3,6]

The intensive research during the 1970's provided valuable insights into the fates of nutrients and toxicants in ecosystems, but the inadequate understanding and the use of inappropriate sludges resulted in adversely-affected soils and a generally negative view of sewage sludges as a soil amendment.[6] The need for some form of disposal encouraged additional research and salesmanship which, in turn, strengthened the resolve of opponents and there was almost no truly objective information on which to base a realistic hazard evaluation. Many of the misinterpretations on both sides were fueled by the following limitations:[6]

1. Failure to recognize the complexity and extreme variability of individual digested sludges.
2. Limited directly applicable experimental data.
3. Reliance upon more manageable greenhouse studies rather than realistic field studies.

Table 5 Average Annual Concentrations of Selected Elements in Digested Sewage Sludges from Chicago Stickney and Calumet Wastewater Treatment Plants

Element	Average annual concentration (mg/kg dry weight)[a]				
	1970	1972	1974	1976	1992[b]
Cd	428	176	312	267	46
Cr	5038	1216	3008	3042	689
Cu	1913	743	1676	1764	450
Mn	322	473	430	556	390
Ni	418	97	347	446	76
Pb	1894	1036	931	917	200
Zn	8087	4324	5143	3792	1664
	Six-year mean ± S.D. concentration (mg/kg dry weight)				
As	19.8 ± 2.3				1
Hg	4.4 ± 2.0				1.2
Sb	1.3 ± 0.5				—
Se	5.7 ± 4.9				<18

[a] Calculated from Table 2 (p. 40) in Reference 5 by: 1000 × [Element loading (kg/ha)]/[annual mt/ha sludge solids] in order to provide a composite average of all sludges used in a given year.

[b] See Table 6.

From Hinesly, T.D., and Hansen, L.G., Progress Report (1971–1977) to Metropolitan Sanitary District of Greater Chicago, Agronomy Department and College of Veterinary Medicine, University of Illinois at Urbana-Champaign, 1979. With permission.

4. Attempts to simply add up the effects of single chemicals or simple mixtures (e.g., metal salts, technical PCBs) to predict the effects of complex sewage sludges, thus ignoring:
   a. The reduced bioavailability and antagonistic actions of many sludge components.
   b. Microflora, plant and animal "barriers" or attenuators.
5. Heavy reliance on results from studies which were purposely designed as worst-case situations (i.e., a very undesirable sludge grossly over-applied).
6. Insistence on interpretations within the limits of classical linear dose-response relationships.

A more complete view of sewage sludge as a net beneficial soil amendment during the next decade led to more rational limits and stationary targets of acceptability for which to aim (Tables 1 and 2).[3] For example, the University of Illinois conducted a large series of field and animal studies in the early 1970s with sewage sludges from the Metropolitan Sanitary District of Greater Chicago containing generally 100 to 400 ppm Cd (Table 5). Current concentrations of many toxicants are much lower due to input and process improvements (Tables 6 and 7).

Other organics commonly found in sewage sludges are those that occur in the waste water to begin with. In the 1970s, the most significant organics included naphthalene, phthalates, chlorobenzenes, chlorophenols, and polychlorinated biphe-

Table 6 Trends in the Concentrations of Selected Elements in Sludges from Two Sewage Treatment Plant Digesters[a] for the Metropolitan Sanitary District of Greater Chicago[8]

Element	Mean concentration (mg/kg) in composite drawoff							
	1988		1990		1991		1992	
	Stk	Cal	Stk	Cal	Stk	Cal	Stk	Cal
No[b]	23	23	19	19	23	23	21	21
Cd	98	77	68	61	47	72	41	52
Cr	1311	493	1018	313	904	338	1079	299
Cu	863	357	595	252	528	295	575	324
Hg	1.5	0.6	1.8	0.6	2.0	0.5	1.7	0.8
Mn	634	361	375	284	538	279	539	240
Ni	185	74	133	43	124	46	111	42
Pb	342	215	279	186	273	154	261	140
Zn	2127	2321	1827	2133	1643	1860	1835	1492

[a] Stickney (Stk) and Calumet (Cal) plants.

[b] Number of composite samples analyzed (generally 2/month).

Table 7 Selected Examples of Chlorinated Pesticides in Sewage Sludges

	Mean concentration (mg/kg dry weight)		
		From 1988[10]	
Pesticide	From 1970s[9]	Positive (%)	Mean of pos.
Aldrin	16.2 (n = 1)	3	0.02
Chlordane	16.0	0.5	0.02
Dieldrin	0.61	4	0.003
DDD	0.25	1	0.06
DDE	NR	1	0.06
DDT	0.35	2	0.03
Lindane	0.075	0.5	0.004
2,4-D	0.12	—	NR[a]

[a] Not reported.

nyls (PCBs).[6-11] Polychlorinated dibenzo-p-dioxins, especially 2,3,7,8-TCDD, and polycyclic aromatic hydrocarbons (PAHs) are of considerable concern because of the high toxicity of TCDD and the carcinogenic potential of PAHs. These toxicants are generally associated with waste incineration and fossil fuel burning. Sewage sludges would make negligible contributions in most cases (Table 8), and PAHs are reasonably well degraded even in sandy soils when sludge applications meet regulatory guidelines.[12] N-Nitrosamines are found in various municipal sludges up to about 0.05 mg/kg if selective analytical procedures are used.[7] Little significance is attached to these low levels.

2. Manures and Composts

Animal manures have long been used as soil amendments and provide valuable macronutrients (Table 9) as well as trace elements (Table 10). As with many soil

Table 8 Concentrations of Polycyclic Aromatic Hydrocarbons (PAHs) in Various Soils and Potential Soil Amendments

PAH	Median (mg/kg)			
	Soils		Fly Ash[13]	Sludge[9,12]
	Agricultural	Urban		
Acenaphthalene	0.006	NR[a]	1.75	NR
Anthracene	0.012	NR	0.25	NR
Benzo(a)pyrene	0.45	0.19	0.20	0.21
Benzo(e)pyrene	0.94	7.0	NR	0.05
Benzoperylene	0.066	23.4	0.95	0.12
Fluoranthene	0.021	83.1	3.25	0.91 (0.42)[b]
Naphthalene	NR	NR	4.78	(1.12)[b]
Phenanthrene	0.94	NR	3.80	(1.82)[b]
Pyrene	0.12	73.6	2.70	(0.93)[b]

[a] Not reported.

[b] n = one industrially contaminated sewage sludge.

Table 9 Macronutrients in Manures and Composts

	Feedlot (dry wt.)		Liquid manures[16]			Bagged cow Manure[4,7]	Composts[17]
	Kansas[16]	Texas[18]	Poultry	Swine[a]	Dairy	(dry wt.)	(dry wt.)[b]
				Percent			
Ca	0.78	1.30	0.77	0.32	0.47	2.6, 2.8	3.2
Fe	0.88	0.21	<0.1	<0.1	<0.1	1.6, 1.6	1.2
K	1.09	1.80	2.9	6.7	2.2	3.3, 1.4	0.9
Mg	0.39	0.50	0.2	0.4	0.3	0.7, 0.8	1.1
N	1.04	1.34	2.0	2.9	2.8	1.2, 1.2	NR[c]
Na	0.22	0.74	NR	NR	NR	0.5, 0.4	NR
P	0.42	1.22	0.2	0.6	0.2	0.9, 0.7	0.2
				mg/kg			
Cu	NR	NR	2	36	3	62, 55	17
Mn	NR	NR	16	76	25	286, 340	620
Zn	66	90	20	790	34	71, 298	160

[a] Averages of samples with 0.6 and 2.6% dry matter.

[b] From 44 samples (11 sites × 4 seasons); mature compost.

[c] Not reported.

amendments, this provides a waste disposal method as well as benefits to agricultural soils. The composition of manures varies widely with species, diet, and presence of bedding materials; an understanding of some of the predictable differences (e.g., poultry vs. ruminant) is essential to gain maximum benefit with minimal harm.

Heavy metals and organics are generally not present near the concentrations found in municipal sludges and ashes, but some unique problems may occur. Feeding and refeeding of animal manures may cause spiraling trace element accumulation.[14] Although transient in nature, the presence of certain biocides and other medications in animal manures may render them temporarily unsuitable for land disposal unless well-diluted with manures from untreated animals. For example, the popular

Table 10 Other Trace Elements in Manures and Composts

Element	mg/kg dry weight		
	Cow manure[4]	Composts[17]	
		Raw	Mature
Ba	268	70	80
Cd	0.8	NR[a]	NR
Co	5.9	6.1	3.4
Cr	56	24	29
Hg	0.2	0.07	0.07
Ni	29	16	18
Pb	16	19	27
V	43	30	40

[a] Not reported.

anthelminthic, ivermectin, is heavily used in horses. For 2 to 4 days after the 200 mg/kg dose, the feces are toxic to certain dipteran larvae responsible for decomposition. Fecal fouling of pastures has been reported, but not demonstrated in controlled studies.[15]

Municipal and other large scale composting of yard and orchard waste have become popular means of diverting large quantities of residue from landfills. Small scale composting is also heavily publicized as a recycling method and home-owner soil amendment. Composting is one of the more natural means of recycling nutrients and is considered a major factor in organic gardening. Regardless of the naturalness, compost is not as ultrapure as is popularly perceived. As with sludges, ashes, and manures, concentrations of potential toxicants in the final product depend partly on amounts in the raw materials. Several pesticides are retained in composts at relatively low, but significant, concentrations (Table 11). These pesticide concentrations are generally higher than those found in sewage sludges and much higher than those found in incinerator ashes.

Disease microbes and parasite ova are special problems in manures, minor problems in sewage sludges, and essentially non-existent in ashes. If composts include animal wastes, the potential for survival of certain disease organisms depends largely on the process and the heat generated. Spores from a few species of fungi and/or bacteria that cause plant diseases may survive even the most rigorous composting, so it is unlikely that mixed compost from residential sources will ever find use in large-scale commercial gardening or growing.

3. Liming Materials, Ash, and Residues from Burning

Agricultural lime is usually quarried and crushed along with limestone to be used for other purposes such as roads. Calcium and magnesium are by far the most abundant elements in all limestones. The content of other nutrient elements and trace elements varies considerably, even within the same formations, due to small veins and inclusions of other minerals.[16,19,20] Manganese, potassium and, especially, iron are frequently adequately high to be of nutrient value if bioavailable (Table 12).

Table 11 Average Pesticide Contents in Composts from 11 Illinois Large-Scale Facilities Sampled During All Four Seasons[17]

Pesticide	Raw compost		Mature compost	
	No. pos.	mg/kg	No. pos.	mg/kg
Chlordane	44	0.53	43	0.40
DDD	43	0.06	43	0.05
DDE	34	0.05	32	0.08
Dieldrin	44	0.01	44	0.008
Heptach. epoxide	44	0.02	44	0.015
Lindane	42	0.50	42	0.31
Methoxychlor	28	0.31	20	0.51
Chlorpyrifos	39	0.01	40	0.08
Diazinon	25	0.99	17	0.59
Fonofos	20	0.01	9	0.05
Malathion	36	0.31	27	0.17
Parathion	30	0.24	26	0.10
Carbaryl	38	22	41	11
Alachlor	40	0.75	39	0.30
Atrazine	35	4.6	25	3.0
2,4-D	44	36	44	56
2,4,5,-T	35	0.79	38	1.2
Dichlobenil	41	0.01	37	0.01
Metalachlor	37	1.1	34	0.97
Trifluralin	40	0.14	30	0.16

Table 12 Trace Elements in Limestones

Element	Concentration (mg/kg)				
	2 Deposits,[19] 70 samples	35 Quarries,[20] 92 samples	Averages		
			Ref. 1	Ref. 2	Ref. 3
As	ND[a]	NR[b]	—	—	—
B	2–10	1–200	4.5	18	NR
Ca	NR	NR	—	—	314000
Cd	NR	NR	—	—	0.7
Cu	3–16	4–70	6.3	18	10
Cr	NR	3–61	NR	11	6
Fe	2700–5100	3200–46000	3600	11300	2900
K	270–920	300–7500	490	1600	1300
Mg	NR	NR	—	—	50900
Mn	840–1700	400–3700	1200	1400	500
Mo	ND	ND-20	ND	1	NR
Na	70–500	10–3300	220	700	700
Ni	8–14	ND-70	9.7	15	20
P	NR	NR	—	—	600
Pb	2–5	6–100	2.6	26	55
Sr	110–320	240–810	164	490	NR
Zn	ND	ND-700	ND	40	113

[a] Not detected.

[b] Not reported.

Strontium is quite high due to its chemical similarity to calcium, but it is unlikely that adverse effects would result from the trace elements present in most agricultural limes by themselves. However, if an unusually high lead or zinc lime, for example,

Table 13 Ranges of Concentrations of Macroelements and Trace Elements in Various Ashes

Element	Concentration (mg/kg)			
	Fly ash[13]	Bottom ash[13]	Wood ash	
			Indust.[21]	Fireplace[a]
Al	5300–26600	5400–53400	15900–32000	3920
As	15–75	1–25	NR[b]–10	<15
Ca	13960–270000	5900–69500	73500–331400	234000
Cd	<5–2210	1–46	3–26	<1
Cr	21–1900	13–520	14–92	<4
Co	2–1670	3–62	NR–20	<1
Cu	187–2380	80–10700	40–140	<2
Fe	900–87000	1000–133500	3300–21000	2540
Hg	1–35	ND[c]–2	<0.1–<5	<0.005
K	11000–65800	920–13133	16600–41700	147000
Mg	2150–21000	880–10100	7100–224000	21200
Mn	171–8500	50–3130	3300–21000	166
Mo	9–700	29	3–123	<3
Na	9780–49500	1800–33300	1500–5400	<700
Ni	10–1966	9–226	12–50	23
P	2900–9300	3400–17800	3300–13600	16200
Pb	200–26600	110–5000	38–127	<8
Sb	139–760	NR	NR–<2	<7
Se	0.5–16	0.1–50	4–20	<16
Sn	300–12500	40–800	NR	<17
Zn	2800–152000	200–12400	200–794	170

[a] Two composite samples of ash from iron heat stove, mulberry (80%), and oak (20%) fuel (S.G. Wood and L.G. Hansen, unpublished results, 1993).

[b] Not reported.

[c] Not detected.

were applied to raise the pH of a soil receiving sewage sludge also containing lead or zinc near the ceiling concentration, overloading of these trace elements could occur.

Another means of providing alkalinity and trace elements is through the use of industrial wood ash. Wood residue is burned in large quantities for biomass-derived steam, mostly at saw mills, paper mills and some electrical generating plants. The ash produced (1.5 to 3.0 million tons per year) can amount to 100 tons per day at some sites and disposal can be quite expensive.[21] The nutritive value of wood ash is emphasized by the fact that it was at one time produced purposely for the production of potash fertilizer. Although much more alkaline than lime (pH generally 11 to 13), the total neutralizing value may vary from 35 to 116% to that of agricultural lime.[21] If the lime equivalent is determined, wood ash can be mixed with or substituted for lime with the added benefit of providing potassium, some phosphorus, and other nutrient elements (Table 13). The high zinc content may be a limitation due to potential phytotoxicity, but the pH considerations will usually limit application to safe zinc levels.[21]

Wood composition is highly dependent on tree species and the portions of the tree used as well as soil type and climate. With biomass-derived energy increasing, home owners and gardeners may already be well aware of the value of wood ash from fireplaces and burnt prunings. As a comparison to published values, the authors composited four grab samples of ash from a woodburning stove for analysis. This particular ash compared even more favorably to limestone than did the industrial ashes considered valuable soil amendments in the northeastern U.S. (Table 13).

Ash from coal and municipal waste incineration is generally much higher in heavy metals (Table 13) and organics (Table 8) and has a lower alkalinity than wood ash. These ashes are generally unsuitable for land application, but may certainly find other uses. The efficient removal of heavy metals from landfill leachate by wood ash[22] and the potential for engineering a concrete substitute from incinerator ash suggest that, rather than a disposal problem, incinerator ash may be useful for lining and covering landfills.

II. BENEFICIAL EFFECTS OF AMENDMENTS

A. REDUCING, RECOVERING, AND RECYCLING

As a nation and, increasingly, as a world we are slowly coming to the realization that resources and the capacity of the ecosystem to store and/or assimilate concentrated waste are not infinite. As one medium (e.g., wastewater) is cleaned for re-use, other media (e.g., sewage sludge) accept the substances removed. When biomass or fossil fuels are converted to energy, emissions and residues must be contained. The efficiencies gained by intensified and more highly localized food production create fecal waste amounts and other imbalances not readily assimilated by the environment. Increased understanding, rational balancing, careful planning, and reasonable control can manage burdensome and potentially devastating waste problems and use them for beneficial purposes.

A rational balance involves thoughtful consideration of the net disposition of toxic, potentiating, and antagonizing substances, not lamenting the presence of carbon dioxide and acetaldehyde in exhaled air!

Raw and processed animal wastes and quarried minerals have long been used as soil amendments. Appropriate ashes, composts, water treatment chemicals, and municipal sewage sludges can also benefit agricultural as well as severely distressed soils. By utilizing their nutrient and neutralizing potential, energy and resources can be saved and the waste products associated with manufacturing refined soil amendments can be avoided. Energy and resources spent trying to contain these potentially valuable wastes can also be saved. The immense amounts of manures, sludges and ashes generated each year make recycling more than a trivial consideration. For agriculture, reassimilation of wastes into the soil has another advantage: most landfills are sited in rural locations.

Successful recycling of appropriate wastes as soil amendments, however, requires a breadth and depth of knowledge and experience not easily achieved. For this we must depend on the generators, regulators, and other specialists such as extension agents to know the limitations of each product as well as the appropriate uses.

B. SOURCES OF NUTRIENTS AND ENHANCED NUTRITION

Each specific soil amendment is used for rather specific properties. Lime and ash are frequently required to economically raise or maintain the pH of the soil, but neither contributes significant available nitrogen. Limestone is also a good source of magnesium while ashes generally have high potassium and moderate phosphate levels. Manures, composts, and sewage sludges generally have more balanced macronutrients since they derive more directly from living organisms. They also provide important organic matter which increases microbial activity, tilth, and moisture holding capacity in the soil.

The main purpose of soil amendment is to increase or retain productivity. Refined soil amendments target specific deficiencies and, as such, are of rather limited composition. Trace elements and organic matter are frequently assumed to be provided by natural cycles; however, continual harvesting can deplete trace nutrients as well as macronutrients. Totally balanced and customized soil amendments would be considerably less cost effective than using a well-characterized and understood waste that provides such nutrients.

Nutrient elements are frequently higher in sludge-fertilized grains than in those fertilized with standard N-P-K formulations.[5,6] Common vegetables grown in sewage sludge-amended soils had consistently higher protein and lower fat levels than commercially fertilized controls.[23] Animals consuming crops fertilized with moderate levels of sewage sludge are often in more optimum states of trace mineral nutrition and frequently have a slight performance advantage over animals consuming N-P-K fertilized feedstuffs.[5,6,23,24] Products from these animals can enhance the trace mineral status of consuming animals as evidenced in improved hematopoesis.[25]

C. COMPLEXING AND/OR DESTRUCTION OF TOXICANTS

Organic matter and colloidal inorganic matter contribute the main source of ion exchange properties in soils. Soils with a high cation exchange capacity are better able to retain nutrients against leaching and retard the accumulation of potentially toxic metals in plants. The organic matter added in manures, composts, and sludges greatly increases cation exchange capacity, especially in distressed soils; however, the organic content reaches an equilibrium in a couple years because of the increased microbial activity while the cations continue to be added.[6]

Increasing organic carbon renders many pesticides less available to plants[26,27] and less subject to leaching.[28] The enhanced microbial activity can accelerate the destruction of organic compounds present in the soil, the soil amendment, or added as a specific biocide. Organic carbon from manure and sewage sludge seemed to accelerate the degradation of alachlor and atrazine; activated carbon was more effective at binding the herbicides, but inhibited degradation.[28] Even if the organic is bioactivated by the microbes, the reactive products bind in the microbes themselves or intervening plants, losing their potential for further harm, and do not persist in the food chain.[29] A disadvantage of organic carbon may be in reducing biocide efficacy, but the net effect or balance by protection from leaching, binding, and degradation has not been determined.

Anions such as inorganic phosphate can also greatly restrict the mobility and bioavailability of potentially toxic elements such as lead and aluminum. Other cations such as calcium, zinc, and iron can successfully retard the absorption of toxic heavy metals. The presence of these substances and organic matter in soil amendments are useful in protecting the food chain from heavy metals which may already be in certain soils.

D. ANTAGONISMS OF TOXIC ACTIONS

Not only do major constituents of recycled soil amendments such as calcium, zinc, and iron restrict the bioavailability of potentially toxic elements, but they also intercede directly in counteracting toxic actions. High dietary zinc induces intestinal metallothionien which helps to trap cadmium and mercury in the cells which are routinely sloughed, excreted, and replaced.[30] Zinc also competes effectively, protecting from cadmium, lead, and cobalt toxicity to the testes.[31,32]

Iron and copper are intimately involved in heme metabolism. Many toxic and potentially toxic substances such as lead, cadmium, aluminum, cobalt, zinc, and halogenated aromatic hydrocarbons interfere with heme synthesis and/or enhance heme destruction. The elevated copper and iron in most recycled soil amendments generally improve the iron/copper status of plants, animals consuming these plants, and animals consuming these animals as evidenced by tissue levels and red blood cell populations (see above and Section IV). There are, however, nearly as many instances (e.g., direct sludge feeding or exceeding recommended limits) wherein iron/copper status is compromised. In most animal studies, there has been a nonlinear dose relationship between exposure and iron/hemoglobin status; thus, these relationships frequently went undetected because of the attempts to apply linear statistical models.

In general, moderate (recommended) and balanced use of recycled soil amendments improves trace element nutrition and general health. This, in turn, has direct and indirect attenuating effects on toxicants from other sources as well as those present in the soil amendments. Excessive use of recycled soil amendments, improper attention to sound agricultural practices (e.g., pH, balanced nutrients), and use of highly imbalanced soil amendments can exacerbate nutritional deficiencies and/or enhance the susceptibilities of animals to the toxic effects of other substances.

E. RESEARCH AND UNDERSTANDING OF ENVIRONMENTAL MIXTURES

Trace element interactions in animal[34] and plant[35] nutrition and toxicity have been investigated to the point where many phenomena are well-understood.[36] The complexities and controversies surrounding agricultural soil amendments, particularly sewage sludges, intensified the interest and broadened the scope of investigation.[36]

The potential immensity of the impact on the environment and food chain forced regulatory attention which encouraged further research and provided direction.[3]

Through the myriad of research came an understanding of the behaviors of nutrients, heavy metals, and toxic organics in ecosystems. This understanding was necessarily developed within the context of mixtures and interactions, thus providing a basis for the more recent emphasis on hazard assessment of complex mixtures.[37]

III. AMENDMENT CHARACTERISTICS RESPONSIBLE FOR ADVERSE EFFECTS

A. COMPLEXITY: RESEARCH AND REGULATORY EFFORT AND EXPENSE

The benefits of using waste materials as soil amendments justify the extensive investigation required to determine reasonable limits for the cumulative adventitious addition of potential toxicants and the close scrutiny required to assure that these limits are not exceeded. The compositions of soil amendments of similar types may:

1. Vary considerably in many components (e.g., municipal sewage sludges, composts)
2. Contain predictably high residues of components such as PAHs and certain heavy metals (ashes)
3. Be relatively similar with occasional elevated concentrations of a few elements (agricultural lime)
4. Be formulated to contain consistent amounts of active ingredients (biocides, fertilizers)

The more variable the amendment material, the more complex is the task of determining suitable application rates and maximum cumulative applications. In spite of these complications, the wealth of data on municipal sewage sludge composition highlights most components of the greatest concern. Infrequent but potentially hazardous exceptions are generally monitored because of their known potential or are detected in routine stringent monitoring programs. Guidance is provided by U.S. EPA regulations[3] and a desire by the generators not to destroy the market for economic disposal of their products.

B. DISEASE TRANSMISSION

Ashes, lime, and formulated biocides and fertilizers are generally quite sterile. Manures almost always contain some pathological microbes or invertebrate eggs and proper sanitation and handling is essential. If animals receive antibiotic feed additives continuously or even sporadically, the very real possibility of shedding gram negative bacteria with transmissible multiple drug resistance must also be considered.[38] Sewage sludges may contain viable pathogens, but infection has rarely been demonstrated; most are apparently destroyed in the digestion processes. Composts certainly contain microbes, invertebrates, and fungi, most of which are destroyed if adequate heat is generated; however, certain plant pathogens can survive rather high temperatures and mixed municipal compost should be used with caution by the commercial grower.

C. TOXICANTS

Every substance is potentially toxic, and it is the dose that determines actual toxicity. With exponentially improved analytical sensitivity over the last 2 decades, extremely toxic substances are found in nearly every matrix examined. The press, self-serving public officials, and some toxicologists are guilty of using names and phrases the public has come to fear in order to generate concern or attention. The public has been conditioned to reject items or activities associated with terms such as "cancer-causing", "mutagen", "dioxin", and "cholesterol". Although the motives may be quite noble, improper and unqualified associations have made it difficult to effectively communicate actual risk to the more critical thinking and/or more highly educated population.

There are many highly toxic substances in the various natural, as well as formulated, soil amendments. These substances are also present in the soils, naturally or through agricultural, industrial, or other human activities such as burning fossil fuels.

For incinerator ash, the elements of greatest concern include lead, cadmium, and mercury. Organics in fly ash receiving the most attention are PAHs, PCDFs (chlorodibenzofurans), and PCDDs (chlorodibenzodioxins). For PCDFs and PCDDs, the highly toxic 2,3,7,8-substituted congeners generally constitute less than 0.4% of ash residues;[39] however, careful analysis of atmospheric emissions indicate that the small contribution of less potent congeners brings the TCDD toxic equivalency up to 0.64%.[40] It is likely that these ratios in emissions would also hold for fly ash from incineration of mixed waste, but the levels in wood ash would be much lower unless treated or otherwise exposed wood were burned. Any burning activity would contribute PAHs, but they are more likely to escape into the atmosphere during field and fireplace burning; industrial wood ash may contain higher amounts due to greater emission control requirements. Agricultural lime may occasionally contain high levels of lead or strontium, but the calcium content is adequate to counteract these elements. Soft lime, calcium hydroxide used for water treatment, is frequently used in place of agricultural lime. Mainly calcium carbonate and magnesium hydroxide, it may include other trace elements, depending on the nature of the water. Superphosphate fertilizers frequently contain relatively high trace concentrations of cadmium, but cadmium has not been shown to accumulate in long-term studies.[41]

In manures and composts, the substances of greatest concern are medications and feed additives (manures) or previously applied pesticides (composts). With manures, the medications would most likely occur in pulses and be diluted with manure containing lower levels; with composts, the pesticide content is somewhat seasonal, but still relatively constant. The main element of concern with manures is copper from nutritional supplements.

Digested municipal sewage sludges have been extensively characterized and there have been extensive animal as well as plant and soil studies conducted. By far, the

Table 14 Plant Accumulation of Important Elements in Soil Amendments Compared to Maximum Tolerable Dietary Intake by Domestic Animals

Element	Plant foliage[a]		Maximum levels chronically tolerated[a]			
	Normal	Phytotoxic	Cattle	Sheep	Swine	Chicken
Cd[b]	0.1–1	5–700	0.5	0.5	0.5	0.5
Co	0.01–0.3	25–100	10	10	10	10
Cu	3–20	25–40	100	25	250	300
Fe	30–300	—	1000	500	3000	1000
Mn	15–150	400–2000	1000	1000	400	2000
Mo	0.1–3	100	10	10	20	100
Ni	0.1–5	50–100	50	50[c]	100[c]	300[c]
Pb[b]	2–5	—	30	30	30	30
Se	0.1–2	100	2[c]	2[c]	2	2
Zn	15–150	500–1500	500	300	1000	1000

[a] mg/kg dry foliage or dry diet.

[b] Maximum tolerated levels based on limited absorption and human food residue considerations; most likely to be decreased for Pb.

[c] Interspecies extrapolation by National Research Council.

From Chaney, R.L., *Land Treatment of Hazardous Wastes,* Parr, J.F., Marsh, P.B., and Kla, J.M., Eds., Noyes Data Corp., Park Ridge, NJ, 1983, p. 426. With permission.

element of most concern has been cadmium. The organics monitored most closely have been polychlorinated biphenyls (PCBs). These and the contaminants in other soil amendments will be considered in more detail in Section IV, "Controlled Exposure Studies".

D. NUTRIENT IMBALANCE AND TISSUE COMPOSITION

The highest concentrations of minerals accumulated in plants are most often in the foliage rather than the grain. For examining the most troublesome elements in soil amendments, it is worthwhile to reiterate some of the relationships developed by Chaney[6,36] in comparing foliage accumulation, phytotoxicity (which limits accumulation), and maximum tolerated levels in animals (Table 14). It should be re-emphasized that amended soil ingestion, amendments adhering to forage, and dust from amended soil adhering to foliage can frequently supply more nutrient/toxicant than the foliage itself.[36]

Concentrations of cadmium, cobalt, molybdenum, and selenium potentially toxic to livestock could be accumulated in plant foliage if soil levels were high.

The two most important factors in attenuating plant uptake of toxic metals are:

1. Maintain the least water soluble form; for most heavy metals, the oxide
2. Carefully control the pH; dramatic decreases in cadmium uptake have been noted with increasing pH, but some amphoteric salts or oxides such as those of Pb, Cr, and Cu become markedly soluble again above pH 10 to 12

Because of their physiology, neutral or higher pHs to reduce gastric heavy metal absorption present a problem for monogastric animals. Soil lead is adequately bioavailable to cause neurological deficits,[42] but even sludge-amended soils are very minor contributors to dietary lead intake.[42] Nevertheless, other factors in soil amendments can help to reduce the bioavailability of heavy metals. High calcium, zinc,

Table 15 Examples of Nutrient and Trace Element Interactions Important for Components of Agricultural Soil Amendments

Element	Element(s) that interfere with: Absorption/utilization	Reciprocity
Cu	Ca, Cd, Fe, Mo, Zn, P	Fe, Mo, Zn
Fe	Cd, Cu, Mn, Zn, P	Cu, Mn, Zn, P
Mg	High K:(Ca + Mg) ratio (grass tetany)	
Mo	Cu, S	Cu
Se	Sludge and some soils low, Ash good source	
Zn	Ca, Cd, Ca + Phytate	Cd, Co

iron, and phosphate can notably impede the absorption of both lead and cadmium.[36] Sewage sludges and, especially, ashes are usually high in more than one of these factors; incinerator ashes generally contain high levels of lead, but the low lead and cadmium and the very high calcium, phosphorus, and iron in wood ash[21] favor its use as a co-amendment over limestone which generally has a low phosphorus content and may contain 100 mg/kg lead (Tables 14 and 15).

Manipulations of pH and cation exchange capacity to decrease toxic element uptake by plants may also exacerbate deficiencies in trace elements essential for animals but not for plants such as fluoride, nickel and, especially due to human deficiencies, chromium.[36] Although selenium and vanadium also fall in this category, vanadium is adequate in most diets. Selenium, with a very narrow range between deficiency and toxicity, is highly dependent on soil concentrations and readily accumulated by plants.

Arsenic, mercury, selenium, and tin are generally lower in digested sewage sludges than in ash due to biomethylation and volatilization. The biomethylated heavy metals, however, are very much more bioavailable and toxic. Thus, trapping them as oxides in the abiotic and alkaline matrix of ashes seems to be highly preferable over permitting less controlled biomethylation in sludges, sediments, and landfills.

Nutrient, trace element, and heavy metal interactions are extensive and complex; adverse effects from soil amendments may frequently be due to micronutrient imbalances rather than single toxic elements. Several interactions have been considered in relationship to sewage sludge soil amendments[6] and have been reviewed in detail for soil-plant-animal interactions.[36] Calcium and zinc protection from heavy metal, especially cadmium, toxicity are well known.[30-36] Iron, copper, zinc, and molybdenum interact extensively in plants as well as animals and deficiencies can also be exacerbated by cadmium (Table 15). Although often very subtle and not a significant health threat, copper and/or iron imbalances are frequently seen in animals over-exposed to sewage sludges, sludge-amended soils, or crops grown on sludge-amended soils.[5,6,24,36,43-45]

E. MORE SUBTLE HAZARDS

Organic contaminants also interact with each other to a great extent and also with trace elements. Most halogenated aromatics such as DDT, PCBs, cyclodienes, and halobenzenes induce one or more forms of cytochrome P-450 monoxygenases as well as certain conjugating enzymes. However, there are highly specific differences among isomeric forms of very similar chemicals. For example, P-450s 1A1 and 1A2 are induced by TCDD, coplanar PCBs, and some polyaromatic hydrocarbons, while PCBs with 2 to 3 ortho chlorines and phenobarbital induce mainly P-450s 2B1 and 2B2.[46] Other PCBs induce both 1A and 2B forms of P-450 and a few PCBs (or metabolites) can inhibit or depress P-450 activity.[46-48]

This divergent and frequently antagonistic activity also extends to estrogenicity, wherein TCDD and coplanar PCBs act as antiestrogens[49,50] and other PCBs as well as DDT and DDE are estrogenic.[50-52] In fact, if the P-450 1A1-inducing activity of PCB mixtures expected to be estrogenic is removed by activated charcoal, the estrogenicity is enhanced.[52]

Heavy metals, trace nutrients, and phytochemicals also interact strongly with the microsomal monooxygenases and steroid hormone function. Cytochromes P-450, as well as hemoglobin, are heme-containing enzymes. Any coincidental monooxygenase inducers present in experimental diets might be augmented by the increased iron and copper status associated with sewage sludge amendments; on the other hand, heme is also destroyed by heme oxygenase which is induced by various heavy metals and other trace elements. The bioavailability of aluminum, a potential neurotoxicant and major component of most sludges, soils, and ashes, is highly dependent on anions present; however, aluminum lactate is readily absorbed causing highly correlated increases in heme oxygenase, decreases in cytochrome P-450, and a mild anemia.[53]

Thus, effects on blood cells and monooxygenases might change in concert, depending on the total composition of the diet. Attempts to correlate monooxygenase induction with low PCB levels in a cruciferous vegetable[54] should have been examined in light of a co-author's previous discovery of monooxygenase induction by cruciferous vegetables which was probably augmented by increased trace mineral status.[55] Various unreconciled reports of stimulation and inhibition of the monooxygenases by metals and organics of concern in soil amendments and soil amendments, themselves, have been addressed previously.[6]

Many monooxygenase and other lipophile metabolizing enzyme alterations can have profound effects on lipophilic hormone homeostasis. Disruption of endocrine function, a subject of concern and study for decades, has been recently recognized as even more critically important and more widespread than previously believed. The many tissues involved in steroid hormone homeostasis as well as thyroid function all seem to be vulnerable to some degree to phytochemicals, heavy metals, and chlorinated hydrocarbons.[56] Most toxicants are much more potent prenatally and perinatally than when administered to adult animals, but effects such as demasculinization, cancer, and learning and behavioral deficits may not be manifest until later in life.

As if the potential interactions among several organics and elements affecting hormone metabolism, receptor activation, receptor blocking, changes in numbers of receptors, and competition for transport binding sites were not enough, the disruption of one endocrine system (e.g., thyroid) may have biphasic effects on other systems (brain, testes). For example, inducing hypothyroidism in newborn male rats results in transient growth depression and reduced testicle size; near adolescence, the testes overdevelop and greatly surpass the size (and sperm production) of testes from untreated litter mates.[57] Chemicals significant in agricultural soil amendments which cause endocrine disruption include: heavy metals (cadmium, lead, and mercury); herbicides such as alachlor, atrazine, 2,4-D, and nitrofen; fungicides such as benomyl, HCB, and tributyltin; insecticides, especially DDT and other chlorinated aromatics, and cyclodienes, carbaryl, parathion, and synthetic pyrethroids; and industrial chemicals such as, especially, TCDD and PCBs as well as phthalates, styrenes, and chlorophenols.[56]

Again, similar chemicals such as PCB congeners have very specific structural requirements for monooxygenase induction,[46] estrogenicity,[50] and thyrotoxicity.[58] In addition, some mechanisms, such as competition for transthyretin binding sites, can disrupt vitamin A as well as thyroid hormone homeostasis.[59] Feeding 20% sewage sludge to broiler chicks dramatically reduced liver vitamin A levels (8% of controls). The same sewage sludge reduced steer liver vitamin A to 40% of controls,[43] and a lyophilized liver/kidney mix from swine foraging on sludge-amended soils[5,24] contained only 30 to 60% of the vitamin A in the mix from matched controls.[25] Several mechanisms and interactions are possible,[43,59] but, as with the steroid hormones and monooxygenases, a combination of several interacting factors must be considered.

Unfortunately, when the early and more extensive investigations of soil amendment effects on animal health were conducted, nutrient element interactions were fairly well established but mechanisms for the relationships among metals and organics in biotransformation, hormone disruption and vitamin status were poorly understood and the subtle or delayed effects were not examined. Lack of adverse effect or even improved performace with moderate use of soil amendments, however, indicate extensive antagonism or neutralization of potential subtle toxic effects.

IV. ADVERSE EFFECTS IN CONTROLLED EXPOSURE STUDIES

Several rather extensive investigations of the effects on animals of relatively long-term exposure to soil amendments were conducted during the 1970s and early 1980s. Most of these dealt specifically with digested municipal sewage sludges, but the results can be extrapolated to other soil amendments by integrating the substances and interactions of concern. It is important to remember that the sewage sludges used in these studies contained considerably higher concentrations of potentially toxic substances than most modern municipal sludges (see Section I) and application rates were considerably greater than current regulations permit; thus, use of modern sludges under carefully regulated conditions would present attenuated hazards.

A. DIRECT FEEDING STUDIES

Some attempts were made to feed sewage sludge directly to animals[6, 60-62] but the food quality was poor and potentially toxic substances accumulated in some animal tissues. Generally, decreases in rate of gain in cattle and swine could not be attributed directly to the sludges,[61,62] but were more likely due to the lower nutrient quality of the sludge-containing diets.[62] On the other hand, inclusion of 20% dried cattle manure in feeder cattle rations resulted in improved performance over controls and no toxic effects could be demonstrated.[63]

In cattle and sheep on poor quality or dormant forages such as semi-desert rangeland, undigested sewage sludge supplements were as effective as cottonseed meal supplements in improving performance and reproduction.[64] Copper, iron, manganese, and zinc nutrition were improved and there were no increases in heavy metals or refractory organics above concentrations encountered in commerce. In sheep fed 62% sewage sludge (mixed with milo, molasses, urea, and bentonite) as the total diet for 3 months, body weight gains were not adversely affected, but the sheep suffered from excessive urination, glucosuria, and slightly enlarged livers.[64]

Extensive long-term feeding of sludge at 10 or 20% of the diet to swine resulted in reduced rate of gain, reduced feed efficiency, and reduced reproductive performance.[61] The adverse effects could readily be attributed to the reduced nutrient quality of the diets. Addition of cadmium to the diets at the levels found in the sludge produced anemia due to iron deficiency, but this effect was not seen in swine receiving sludge diets.[61]

Similarly, broiler chicks receiving the iron equivalent found in 6% sewage sludge-diluted diets exhibited significant depression of body weight; however, neither broilers nor layers exhibited any decreases in performance parameters when digested Chicago sewage sludge was included in their diets at 6 or 7%.[61] Interference in iron absorption and metabolism by cadmium[34] and other elements in soil amendments[6] are well known. The absolute demonstration of more severe effects from added cadmium salts (swine) or iron salts (chicks) stresses the importance of considering interactions in soil amendments.

B. SLUDGES IN SOILS AND ON FORAGE CROPS

Even though plants may act as barriers and buffers for many toxic substances in soil amendments (see Section III. D), some soil is ingested along with forage plants and dust as well as soil amendments may adhere to forages. Thus, direct ingestion of soil amendments cannot be avoided in foraging animals.

1. Cattle

In steers grazing on control pastures, pastures sprayed with 1.5 acre inches of liquid Pensacola (FL) sludge during grazing or with 3 acre inches before planting and an additional 3 acre inches during grazing, concentrations of copper and zinc were decreased in the liver. The increased lead and cadmium seen in livers and kidneys of cattle fed the sludge directly were not seen in cattle foraging on sludge amended pastures.[61]

After foraging for the full 6 years on winter wheat grown on a 6-year Denver sewage sludge disposal site, range cattle showed no health effects and no accumulation of heavy metals or chlorinated hydrocarbons.[62] As with the Florida study, the sludge-exposed cattle had lower concentrations of copper in liver tissue than would be expected, except for dietary deficiency, but the forages had slightly elevated copper concentrations.

In a long-term exposure study, initially designed to determine if cattle foraging on sewage sludge amended soils had increased parasitism, adverse health effects were not seen. With a few modifications, the cattle were matched with cattle rotated through control pastures and comparative accumulation of potential toxicants were determined.[44,65] Over a 5-year period, cows having foraged on sludge-amended pastures (n ≈ 60) were compared to previously matched cows having foraged on control pastures (n ≈ 30). For the treatment group, cadmium was about sevenfold higher in liver, fivefold in kidney, and threefold in heart muscle (but not significantly increased in diaphragm muscle). Lead was about threefold higher in bone and, in this case, copper was about twofold higher in livers.

Cadmium levels in calf livers and kidneys were not significantly elevated up to 120 days of age; at 300 days of age, cadmium in livers and kidneys were at about the same ratio as seen for the control and treated cows, although the total concentrations were lower. As with the cows, copper levels in livers and, occasionally, heart muscle were generally elevated over the controls which appeared to be somewhat copper deficient. Milk samples, after 2 years exposure, showed slight, but not significant due to high variability, increases in cadmium, copper, chromium, nickel, lead, and zinc, but not in mercury. The largest difference in blood metals for 80 experimental and 20 control samples was in lead (0.086 mg/kg vs. 0.022 mg/kg).[44] Although mean tissue lead concentrations for necropsied calves were generally not significantly elevated, rib biopsies of nursing calves (90 days old) averaged 2.3 mg/kg lead in five controls and 14.9 mg/kg lead in ten treated animals; bone lead had decreased in the same calves 1 year later to 2.5 and 4.0 mg/kg, respectively.

These cows[44] foraged for at least 5 years on pastures heavily fertilized with liquid-digested sludges from Chicago, which contained relatively high levels of heavy metals during the mid-1970s (Table 5). There was also accumulation of 0.2 to 0.7 mg/kg chlorinated aromatics reported as PCBs in visceral fat from treated cows and calves compared to non-detectable to 0.03 mg/kg in control cows and calves. Under these conditions, there were no discernible adverse health effects or effects on reproduction which could be correlated with sewage sludge exposure.[44,65]

2. *Swine*

Fitzgerald also conducted extensive parasite transmission studies with swine exposed to pens having received 0, 12, 26, 50, and 150 metric tons Chicago sewage sludge per hectare. Again, there were no adverse health effects from the sludge applications.[44,65] Residues of heavy metals were reported rather sporadically, the 150 mt/ha[44] group not being included in published reports.[65] If the 50 to 54 mt/ha group for which n = 2 is excluded, more definite trends can be seen from the data. Zinc and copper concentrations were sporadic in all tissues (Table 16) and chromium concentrations increased consistently only in muscle and bone. Cadmium concentrations increased with exposure in liver and kidney, but not in muscle, heart or bone.

Table 16 Concentrations of Total Heavy Metals in Sludge, Soil, Corn Grain, Swine Feces, and Swine Tissues

Overwintered sows (1976-1977)[5,24,66]							Corn-fed gilts (1974)[5,66]						Penned gilts[44,65]				
Group[a]	Cd	Fe	Mn	Zn	Cu	Pb	Group[a]	Cd	Fe	Mn	Zn	Cu	Group[a]	Cd	Fe	Zn	Cu
Sludge (total amount applied, kg/ha dry weight)							**total amount applied, kg/ha dry weight**						**mg/l in liquid sludge[b]**				
504	50.7	22476	203	2716	644	570	417	27.7	18189	151	2358	515	—	13.7	—	188.3	89.8
Soil (mg/kg dry weight)																	
0	0.68	19300	1617	82	18	38	0	0.9	23000	2320	78	23	0	2.1	20618	52.8	23.0
126	4.59	18500	1291	159	41	56	—	—	—	—	—	—	>0	31.5	19113	490.4	212.7
252	9.90	19500	1406	265	72	82	—	—	—	—	—	—	—	—	—	—	—
504	19.43	21400	1535	435	122	131	417	15.6	20000	1663	460	100	—	—	—	—	—
Corn grain mg/kg dry weight)[c]																	
—	—	—	—	—	—	—	0	0.10	14.2	7.40	26.8	1.99	—	—	—	—	—
—	—	—	—	—	—	—	417	0.56	18.5	6.98	51.9	2.22	—	—	—	—	—
Feces (mg/kg fresh weight; March, on plots)																	
0	1.16	—	250	209	19	7.9	—	—	—	—	—	—	—	—	—	—	—
126	1.23	—	—	—	—	—	—	—	—	—	—	—	—	—	—	—	—
252	2.33	—	176	315	36	11.4	—	—	—	—	—	—	—	—	—	—	—
504	9.01	—	253	336	64	41.7	—	—	—	—	—	—	—	—	—	—	—
Kidney (mg/kg fresh weight)																	
0	0.35	69	1.03	27	5.33	0.25	control	0.66	168	7.7	127	22.6	0	0.54	195	100	45
126	1.90	110	1.01	28	3.70	0.26	14 day	0.78	144	6.8	136	26.7	13	1.32	194	128	45
252	3.69	78	0.96	30	5.51	0.28	56 day	1.41	138	6.4	126	26.7	27	4.38	166	116	32
504	4.71	66	0.90	31	6.36	0.40	—	—	—	—	—	—	150	11.00	159	138	39
Muscle (mg/kg fresh weight)																	
0	<0.02	14	0.09	36	0.66	<0.18	control	0.09	40.0	0.49	60	2.9	0	0.03	109	93	4
126	<0.02	16	0.07	36	0.66	0.25	14 day	0.08	33.8	0.72	53	2.8	13	0.05	81	79	4
252	0.02	15	0.11	36	0.59	0.53	56 day	0.08	34.7	0.47	60	3.1	27	0.37	105	82	6
504	0.03	15	0.12	36	0.63	0.28	—	—	—	—	—	—	150	0.07	89	131	8
Liver (mg/kg fresh weight)																	
0	0.03	150	1.87	57	6.15	0.25	control	0.19	168	12.5	198	13.9	0	0.07	464	234	17
126	0.33	211	1.83	58	13.19	0.26	14 day	0.27	596	13.1	217	13.6	13	0.37	607	367	38
252	0.42	202	1.96	64	3.53	0.28	56 day	0.32	409	13.0	185	12.3	27	0.82	517	247	18
504	0.83	158	1.57	52	5.35	0.40	—	—	—	—	—	—	150	1.30	159	254	31

Table 16 (Continued) Concentrations of Total Heavy Metals in Sludge, Soil, Corn Grain, Swine Feces, and Swine Tissues

	Overwintered sows (1976-1977)[5,24,66]						Corn-fed gilts (1974)[5,66]						Penned gilts[44,65]				
Group[a]	Cd	Fe	Mn	Zn	Cu	Pb	Group[a]	Cd	Fe	Mn	Zn	Cu	Group[a]	Cd	Fe	Zn	Cu
Spleen (mg/kg fresh weight)																	
0	<0.02	400	0.21	27	0.74	0.29	control	0.15	614	1.36	117	4.6	—	—	—	—	—
126	<0.02	503	0.54	28	0.67	<0.18	14 day	0.14	507	1.13	111	4.0	—	—	—	—	—
252	0.04	595	0.28	27	0.83	0.37	56 day	0.22	562	1.33	114	4.8	—	—	—	—	—
504	0.10	507	0.23	26	1.48	0.23	—	—	—	—	—	—	150	—	—	—	—

[a] Total sludge application (mt/ha dry weight) or time on study.

[b] Analysis of anaerobically digested sludge applied to cropland during July, 1978.[65]

[c] Levels in sludge-corn and control-corn differed significantly ($P = 0.01$).

Iron concentrations declined with increasing sludge exposure in heart, kidney, and bone but increased in livers from swine exposed for 4 months to the intermediate sludge levels, falling back to control levels at the maximum sludge treatment (Table 16).[44] The same relationship between sludge application and tissue iron was seen in swine foraging intermittently on more heavily sludge-amended plots (Table 16)[24,66] and has been discussed relative to bioavailability of iron and interference from other elements in the sludges.[6]

In the intermittent swine foraging studies, weanling gilts were placed on corn plots which had been grossly over-amended with digested Chicago sewage sludge (totals of 0, 126, 252, and 504 metric tons of dry solids per hectare) over a 9-year period (Table 16).[5,24,66] The gilts were first on the plots for 4 months, during which time they rooted and consumed sludge-fertilized corn leaves, stalks, and cobs which had been returned to respective plots; they were then removed, bred, farrowed, and weaned during the subsequent growing season. The following fall, they were returned to the same plots following the corn harvest. Here they were bred so that they could be removed in the spring for farrowing.

Although a number of parameters were assessed in these heavily exposed swine, the most consistent were dose-dependent accumulation of cadmium in kidney, liver, and spleen (Table 16) and PCBs and DDE in backfat (Table 17).[24] Although dose-dependent, the maximum cadmium level was still less than 6 mg/kg in kidney. At these low levels, cadmium-metallothionein was not readily detectable using spectrophotometric methods. Nevertheless, graphite furnace atomic absorption spectroscopy of Sephadex fractions of soluble proteins indicated about an equal distribution of cadmium between albumen fractions and metallothionein in the livers; a similar relationship was seen in the soluble proteins from kidneys of sows from low and intermediate sludge plots, but adequate metallothionein was induced in maximum-treated sows to sequester all of the cadmium in the soluble kidney proteins.[24]

Interestingly, iron in most tissues, copper in some, blood hemoglobin content, sow weight gain, and reproductive performance showed non-linear relationships: the sows foraging on plots having received low and/or intermediate sewage-sludge treatments outperformed controls, but the maximum treatment underperformed controls.[5,6,24,66]

These results are comparable to 31- and 33-sow studies wherein the diets were diluted in digestible nutrients, metabolizable energy, and nitrogen retained with 10 and 20% dried Florida sewage sludge.[61,67,68] When these levels of sludge were fed directly to sows through two pregnancies, the 20% sludge group farrowed and weaned more live pigs than the control group, although the pigs were smaller at weaning.[68] Kidney cadmium in the sows was 4, 17, and 24 mg/kg for 0, 10, and 20% dried sludge, respectively.[61]

Although the exact nature of the above study is unclear, a subsequent study restricted feed intake except during lactation and confined the pigs to "outside dirt lots with no access to pasture".[61] In this study, the first litters showed little difference except that the control sows had 2 times more stillbirths than the 20% sludge group and the 20% sludge group had 40% more lost pigs during nursing than the control group. There was a tendency for the 20% sludge sows to weigh less at breeding and farrowing. Second generation gilts from the above sows were maintained on the

Table 17 Tissue Levels of Chlorinated Hydrocarbons in Fatty Tissues from Sows Overwintered Two Seasons on Sewage Sludge Amended Plots

8-Year sludge treatment	Back fat[a]	Marrow[a]
Control		
Lindane	3.0 ± 2	5 ± 2
Heptachlor epoxide	7.0 ± 9	4 ± 3
Dieldrin	11.0 ± 8	6 ± 4
p,p′-DDE	12.0 ± 3	10 ± 5
DDE (fat basis)	27.0 ± 5	—
PCB (fat basis)	36.0 ± 9	—
2,4,5,2′,4′,5′-PCB[b]	4.1 ± 3.2	—
2,3,4,2′,4′,5′-PCB	5.5 ± 4.0	—
126 t/ha		
LIndane	2.0 ± 1	3 ± 3
Heptachlor epoxide	2.0 ± 2	8 ± 6
Dieldrin	14.0 ± 10	11 ± 3
p,p′-DDE	21.0 ± 15	15 ± 3
DDE (fat basis)	41.0 ± 18	—
PCB (fat basis)	106.0 ± 64	—
2,4,5,2′,4′,5′-PCB	23.8 ± 13.8	—
2,3,4,2′,4′,5′-PCB	27.0 ± 17.6	—
252 t/ha		
Lindane	<1.0	5 ± 5
Heptachlor epoxide	5.0 ± 8	4 ± 4
Dieldrin	25.0 ± 35	8 ± 6
p,p′-DDE	30.0 ± 16	32 ± 22
DDE (fat basis)	51.0 ± 17	—
PCB (fat basis)	191.0 ± 97	—
2,4,5,2′,4′,5′-PCB	31.2 ± 16.7	—
2,3,4,2′,4′,5′-PCB	36.8 ± 19.5	—
504 t/ha		
Lindane	<1.0	5 ± 4
Heptachlor epoxide	<1.0	<1.0
Dieldrin	12.0 ± 12	7 ± 5
p,p′-DDE	101.0 ± 18	58 ± 4
DDE (fat basis)	97.0 ± 6	—
PCB (fat basis)	389.0 ± 118	—
2,4,5,2′,4′,5′-PCB	89.3 ± 23.4	—
2,3,4,2′,4′,5′-PCB	110.3 ± 34.9	—

[a] n = 3 or 4.

[b] Major congeners; see Hansen et al.[24] for other congeners.

From Hansen, L.G., Washko, P.W., Tuinstra, L.G.M.Th., Dorn, S.B., and Hinesly, T.D., *J. Agric. Food Chem.*, 29, 1012, 1981. With permission.

respective diets through two breeding cycles. By the second litter, the 20% sludge-fed sows were significantly smaller than the control or 10% sludge-fed sows.[61] The 10% sludge-fed sows had fewer stillborn pigs and live pigs were slightly larger than control and 20% sows; however, more pigs were lost during nursing.

In the swine foraging study, all sludges applied exceeded the current ceiling concentrations of cadmium and the soil concentrations of cadmium were at the cumulative loading limit (Table 1) prior to the second season application. In addition, the pigs consumed nearly 100% of the sludge-grown leaves, stalks, and cobs returned to the respective plots. Thus, the swine foraging study [5,24,66] was a sound test of current limits. At the two intermediate application rates, the pigs outperformed the controls even though diets were complete as recommended for each stage of growth. In the sludge feeding studies, when pigs were allowed to supplement diets diluted with dried sludge,[61,68] they outperformed controls. When sludge-diluted diets were restricted,[61,67] they underperformed controls in some aspects. An unambiguous measure of toxicity would be fetotoxicity; however, in none of the above cases did intermediate plot treatment levels or 20% sludge-diluted diets result in increased fetal death or stillborn full-term piglets; in fact, all treatment levels decreased stillborns when compared to controls and lower treatment levels, except in the second litter of the maximum plot treated sows. Therefore, if guidelines for sludge application are followed and modern sludges are used, it appears that there would be little hazard involved in allowing swine to forage on these fields. Potential hazards from tissue residues (Table 16) will be discussed later.

Complete blood cell counts and hemoglobin were determined in the foraging sows during gestation and within 3 days of weaning their second litter. The stress of farrowing and nursing resulted in red blood cell (RBC) decreases in all animals, but the decrease was most pronounced in the controls and least pronounced in the sows from the maximum sludge-treated plots.[5] The sows from plots treated with 1/4th the maximum sludge amount, previously shown to be in the most optimum state of iron and copper nutrition,[5,66] also had the greatest numbers of RBC, highest hemoglobin levels, and the greatest packed cell volume (PCV, hematocrit); their piglets also had the highest mean PCV by litter.[5] The control sows were invariably lowest in these three parameters and piglet PCV, even though the diets were at recommended nutritional levels and the piglets had received iron injections during nursing. In addition, the control sows had a tendency toward larger RBCs (mean corpuscular volume), indicating response to a minimal anemia.

White blood cell counts were well within normal ranges except in three sows (one each from 0, 1/4 and 1/2 groups) believed to have undiagnosed microbial disease.

There were no treatment-related differences in major organ weights of sows or piglets from the swine foraging study; however, individual animals which had the highest hemoglobin levels had the lowest heart weight relative to body weight, further suggesting that the elevated hemoglobin was not a transient phenomenon.

Gross observations and histopathology of several tissues revealed some degree of congestion in all sows and their age was reflected in other microscopic changes. All kidneys had glomerular disease, but the maximum treated sows had subtle but greater kidney involvement (microscopic pathology was totally "blind").[5] Ten different clinical chemistry parameters were measured when the pigs as gilts were first placed on the plots, just prior to removing from the plots during the first season, and when the animals were terminated. There were minor age-related changes, but values were almost identical for all treatment groups (e.g., increasing total serum protein, decreasing serum phosphate, glucose, and alkaline phosphatase).[5] Thus, there was no functional evidence of kidney damage in spite of the slightly greater mild renal disease in the maximum-treated sows.

3. Dogs and Cats

Livestock are generally restrained and restricted from access to areas where agricultural soil amendments are applied, especially biocides. Other domestic animals with minimal restrictions to movement on most farms, may readily contact freshly applied amendments. Cats access toxicants through grooming and hunting as well as direct ingestion, but are more selective than dogs in what they consume. Dogs consume surprisingly large amounts of forage, especially fine grasses, and may suffer related toxicities. Herbicide poisoning, especially by 2,4-dichlorophenoxy acetic acid (2,4-D), is not rare. Nevertheless, dogs penned on freshly treated (2,4-D) grass plots failed to develop any signs of poisoning, even though stomach contents revealed massive grass consumption.[69] Pets are more likely to become intoxicated from access to concentrates and discarded containers than from walking, hunting, and foraging on amended fields.

C. FORAGES AND GRAINS GROWN ON AMENDED SOILS

Plants tend to attenuate soil extremes in potentially toxic substances and other imbalances; the alternative is poor growth or death (see Section III, Table 14). Nevertheless, some toxicants have low phytotoxicity and greater potential for harm to animals. Grains, the means of species survival, are generally more protected from accumulations of toxic elements than are leaves and stalks; therefore, harvested forage crops generally have a greater potential for toxicant accumulation than do grains and fruits. Forages and grains grown on soils amended with sewage sludges and ash have been investigated for adverse effects to animals. Again, specific biocides have been carefully studied and should present no problems if label directions are followed and animals are not permitted to contact concentrates.

1. Cattle

Most studies with cattle related to soil amendments have been free-foraging (see Section IV. A). The lack of notable adverse effects in these studies indicates that the lower levels of toxicants in sewage sludge-fertilized grains and forages than in the sludge, itself, should present little hazard. Heavy metals (e.g., cadmium, lead) and refractory organics accumulate in cattle as a function of diet and age and surveyed "control" cattle indicate a propensity for copper deficiency.[62] These conditions could be exacerbated by unbalanced element contents of feeds. "Background" concentrations of cadmium and lead in the livers and kidneys of cattle seem to be higher than those in swine and poultry, but the subsequent accumulation from added dietary sources is lower than in swine and poultry.[62,70]

2. *Sheep and Goats*

a. *Fly Ash*

Fly ash precipitated from coal and municipal waste incinerators generally contains high levels of many heavy metals and some toxic organics, especially PAHs (Table 13). Although not considered suitable as a soil amendment, white sweet clover volunteering on a deep pile of fly ash was fed as 23.5% of a dry-pelleted ration to lambs and pregnant goats for up to 173 days with no demonstrable adverse effects.[71] There were, however, high concentrations of selenium in several tissues and elevated molybdenum, strontium, bromine, and rubidium in some tissues. Since wood ash contains selenium in levels comparable to those in fly ash (Table 13), it would be prudent to avoid applying ash as an amendment in high-selenium soils; however, even high-selenium soils would be unlikely to compete with the pure fly ash on which the clover was grown. The lack of toxicity in this very extreme situation further suggests that wood ash is a safe soil amendment when used under reasonable conditions.[21]

b. *Sewage Sludge*

Also grown on the pure amendment rather than amended soil, cabbage grown on 1.2 m of sewage sludge was fed as 30% of the diet to sheep.[72] Some increase in liver weight and ultrastructural changes in liver endoplasmic reticulum and mitochondria were suggestive of PCB effects, but the cabbage PCB content was less than 0.5 mg/kg so that the changes observed were probably due to high levels of trace elements such as cadmium and copper or the enhancement of microsomal induction by the cabbage itself.[6] Corn silage grown on heavily sludge-amended soil resulted in cadmium accumulation, but no gross adverse effects were observed when fed to sheep for 274 days;[73] however, there were, again, ultrastructural changes in the liver suggestive of toxicant injury.

With goats and sheep fed sludge-fertilized silage for 3 years, there were no effects on performance.[45,74,75] Liver weights for lambs fed the sludge fertilized silage tended to be greater than for lambs fed control silage; in goats fed the sludge-fertilized silage, there was a tendency for absolute liver and kidney weights to be less than those of the controls. Liver and kidney cadmium increased and copper and iron decreased with increasing sludge applications. Copper and iron levels in the heart muscle tended to increase with increasing sludge applications and the PCVs and blood hemoglobin were, again, greatest in the highest treatment group, but the relationship was quadratic rather than linear.

c. Manures

Interrelationships among copper, molybdenum, and sulfate and the sensitivity of sheep to copper toxicity is generally recognized.[76] Elevated copper in sewage sludges and, especially, wastes from animals receiving copper supplements can be anticipated to produce copper toxicity in sheep following chronic exposure; however, antagonistic elements (Table 15) which are also high in these potential soil amendments are generally protective[77] and anticipated toxicoses are generally not seen.[36] Dried poultry manure included at 30 to 60% of the diet causes severe ascites and other effects not related to copper toxicity.[78] On the other hand, pastures fertilized with chicken litter[79] or swine manure[80] caused typical severe copper poisoning in sheep foraging on these pastures. Since sewage sludge feeding and amendments frequently cause decreased copper levels, it can be assumed that the antagonistic elements protect sheep from copper; however, the situation with direct feeding of manures or foraging on fields heavily fertilized with manures appears to be somewhat different.

3. Swine

Most feeding studies in swine involved direct feeding of sewage sludge or, taking advantage of their rooting and foraging, penned studies on sludge-amended soils. A feeding study in which a balanced ration was prepared from sludge-fertilized corn was thorough, but of relatively short duration.[81] Corn harvested from plots having received over 400 mt/ha Chicago sewage sludge was found to contain significantly higher levels of nutrient as well as trace elements when compared to corn harvested from control plots (Table 16). This resulted in a higher protein content (17.7 vs. 16.5%) for the diet formulated from the sludge-fertilized corn.

Three groups of three weanling gilts were fed sludge-fertilized diets for 0, 14, or 56 days and all were terminated on day 56. The gilts were not treated for parasites (to avoid possible interactions with potential sludge toxicants) and they were limit-fed. Under these conditions, the 56-day SF group (220.5% $\pm$ 24.7% gain) grew slightly better than control or 14-day SF pigs (204 and 206%, respectively). In addition, all pigs had red blood cell parameters indicating a mild microcytic anemia from parasitism, but the controls had significantly lower mean corpuscular volume and mean corpuscular hemoglobin, indicating a more intense response.[5,66,81] There were no consistent differences in body fat content, electrocardiograms, electroencephalograms, major organ weights, gross pathology, or histopathology. The 56-day SF group had slightly larger livers and significantly greater hepatic microsomal protein levels; the microsomal oxidase activity for this group was incorrectly reported as elevated,[81] but was actually significantly lower due to the elevated protein.[5,66] Total serum protein and phosphate were higher in both SF groups at the end of sludge-corn feeding (14 and 56 days), but the 14-day group returned to control values by 56 days. At termination, the 56-day SF group also had significantly higher serum glucose and blood urea nitrogen values than the other two groups. All of these changes can be related to the elevated nutrient content in the sludge-fertilized corn.

There were significant increases in cadmium concentrations in the livers and kidneys of swine receiving sludge-fertilized corn (Table 16).

Rats were fed a similar diet with 15% sludge fertilized soybeans as well as 79% SF corn for 20 weeks. The only effect seen was increased liver and kidney cadmium at 20 weeks, but not at 6 weeks.[5,25]

4. Poultry

An early study[82] compared the effects on penned pheasants of feeding corn grain fertilized with zero or three progressive levels of sewage sludge. The sludge-fertilized corn had significant progressive increases in cadmium, zinc, nickel, potassium, and phosphorus; manganese was significantly lower in the sludge-fertilized corn. The pheasants were fed for 100 days and those receiving the highest sludge corn recovered most rapidly from initial weight loss due to confinement. Subsequently, there was no demonstrable difference in weight gain or feed consumption. Cadmium accumulated in liver, kidney, and duodenum with increasing dose, but percent retention decreased as the dietary level increased. Copper also tended to increase in the kidneys and iron decreased in livers with increasing dose, but the differences were not significant; as with previously mentioned studies, copper and iron increased in several tissues at low sludge-fertilization rates and returned toward control levels at the highest application.

Neither broiler chicks nor laying hens were adversely affected by including 50 to 100% sludge-fertilized corn in their diets.[61] In one replicate, broiler feed efficiency was reduced by 100% sludge-fertilized corn for 21 days, while the same level resulted in increased weight gain in laying hens after 112 days. No changes in the mineral content of blood, liver, kidney, or muscle were observed.

In another study, laying hens were fed starter, developer, and layer rations formulated from sludge-fertilized corn and soybeans for their entire productive life (7 days through 560 days).[83] Corn and soybean crosses were selected for their ability to accumulate cadmium, grown on sludge-fertilized strip mine soils, harvested, and processed into ground corn and soybean meal to be included in the appropriate medium (0.57 ± 0.11 mg/kg cadmium) or high (0.97 ± 0.14 mg/kg cadmium) ration. The low (0.09 ± 0.05 mg/kg cadmium) rations were formulated from commercial corn and soybean meal. All sludge-fertilized rations contained comparable concentrations of zinc which was higher (37, 38, and 50 mg/kg for starter, developer, and layer rations) than those formulated from commercial corn and soybean meal (25, 25, and 32 mg/kg). Copper was slightly elevated in the sludge-fertilized rations, while iron and manganese were highest in the medium cadmium starter and developer rations, respectively.

There were no differences in rate of gain, feed consumption, or rate of egg lay among the groups, although age differences were seen in all groups.[83] There was no difference among groups in the cadmium content of egg shells, egg whites, or egg yolks during the study.

Table 18 Hepatic[85] and Hematological[86] Parameters in Laying Hens that may be Associated with Iron-Copper Imbalances

Age and dietary cadmium	(n)	Liver (% body)	Fat (% liver)	Liver microsomes Protein[a]	Liver microsomes P-450[b]	Hematology PCV(%)[c]	Hematology Hb[d]	Liver iron (mg/kg)
8 Weeks								
Low	(16)	2.1	4.8	28.2	0.28	32.3	—	444
Medium	(15)	2.2	4.4	24.5	0.24	32.0	—	442
High	(13)	2.4	4.5	22.0	0.18	31.4	—	419
20 Weeks								
Low	(8)	1.8	12.3	10.6	0.19	32.4	—	788
Medium	(8)	2.2	21.6	9.4	0.18	31.1	—	749
High	(8)	2.0	13.9	10.6	0.14	32.5	—	667
50 Weeks								
Low	(10)	2.3	9.9	15.9	0.24	33.1	8.5	337
Medium	(10)	2.7	12.0	17.1	0.18	29.9	8.3	303
High	(10)	2.8	12.8	16.0	0.22	28.0	8.2	257
72 Weeks								
Low	(10)	2.5	13.9	17.7	0.23	31.2	9.6	397
Medium	(10)	2.6	18.1	17.0	0.19	30.0	8.2	374
High	(10)	2.6	15.3	15.4	0.19	30.4	8.9	452
80 Weeks								
Low	(10)	2.0	9.6	15.6	0.20	29.7	9.1	740
Medium	(10)	2.3	11.0	16.5	0.14	28.4	8.5	567
High	(10)	2.2	12.5	16.9	0.14	27.6	8.8	500

[a] Mg microsomal protein/g fresh weight liver.

[b] nMoles cytochrome P-450/mg microsomal protein.

[c] Packed cell volume ("Hematocrit").

[d] Hemoglobin (g/dl).

Four replicate groups of 25 birds were included for each dietary level and random birds were sacrificed from each replicate group at 8, 20, 50, 72, and 80 weeks. Gross observation and microscopic pathology were sporadic and reflected age changes rather than dosage differences.[84] Livers and gizzards tended to be slightly larger (relative to body weight) in the sludge-fertilized groups at weeks 8, 20, and 50; kidneys were slightly larger in the medium and high cadmium groups at week 8, but were identical (± 0.01% body weight) for the rest of the study.[85] Heart weights relative to body weight were slightly higher in the high cadmium group at 20 weeks.

There were no significant patterns in serum clinical chemistry.[86]

Cadmium accumulated in a dose- and time-dependent manner in gizzard, kidney, and liver (Table 18).[83] Cadmium increases in the alimentary canal (crop, proventriculus, duodenum), spleen, and lung were dose-dependent after 50 weeks. By 80 weeks, there were also dose-dependent cadmium increases in femur bone, heart, leg muscle, and pancreas, but not in brain, breast muscle, or feathers.[83]

Copper and iron were significantly lower in the gizzards of 8-week birds, but at 50 weeks copper was higher in these gizzards. Copper was also significantly higher (dose dependent) in crop and duodenum and iron in crop at 50 weeks. At 8, 20, and 80 weeks, iron tended to be higher in kidneys and lower in livers of birds receiving sludge-fertilized diets, but the tendency was weak at 50 weeks and weakly reversed at 72 weeks. Iron in the lung was slightly elevated by sludge exposure at 50 weeks and significantly elevated in the lungs of the aged birds at 80 weeks.

Iron and copper fluctuations are frequently associated with sewage sludge exposure[24,43,66] and may be reflected in hematology and heme-containing monooxygenases. Relative liver weights in confined laying hens are frequently decreased by microsomal-inducing agents because of greater fat catabolism.[87] The minor differences in liver weights in the sludge-feeding study are loosely associated with liver fat content (Table 18). This fat increase is frequently at the expense of microsomal protein, but inversely related to total P-450 estimates (Table 18). Thus, the apparent increase in liver fat and liver weight could actually reflect a decrease in the controls due to greater monooxygenase catabolism of fats.

Hematological changes were very subtle, but also very consistent (Table 18). Although PCVs and hemoglobin (and red blood cell counts, not shown) were not always affected in a linear dose-dependent fashion, layers receiving the two sludge-fertilized diets were always lower than controls. At 72 weeks, the high cadmium group was in a greater state of iron nutrition and it must be remembered that only five windows out of 560 days were sampled. Thus, other fluctuations and interactions with copper, zinc, cadmium, and other factors would be expected to influence the gross hematology. Therefore, it is probable that very minor changes in hematology were associated with incorporating the maximum amount of sludge-borne cadmium possible into the diets; nevertheless, these changes did not affect performance.

D. ANIMAL PRODUCTS FROM SOIL AMENDMENT EXPOSURES

Adverse effects on animal health from municipal sewage sludges have been demonstrated only in direct feeding of sewage sludges or when proposed soil amendment guidelines[3] have been violated and animals are permitted direct contact with soils and forages containing these overapplications. Manures can be injurious when used inappropriately as soil amendments and, again, animals are allowed contact before components are assimilated and/or stabilized. The worst case exposure to fly ash, considered a highly toxic waste, resulted in no demonstrable adverse effects.[71] The authors are unaware of controlled studies investigating adverse animal effects of using agricultural lime, wood ash, or composts.

Even if these soil amendments do not adversely affect the health of domestic animals when used properly, increased residues of potentially toxic substances are found in various tissues. It is appropriate to be concerned with potentiating effects of co-contaminants and bioactivation through the food chain, but most instances indicate plants as barriers and balancers[36] and animal products as filters and further nutritional balancers.[29] In addition, meat products from general commerce frequently contain greater concentrations of toxic elements and chlorinated hydrocarbons than do amendment-exposed animals[62,64,81] and food processing should be examined for contributions of toxic chemicals to the food chain in excess of those from agricultural practices.[81]

In addition to frequent performance improvements in domestic animals fed sludge-fertilized products, laboratory rodents also show improved growth and state of nutrition.[5,23,25] Although nutritional requirements have been investigated and revised for generations, recycled wastes appear to provide a more appropriate balance, perhaps due to the general biological origin of the wastes. If used cautiously and appropriately, they may well be superior to refined soil amendments if not grossly contaminated with specific toxicants.

To emphasize the value of soil amendment vs. direct feeding, mice fed livers and kidneys from cattle or swine fed sewage sludge directly showed elemental imbalances (accumulation of cadmium and lead, decrease in copper) and decreased reproduction performance.[61] Those fed kidney or liver from cattle receiving sludge-fertilized sorghum had a better copper status and improved reproduction performance.[61]

Rats fed beef from cattle which had been fed undigested sewage sludge grew faster and reproduced more efficiently than rats fed a commercial rat chow.[64]

Rats fed liver and kidney from swine foraging on sewage sludge-amended soils accumulated cadmium in their kidneys in a dose-dependent manner. Nevertheless, growth and reproduction (two litters each) were not affected. The female rats having received tissues from swine exposed to any of the three levels of sewage sludge were in a better state of iron nutrition and hematology after their first litter than rats receiving tissues from unexposed swine.[25] Only rat pups from dams which had received tissues from maximum exposed swine (which had been adversely affected, themselves) showed improved hematology. Second litters and terminal dams were indistinguishable in terms of hematology.

V. MINIMIZING ADVERSE EFFECTS AND MAXIMIZING BENEFICIAL EFFECTS

Federal programs, such as those for industrial pretreatment, have had a dramatic effect on improving the quality of municipal sludges. The Chicago sludges Hinesly and Hansen[5,24,66] used in their studies in the early 1970s bear no resemblance to modern sludges. Industrial pretreatment programs instituted by the Metropolitan Water Reclamation District of Greater Chicago have reduced heavy metal (e.g., cadmium, chromium, copper, lead, and zinc) concentrations as much as tenfold since 1978.[88]

After 15 years of development, EPA issued final regulations for use and disposal of wastewater sludge on February 19, 1993. The final regulations do not distinguish between agricultural and nonagricultural application of sludge and sludge products. Specific limits are defined for 11 metals plus total hydrocarbons, and sludges containing 50 ppm or more of polychlorinated biphenyls cannot be land-applied and must be disposed of in accordance with Toxic Substance Control Act (TSCA) regulations. The regulations promulgate ceiling concentrations in the sludge (mg/kg), annual loading rates (kg/ha), cumulative loading rates (kg/ha), and "clean" sludge concentrations (mg/kg)[89,90] that allow sludge use without the strict recordkeeping otherwise required. Thus, adherence to federal regulations will assure that adverse effects are minimized while the many beneficial effects of sludge amendment of soils are maximized.

REFERENCES

1. Bowen, H. J. M., *Trace Elements in Biochemistry,* Academic Press, New York, 1966, pp. 25-41.
2. John, S.F., Kane, D.N., and Hinesly, T.D., Use and disposal of wastewater sludges in Illinois, Illinois Department of Energy and Natural Resources, Springfield, 1991.
3. USEPA, Standards for the use or disposal of sewage sludge (40 CFR Parts 257, 403, and 503), *Fed. Reg.,* 58 (32), 9248, February 19, 1993.
4. Furr, A.K., Lawrence, A.W., Tong, S.S.C., Grandolfo, M.C., Hofstader, R.A., Bache, C.A., Gutenmann, W.H., and Lisk, D.J., Multielement and chlorinated hydrocarbon analysis of municipal sewage sludges of American cities, *Environ. Sci. Technol.*, 10, 683, 1976.
5. Hinesly, T.D., and Hansen, L.G., Agricultural benefits and environmental changes resulting from the use of digested sludge on field crops: Including animal health effects, *Progress Report (1971–1977) to Metropolitan Sanitary District of Greater Chicago*, Agronomy Department and College of Veterinary Medicine, University of Illinois at Urbana-Champaign, 1979.
6. Hansen, L.G., and Chaney, R.L., Environmental and food chain effects of the agricultural use of sewage sludges, *Rev. Environ. Toxicol.*, 1, 103, 1984.
7. Mumma, R.O., Raupach, D.C., Waldman, J.P., Tong, S.S.C., Jacobs, M.L., Babish, J.G., Hotchkiss, J.H., Wszolek, P.C., Gutenmann, W.H., Bache, C.A., and Lisk, D.J., National survey of elements and other constituents in municipal sewage sludges, *Arch. Environ. Contam. Toxicol.*, 13, 75, 1984.
8. Prakasam, T.B.S., Treatment Plant Analytical Data Reports, Metropolitan Water Reclamation District of Greater Chicago, Research and Development Department, Personal Communication, April 29, 1993.
9. Dacre, J.C., Potential health hazards of toxic organic residues in sludge, in *Sludge — Health Risks of Land Application*, G. Bitton, B.L. Damron, G.T. Edds, and J.M. Davidson, Eds., Ann Arbor Science, Ann Arbor, 1980, pp. 85-102.
10. USEPA, National sewage sludge survey: Availability of data and anticipated impacts on proposed regulations, *Fed. Reg.,* 55(218), 47210, Nov. 9, 1990.
11. Jacobs, L.W., and Zabik, M.J., Types and concentrations of organics in municipal sludge, *Proc. Municipal Wastewater Sludge Health Effects Research Planning Workshop*, USEPA, Cincinnati, OH, 3, 12, 1984.
12. Demirjian, Y.A., Joshi, A.M., and Westman, T.R., Fate of organic compounds in land application of contaminated municipal sludge, *J. Wat. Poll. Cont. Fed.*, 59, 32, 1987.
13. USEPA, Characterization of municipal waste combustion ash, ash extracts, and leachates, EPA-530-SW-90-029A, National Technical Information Service, Springfield, VA, 1990.
14. Capar, S.G., Tanner, J.T., Friedman, M.H., and Boyer, K.W., Multielement analysis of animal feed, animal wastes, and sewage sludge, *Environ. Sci. Technol.*, 12, 785,1978.
15. Ewert, K.M., DiPietro, J.A., Danner, C.S., and Lawrence, L.M., Ivermectin treatment of horses: effect on fecal-fouled areas in pastures, *Vet. Rec.*, 129, 140, 1991.

16. Murphy, L.S., and Walsh, L.M., Correction of micronutrient deficiencies with fertilizers, in *Micronutrients in Agriculture*, Mortvedt, J.J., Giordano, P.M., and Lindsay, W.L., Eds., Soil Science Society of America, Inc., Madison, WI, 1972, pp. 347-387.
17. Wood, S.G., personal communication, 1993.
18. Mathers, A.C., Stewart, B.A., Thomas, J.D., and Blair, B. J., Effects of cattle feedlot manure on crop yields and soil conditions, Technical Report No. 11, U.S.DA Southwestern Great Plains Research Center, Bushland, TX, 1973.
19. Lamar, J.E., and Thomson, K.B., Sampling limestone and dolomite deposits for trace and minor elements, Illinois State Geological Survey, Urbana, Circular 221, 1956.
20. Ostrom, M.E., Trace elements in Illinois Pennsylvanian limestones, Illinois State Geological Survey, Urbana, Circlular 243, 1957.
21. Campbell, A.G., Recycling and disposing of wood ash, *Tappi J.*, 73, 141, 1990.
22. Gray, M.N., Rock, C.A., and Pepin, R.G., Pretreating landfill leachate with biomass boiler ash, *J. Environ. Engin.*, 114, 465, 1988.
23. Boyd, J.N., Changes in blood alpha-fetoprotein concentrations in rats fed carcinogens and dietary modifiers of carcinogenesis, Ph.D. Thesis, Cornell University, Ithaca, NY, p. 53, 1981.
24. Hansen, L.G., Washko, P.W., Tuinstra, L.G.M.Th., Dorn, S.B., and Hinesly, T.D., Polychlorinated biphenyl, pesticide, and heavy metal residues in swine foraging on sewage sludge amended soils, *J. Agric. Food Chem.,* 29, 1012, 1981.
25. Lambert, R.L., Tissue Residues and Toxicities of Inorganic and Protein-bound Cadmium in Rats, Ph.D. Thesis, University of Illinois, Urbana, 1983.
26. Beetsman, G.B., Keeney, D.R., and Chesters, G., Dieldrin uptake by corn as affected by soil properties, *Agron. J.*, 61, 247, 1969.
27. Guo, L., Bicki, T.J., Felsot, A.S., and Hinesly, T.D., Phytotoxicity of atrazine and alachlor in soil amended with sludge, manure, and activated carbon, *J. Environ. Sci. Health*, B26, 513, 1991.
28. Guo, L., Bicki, T.J., Hinesly, T.D., and Felsot, A.S., Effect of carbon-rich waste materials on movement and sorption of atrazine in a sandy, coarse-textured soil, *Environ. Toxicol. Chem.*, 10, 1273, 1991.
29. Hansen, L.G., and Lambert, R.L., Transfer of toxic trace substances by way of food animals: Selected examples, *J. Environ. Qual.*, 16, 200, 1983.
30. Cousins, R.J., Metallothionein synthesis and degradation: Relationship to cadmium metabolism, *Environ. Health Perspec.,* 28, 131, 1979.
31. Saxena, D.K., Murthy, R.C., Singh, C., and Chandra, S.V., Zinc protects testicular injury induced by concurrent exposure to cadmium and lead in rats, *Res. Commun. Chem. Pathol. Pharmacol.*, 64, 317, 1989.
32. Anderson, M.B., Lepak, K., Farinas, V., and George, W.J., Protective action of zinc against cobalt-induced testicular damage in the mouse, *Reprod. Toxicol.*, 7, 49, 1993.
34. Bunn, C.R., and Matrone, G., *In vivo* interactions of cadmium, copper, zinc and iron in the mouse and rat, *J. Nutr.*, 90, 395, 1966.
35. Olsen, S.R., Micronutrient interactions, in *Micronutrients in Agriculture*, Mortvedt, J.J., Giordano, P.M., and Lindsay, W.L., Eds., Soil Science Society of America, Madison, WI, 1972, p. 243.
36. Chaney, R.L., Potential effects of waste constituents on the food chain, in *Land Treatment of Hazardous Wastes*, Parr, J.F., Marsh, P.B., and Kla, J.M., Eds., Noyes Data Corp., Park Ridge, NJ, 1983, p. 426.

37. National Research Council, *Complex Mixtures: Methods for In Vivo Toxicity Testing*, National Academy Press, Washington D.C., 1988.
38. Siegel, D., Huber, W.G., and Enloe, F., Continuous nontherapeutic use of antibacterial drugs in feed and drug resistance of the gram-negative enteric florae of food-producing animals, *Antimicrob. Agents Chemother.*, 6, 697, 1974.
39. NUS Corp., Characterization of MWC ashes and leachates from MSW landfills, monofills, and co-disposal sites — Summary. Vol. 1, *EPA/530-SW-87-028A*, Office of Solid Waste, USEPA, Washington, DC, 1987.
40. Lowe, J.A., Methods of estimating toxic equivalents for polychlorinated dibenzodioxins and dibenzofurans, in *Health Effects of Municipal Waste Incineration*, Hattemer-Frey, H.A., and Travis, C., Eds., CRC Press, Boca Raton, FL, 1991, p. 238.
41. Mortvedt, J.J., Cadmium levels in soils and plant tissues from long-term soil fertility experiments in the U.S., Programme XIII Congress of the International Society of Soil Sci., Hamburg, Germany, 1986, p. 870.
42. Davidson, C.I., and Rabinowitz, M. Lead in the environment: from sources to human receptors, in *Human Lead Exposure*, Needleman, H.L., Ed., CRC Press, Boca Raton, FL, p. 79, 1992.
43. Kienholz, E.W., Effect of toxic chemicals present in sewage sludge on animal health, in *Sludge — Health Risks of Land Application*, Bitton, G., Damron, B.L., Edds, G.T., and Davidson, J.M., Eds., Ann Arbor Science, Ann Arbor, MI, 1980, p. 153.
44. Fitzgerald, P.R., Use of anaerobically digested sludge for pasture reclamation: Parasitology and the occurrence of heavy metals in soil, vegetation, and animal tissues, Final Report to the Metropolitan Sanitary District of Greater Chicago, University of Illinois, Urbana, 1981.
45. Bray, B.J., Dowdy, R.H., Goodrich, R.D., and Pamp, D.E., Trace metal accumulations in tissues of goats fed silage produced on sewage sludge-amended soil, *J. Environ. Qual.*, 14, 114, 1985.
46. Parkinson, A., and Safe, S., Mammalian biologic and toxic effects of PCBs, *Environ. Toxin Ser.*, 1, 49, 1987.
47. Hansen, L.G., Environmental toxicology of PCBs, *Environ. Toxin Ser.*, 1, 15, 1987.
48. Hansen, L.G., Selective accumulation and depletion of PCB components: Food animal implications, *Ann. NY Acad. Sci.*, 320, 238, 1979.
49. Krishnan, V., and Safe, S., PCBs, PCDDs and PCDFs as antiestrogens in MCF-7 human breast cancer cells: Quantitative structure-activity relationships, *Toxicol. Appl. Pharmacol.*, 120, 55, 1993.
50. Jansen, H.T., Cooke, P.S., Porcelli, J., Liu, T.-C., and Hansen, L.G., Estrogenic and antiestrogenic actions of PCBs in the female rat: *in vitro* and *in vivo* studies, *Reprod. Toxicol.*, 7, 237, 1993.
51. Bitman, J., and Cecil, H.C., Estrogenic activity of DDT analogs and PCBs, *J. Agric. Food Chem.*, 1, 1108, 1070.
52. Li, M.-H., Porcelli, J., and Hansen, L.G., Estrogenic and enzyme induction activities of PCB congeners in prepubertal rats, *The Toxicologist*, 13, 357, 1993.
53. Fulton, B., and Jeffery, E.H., Heme oxygenase induction: A possible factor in aluminum-associated anemia, *Biol. Trace Element Res.*, 40, 9, 1994.
54. Babish, J.G., Stoewsand, G.S., Furr, A.K., Parkinson, T.F., Bache, C.A., Dugenmann, W.H., Wszolek, P.C., and Lisk, D.J., Elemental and polychlorinated biphenyl content of tissues and intestinal aryl hydrocarbon hydroxylase activity of guinea pigs fed cabbage grown on municipal sewage sludge, *J. Agric. Food Chem.*, 27, 399, 1979.

55. Babish, J.G., and Stoewsand, G.S., Hepatic microsomal enzyme induction in rats fed varietal cauliflower leaves, *J. Nutr.*, 105, 1592, 1975.
56. Hileman, B., Concerns broaden over chlorine and chlorinated hydrocarbons, *Chem. Eng. News*, April 19, 1993, 11.
57. Cooke, P.S., Thyroid hormones and testis development: A model system for increasing testis growth and sperm production, *Ann. NY Acad. Sci.*, 637, 122, 1991.
58. Ness, D.K., Schantz, S.L., Moshtaghian, J., and Hansen, L.G., Effects of perinatal exposure to specific PCB congeners on thyroid hormone concentrations and thyroid histology in the rat, *Toxicol. Lett.*, 68, 311, 1993.
59. Brouwer, A., Reijnders, P.J.H., and Koeman, J.H., PCB-contaminated fish induces vitamin A and thyroid hormone deficiency in the common seal *(Phoca vitulina), Aquat. Toxicol.*, 15, 99, 1989.
60. Cheeke, P.R., and Myer, R.O., Evaluation of the nutritive value of activated sewage sludge with rats and Japanese quail, *Nutr. Rep. Intern.,* 8, 383, 1973.
61. Edds, G.T., Osuna, O., and Simpson, C.F., Health effects of sewage sludge for plant production or direct feeding to cattle, swine, poultry, or animal tissue to mice, in *Sludge — Health Risks of Land Application*, Bitton, G., Damron, B.L., Edds, G.T., and Davidson, J.M., Eds., Ann Arbor Science, Ann Arbor, MI, 1980, chap. 14.
62. Baxter, J.C., Johnson, D.E., and Kienholz, E.W., Uptake of trace metals and persistent organics into bovine tissues from sewage sludge — Denver Project, in *Sludge — Health Risks of Land Application*, Bitton, G., Damron, B.L., Edds, G.T., and Davidson, J.M., Eds., Ann Arbor Science, Ann Arbor, MI, 1980, p. 285.
63. Richter, M.F., Health effects of recycled cattle manure, in *Sludge — Health Risks of Land Application*, Bitton, G., Damron, B.L., Edds, G.T., and Davidson, J.M., Eds., Ann Arbor Science, Ann Arbor, MI, 1980, p. 337.
64. Smith, G.S., Kiesling, H.E., Ray, E.E., Hallford, D.M., and Herbel, C.H., Sewage solids as supplemental feed for ruminants: Bioassays of benefits and risks, in *Sludge — Health Risks of Land Application*, Bitton, G., Damron, B.L., Edds, G.T., and Davidson, J.M., Eds., Ann Arbor Science, Ann Arbor, MI, 1980, p. 357.
65. Fitzgerald, P.R., Observations on the health of some animals exposed to anaerobically digested sludge originating in the Metropolitan Sanitary District of Greater Chicago system, in *Sludge — Health Risks of Land Application*, Bitton, G., Damron, B.L., Edds, G.T., and Davidson, J.M., Eds., Ann Arbor Science, Ann Arbor, MI, 1980, p. 267.
66. Hansen, L. G. and Hinesly, T. D., Cadmium from soil amended with sewage sludge: effects and residues in swine, *Environ. Health Perspect.,* 28, 51, 1981.
67. White, C.E., Hammell, D.L., and Osuna, O., Effect of feeding digested sewage sludge on long-term sow reproductive performance, in *Sludge — Health Risks of Land Application*, Bitton, G., Damron, B.L., Edds, G.T., and Davidson, J.M., Eds., Ann Arbor Science, Ann Arbor, MI, 1980, pp. 339-340.
68. Beaudouin, J., Shirley, R.L., and Hammell, D.L., Effect of sewage sludge diets fed swine on nutrient digestibility, reproduction, growth, and minerals in swine, in *Sludge — Health Risks of Land Application*, Bitton, G., Damron, B.L., Edds, G.T., and Davidson, J.M., Eds., Ann Arbor Science, Ann Arbor, MI, 1980, p. 341.
69. Arnold, E.K., Lovell, R.A., Beasley, V.R., Parker, A.J., and Stedelin, J.R., 2,4-D toxicosis III. An attempt to produce 2,4-D toxicosis in dogs on treated grass plots, *Vet. Human Toxicol.,* 33, 457, 1991.
70. Sharma, R.P., and Street, J.C., Public health aspects of toxic heavy metals in animal feeds, *J. Am. Vet. Med. Assoc.*, 177, 149, 1980.

71. Furr, A.K., Parkinson, T.F., Heffron, C.L., Reid, J.T., Haschek, W.M., Gutenmann, W.H., Bache, C.A., St. John, Jr., L.E., and Lisk, D.J., Elemental content of tissues and excreta of lambs, goats and kids fed white sweet clover growing on fly ash, *J. Agric. Food Chem.*, 26, 847, 1978.
72. Haschek, W.M., Furr, A.K., Parkinson, T.F., Heffron, C.L., Reid, J.T., Bache, C.A., Wszolek, P.C., Gutenmann, W.H., and Lisk, D.J., Element and polychlorinated biphenyl disposition and effects in sheep fed cabbage grown on municipal sewage sludge, *Cornell Vet.*, 69, 302, 1979.
73. Heffron, C.L., Reid, J.T., Elfving, D.C., Stoewsand, G.S., Hashek, W.M., Telford, J.N., Furr, A.K., Parkinson, T.F., Bache, C.A., Gutenmann, W.H., Wszolek, P.C., and Lisk, D.J., Cadmium and zinc in growing sheep fed silage corn grown on municpal sludge amended soil, *J. Agric. Food Chem.*, 28, 58, 1980.
74. Dowdy, R.H., Bray, B.J., Goodrich, R.D., Marten, G.C., Pamp, D., and Larson, W.E., Performance of goats and lambs fed corn silage produced on sludge-amended soil, *J. Environ. Qual.*, 12, 467, 1983.
75. Dowdy, R.H., Goodrich, R.D., Larson, W.E., Bray, B.J., and Pamp, D., Effects of sewage sludge on corn silage and animal products, USEPA Rep. 600/S2-84-075, Cincinatti, OH.
76. Osweiler, G.D., Carson, T.L., Buck, W.B., and Van Gelder, G.A., *Clinical and Diagnostic Veterinary Toxicology*, 3rd Edition, Kendall/Hunt, Dubuque, IA, 1985, p. 87.
77. Bremner, I., Young, B.W., and Mills, C.F., Protective effect of zinc supplementation against copper toxicosis in sheep, *Br. J. Nutr.,* 36, 551, 1976.
78. Angus, K.W., Suttle, N.F., Munro, C.S., and Field, A.C., Adverse effects on health of including high levels of dried poultry waste in the diets of lambs, *J. Comp. Pathol.,* 88, 449, 1978.
79. Miller, S., and Nelson, H.A., Copper poisoning in sheep grazing pastures fertilized with chicken litter, *J. Am. Vet. Med. Assoc.,* 173, 1587, 1978.
80. Kerr, L.A., and McGavin, H.D., Chronic copper poisoning in sheep grazing pastures fertilized with swine manure, *J. Am. Vet. Med. Assoc.,* 198, 99, 1991.
81. Hansen, L.G., Dorner, J.L., Byerly, C.S., Tarara, R.P., and Hinesly, T.D., Effects of sewage sludge-fertilized corn fed to growing swine, *Am. J. Vet. Res.*, 37, 711, 1976.
82. Hinesly, T.D., Ziegler, E.L., and Tyler, J.J., Selected chemical elements in tissues of pheasants fed corn grain from sewage sludge-amended soil, *Agro-Ecosystems*, 3, 11, 1976.
83. Hinesly, T.D., Hansen, L.G., Bray, D.J., and Redborg, K.E., Transfer of sludge-borne cadmium through plants to chickens, *J. Agric. Food Chem.*, 33, 173, 1985.
84. Simon, J., Busch, J., and Hansen, L.G., unpublished results.
85. Sundlof, S.M., and Hansen, L.G., unpublished results, Interim Report to T.D. Hinesly, 1981.
86. Dorner, J.L., and Hansen, L.G., unpublished results, Interim Report to T.D. Hinesly, 1981.
87. Hansen, L.G., Dorn, S.B., Sundlof, S.M., and Vogel, R.S., Toxicity, accumulation, and depletion of hexachlorobenzene in laying chickens, *J. Agric. Food Chem.*, 26, 1369, 1978.
88. Developed from analytical data provided the authors by Dr. Richard I. Pietz, Metropolitan Water Reclamation District of Greater Chicago, 10 May 1993.
89. Chaney, R.L., Twenty years of land application research, *Biocycle*, 31(9), 54, 1990.
90. Chaney, R.L., Public health and sludge utilization, *Biocycle*, 31(10), 68, 1990.

CHAPTER 3

WILDLIFE HABITATS AND POPULATIONS

Jonathan B. Haufler and Henry Campa, III

TABLE OF CONTENTS

I. INTRODUCTION

Soil amendments are typically applied to an area for the purposes of fertilization, thus increasing site productivity. Sometimes a secondary objective is to productively dispose of materials such as sewage sludges. In either case, the amendments should produce responses in the plant community. Wildlife populations can be affected by soil amendments either directly from the application, or indirectly through changes in the plant community. Direct effects of the application process on wildlife populations tend to be very limited. The greatest effects of soil amendments on wildlife are secondary, in response to the changes in the plant community's composition, structure, productivity, or chemical composition. These habitat responses influence

0-87371-860-7/95/$0.00+$.50
© 1995 by CRC Press, Inc.

wildlife populations by causing changes in habitat use, survival rates, and/or reproductive rates. In addition, elements or compounds contained in the amendments could influence wildlife populations through direct ingestion of amendments or by entering wildlife food chains.

Soil amendments have been used as tools for wildlife management. Their primary use has been to increase either the quantity or quality of wildlife foods in an area. Typically, this involves fertilization in conjunction with the planting of herbaceous species. Rarely are natural ecosystems fertilized, as the costs of commercial fertilizer are expensive relative to the returns in population responses by one or two wildlife species. When total wildlife communities are considered, management activities usually focus on ecosystem management approaches, which generally emphasize natural processes rather than supplemental enhancement of primary or secondary productivity. However, application of sewage sludges into natural ecosystems has been used as a means of increasing productivity of the ecosystems while recycling the nutrients contained in the sludge.

This chapter discusses these various wildlife habitat and population responses to soil amendments. Specifically, it discusses wildlife responses to changes in habitat features caused by soil amendments, and also discusses responses of populations to potential contaminants contained in amendments.

II. EFFECTS OF SOIL AMENDMENTS ON WILDLIFE HABITAT

The application of soil amendments to wildlife habitat has received considerable attention by wildlife managers for enhancing the growth and nutritional qualities of vegetation for wildlife use.[1] Past land applications have primarily focused on applying inorganic fertilizers to grasslands and wildlife food plots. Applying sewage sludges and effluents to ecosystems such as forests and a variety of disturbed areas such as reclaimed strip mines is a relatively new habitat management practice. Following secondary or tertiary waste treatment procedures, sludges and effluents have been applied to a diversity of vegetation types by agricultural producers and natural resource managers. While the application of soil amendments to sites has largely shown positive impacts on wildlife habitat components, such as increasing the availability and quality of food or cover, potential long-term negative impacts to wildlife habitat may also occur. For example, disposing of untreated or primary wastewater for extended periods of time has been shown to be detrimental to tree growth.[2] The addition of soil amendments as a wildlife habitat management practice, therefore, should be carefully evaluated prior to implementation.

A. VEGETATION COMPOSITION, STRUCTURE, AND PRODUCTIVITY

Applying soil amendments to forests and grasslands for wildlife can produce changes in the composition, structure, and productivity of plant communities. When changes do occur they are typically attributable to the addition of nutrients to nutrient-limited sites, increased moisture levels, and the introduction of new species from seeds contained in sewage sludges.[3]

Studies investigating the responses of various ecosystems to soil amendments have documented that plant species composition may be altered by nutrient enrichment. Investigations of short-grass prairies,[4] old fields,[5] and tropical forests[6] have shown nutrient enrichment with soil amendments can influence vegetative community composition. Michigan researchers, however, did not document consistent changes in the composition of plant species when municipal sewage sludges were applied to five forest types.[7–10] The lack of consistent shifts in species composition on Michigan sites were partially attributed to sludge being applied to relatively undisturbed forest stands.[3] Sludge applications to more disturbed clear-cut sites in Washington did cause enhanced growth of herbaceous species and a reduction in the regeneration of woody species.[11–12] Impaired regeneration of seedlings in Washington was attributed to increased numbers of Townsend's voles *(Microtus townsendii)* that girdled trees.

Soil amendments to forested ecosystems have also caused alterations in forest stand structure by increasing vertical and horizontal cover.[3] Following the application of municipal sewage sludge to a northern hardwoods stand, red pine *(Pinus resinosa)*/jack pine *(P. banksiana)* plantation, oak (*Quercus* spp.) stand, and a 15-year-old aspen clear-cut in Michigan, percent cover was greater on treated plots than controls. Changes in structure were attributed to annual increased plant productivity.[3]

The application of soil amendments to wildlife habitat typically does enhance annual plant productivity.[7,13–17] Jokela et al.[18] reported that tree growth in a slash pine *(P. elliottii)* plantation increased 170% when municipal garbage composted with sewage sludge was applied. Woodyard[7] documented that the application of municipal sewage sludge to a 4-year-old jack pine clear-cut enhanced the annual productivity of cherries (*Prunus* spp.), jack pine, and brambles (*Rubus* spp.). Total annual production >2 m in height averaged 155% greater on sludge-amended plots than controls for the first 2 years of Woodyard's[7] study. Sopper and Kardos[13] documented similar results for red pine irrigated with wastewater. They noted that irrigation increased both the diameter and height of red pine. Berry[16] evaluated tree growth in response to three types of soil amendment treatments: the application of dried municipal sewage sludge, fertilizer and lime, and control sites which did not receive any amendment. Berry[16] documented that in all experiments, seedling growth was greater on sludge-treated plots than control plots and plots amended with fertilizer and lime. Researchers evaluating pin cherry *(Prunus pensylvanica)* and bramble response to commercial fertilization also documented increased productivity even when in competition with birch (*Betula* spp.) and maple (*Acer* spp.) regeneration.[19] Studies in the Pacific Northwest comparing the growth response of Douglas-fir *(Pseudotsuga menziesii)* to sludge and urea applications found sludge produced growth at least two times greater and lasted at least two times longer than urea.[20]

Significant increases in productivity after soil amendments have been documented for herbaceous species as well as woody species.[17,21–22] Hahn et al.[23] noted that spray irrigating pasture grasses with anaerobically digested, domestic sludge significantly increased yields. Sopper and Kardos[13] observed a similar response of increased annual production of old field vegetation irrigated with municipal wastewater.

They noted that with irrigation, productivity increased an average of 201% annually over a 10-year period. Applying dried, anaerobically digested sewage sludge has been evaluated as a land management technique to enhance the productivity of grasslands in the southwestern U.S. In their study, Fresquez et al.[17] documented that total herbaceous plant yield, especially of blue grama *(Bouteloua gracilis)*, increased two to three times on sludge-treated plots. The authors attributed increases in plant productivity to the nitrogen content of the sludge.

Although it is well documented that many soil amendments can be used as a wildlife habitat management technique to enhance plant productivity, natural resource managers must consider other potential changes that may occur in plant communities as a result of applying nutrients and if these changes are counterproductive in meeting management objectives. For example, as plant productivity increases, plant species composition on nutrient-poor sites may shift due to accelerated growth rates of species which are sensitive to the addition of nutrients. Applying soil amendments to an opening created for wildlife may increase the productivity of herbaceous species on the opening, however, it may also shorten the interval in which the area remains in the desired early successional stage.[7]

B. NUTRITIONAL QUALITY

One goal of applying soil amendments to non-agricultural ecosystems is to use the nutrients available to manipulate and improve wildlife habitat quality. Researchers who have studied the effects of soil amendments on ecosystems have found, in most instances, that the nutritive quality of vegetation was improved.[9,16–18,24–26] The nutrient component which consistently increased with soil amendments was the nitrogen content of plants or crude protein (Table 1).[9,16–18,24–27] Protein levels in vegetation are commonly used as an index of forage quality by natural resource managers.[28–29]

Blessin and Garcia[25] studied the nutrient content of vegetation grown on sludge-treated strip mine lands and found a 2.5% increase in the protein content of vegetation after sludge was applied. Anderson[27] found that sludge application increased crude protein content of Italian rye grass *(Lolium multiflorum)* 160% in December samples and 176% in March samples. Campa[26] found that a spring application of municipal sewage sludge to a 4-year-old jack pine clear-cut in Michigan significantly increased crude protein levels in all six plant species sampled the first summer after application and in all three plant species sampled in late fall. An additional Michigan study investigating the nutritional response of wildlife forages to sludge application documented higher crude protein levels in spring and summer vegetation from aspen plots the first year after treatment (Table 1).[30] The only species exhibiting a significant increase in crude protein of three woody species sampled in the winter was bigtooth aspen *(P. grandidentata)*. On oak plots amended with sludge in Michigan, red maple *(Acer rubrum)* had significant increases in protein levels, with protein levels being greatest in spring and summer samples and lowest during the winter.[10] Increases in protein levels were also observed in sedge (*Carex* spp.), red oak *(Quercus borealis)*, and red maple on the jack pine/red pine plots in Michigan following sludge application.[9]

Table 1 Foliar Nitrogen Content (%) of Plant Species Treated with Various Types of Soil Amendments

Location	Type of amendment	Time of amendment	Sample collection	Plant species	% foliar-N	Ref.
Michigan	Liquid, anaerobically digested, non-industrial sewage sludge	June	August	Red maple	2.57	9
				Red oak	3.53[b]	
South Carolina	Sewage sludge	NG[a]	August	Green ash	1.83[b]	16
				Sweetgum	1.79[b]	
				Loblolly pine	1.51[b]	
South Carolina	Commercial 10-10-10 fertilizer and dolomitic limestone	NG	August	Green ash	1.39	16
				Sweet gum	1.06	
				Loblolly pine	1.09	
New Mexico	Dried, anaerobically digested sewage sludge	June	September	Blue grama	2.80[b]	17
				Galleta	2.40	
				Bottlebush squirreltail	2.90[b]	
Florida	Municipal garbage composted with sewage sludge	NG	NG	Slash pine	1.13[b]	18
Michigan	Liquid, anaerobically digested, non-industrial, municipal sewage sludge	April-May	July-August	Cherry	4.46[b]	26
				Jack pine	1.90[b]	
				Brambles	2.64[b]	
				Orange-hawkweed	3.64[b]	
				Sedge	1.93[b]	
				Panic grass	2.74	
Washington	Sewage sludge	NG	December	Italian ryegrass	5.10[b]	27
Michigan	Liquid, anaerobically digested, non-industrial municipal sewage sludge	October	July-August	Bigtooth aspen	2.78[b]	30
				Quaking aspen	2.64[b]	
				Pin cherry	4.05[b]	
				Wild strawberry	2.03[b]	
				Orange-hawkweed	2.70[b]	
				Panic grass	2.53[b]	
Missouri	Nitrogen-phosphorus-potassium granulated fertilizer	March	May-November	Tall fescue	1.70	33

[a] Information not given.

[b] Significantly greater (at least $P < 0.10$) than controls.

The fourth vegetation type investigated in Michigan was a hardwood study area. Increases in the nutritional qualities of plant species on this site were minimal, probably due to its higher site quality.[10]

In addition to nitrogen, soil amendments can also be a source of phosphorus for vegetation. Dressler and Wood[24] noted that phosphorus levels were significantly greater in plants irrigated with effluent. Campa[26] reported higher levels of phosphorus in orange-hawkweed *(Hieracium aurantiacum),* sedge, and panic grass *(Panicum virgatum)* on sludge-treated plots. None of the woody species sampled by Campa[26] during the summer had significant increases in phosphorus. All of the woody species collected in this study on sludge-treated sites during the winter, however, had greater phosphorus contents than samples from control plots. Berry[16] documented elevated levels of phosphorus and nitrogen in foliage from a diversity of tree species on sludge-amended plots compared with those treated with fertilizer. Campa et al.[30] and Seon[9] reported higher phosphorus levels in herbaceous species sampled in the summer from aspen and pine study areas, respectively, in Michigan. No differences were observed in winter samples of woody species on either of these sites.

Applying supplemental sources of nutrients, such as nitrogen and phosphorus, to wildlife forages may be essential for animals to meet growth, maintenance, and reproduction requirements. For example, Murphy and Coates[31] documented that forage protein content may be the most critical nutrient on some ranges and may account for low production and physical development of white-tailed deer *(Odocoileus virginianus).* They noted body weights and antler developments of yearlings and adult males were retarded by a diet of 7% protein. Researchers who have evaluated the nutritional quality of forages treated with soil amendments documented that many plant species had protein contents greater than 7%.[9,26] Plants on plots not receiving amendments, however, had protein levels below the minimum levels required by deer.[9,26] The wildlife habitat management implications of these results indicate that applying soil amendments to nutrient-poor sites may be an essential component for maintaining productive populations of wild herbivores.

Other nutrients, such as calcium and potassium, have not shown the same consistent or significant increases in plant tissues following soil amendments that have been documented for nitrogen and phosphorus. Bledsoe and Zasoski[32] attribute this response to lower concentrations of these nutrients typically occurring in soil amendments. Berry[16] noted calcium and magnesium were usually higher in foliage of trees from plots amended with fertilizer as opposed to sludge-treated and control plots. Campa,[26] however, analyzed herbaceous and woody species after sludge application and documented consistently greater concentrations of only potassium accumulating in herbaceous species on sludge-amended plots. Fresquez et al.[17] documented similar results after applying dried, anaerobically digested sewage sludge to a degraded semiarid grassland in New Mexico. In their study, potassium concentrations increased linearly with sludge application in blue grama, galleta *(Hilaria jamesii),* and bottlebrush squirreltail *(Sitanion hystrix).*

In addition to protein content, other indicators of forage nutritive quality are fiber constituents (cellulose, hemicellulose, and lignin) and *in vitro* dry matter digestibility. The digestibility for a given forage species, however, is not constant. Factors such as stage of maturity, season, and site conditions may influence digestibility levels. The relationship between digestibility and fiber constituents is also important to examine when attempting to evaluate the effects of land management practices on the nutritional qualities of a forage species. Typically, as plants mature fiber levels increase and digestibility decreases. By determining the fiber constituent content and the level of digestibility of forages, managers will be given an indication of how plants are altered by soil amendments and how well plants may be used by herbivores.

In some cases, the application of soil amendments to nutrient-poor sites has increased the digestibility of plant species,[9,26,30] while in other cases they have not shown influential effects.[33] The *in vitro* digestibility of cherry twigs and leaves, brambles, and panic grass was substantially greater on sludge-treated plots of a young jack pine clear-cut than control plots.[26] Panic grass showed the greatest response in digestibility, a 26% difference, on sludge-amended plots during the summer. Greater digestibility of some species in Campa's[26] study was attributed to consistently lower cellulose and lignin contents of vegetation on sludge-treated plots. Lower fiber levels and thus greater digestibility on sludged plots were attributed to the active amount of plant growth on treated plots. Kozlowski and Keller[34] stated that actively growing vegetation usually has a higher nutritional plane than dormant vegetation. This result may be attributed to nutrients being assimilated into plant cells during production. Therefore, if additional nutrients are available to plants, such as with soil amendments, they will be accumulated, thus increasing their nutrient content.[35] This type of a response was observed by Campa[26] for cherry plant tissues, brambles, and panic grass. Dressler and Wood[24] also documented significantly lower fiber contents of vegetation on effluent irrigated sites, while crude protein, potassium, and phosphorus were significantly higher.

When 54 kg/ha of nitrogen-phosphorus-potassium granulated fertilizer was applied in the spring to an Ozark forest in Missouri the digestibility of tall fescue *(Festuca arundinacea)* decreased 17% from May to September and then increased 14.5% in September.[33] Probasco and Bjugstad[33] attributed the decline in digestibility during the summer to fescue becoming nearly dormant during this season, so that the fertilizer treatment was not felt to be as influential a factor as anticipated. With the decline in digestibility of tall fescue, there was an increase in fiber production. They concluded that an increase in acid-detergent fiber (cellulose and lignin) in tall fescue resulted in the lower digestibility levels.

The timing of the application of soil amendments influences the response of vegetation in terms of nutrient accumulation. Applications of amendments, such as sewage sludges, prior to the growing season have shown to have significant effects on the nutritional quality of wildlife forages. Campa[26] documented significantly greater crude protein contents for all but one species sampled in the summer and fall on sludge-treated plots when municipal sewage sludge was applied in late-April to early-May in northern Michigan. Applications of sludge during the growing season, however, have not demonstrated consistent increases in plant nutritional qualities until the following growing season. Haufler et al.[10] found that sludge application during June to a northern hardwoods site did not produce significant increases in protein levels of plant samples collected during the summer. Protein levels of sugar maple *(Acer saccharum)* on sludge-amended plots of this site, however, were significantly greater the following winter and spring, but declined to control levels by the following summer.

How long plant species in various vegetation types will exhibit elevated levels of nutrients with a single application of a soil amendment may be dependent on the type and amount of amendment applied, its nutrient composition, timing of application, and the specific plant species. Berry[16] recommended that if sludge is used as a soil amendment instead of fertilizer, plant nutrients will be retained on sites because they are released slowly. A consistent source of additional nutrients for vegetation suggests that the nutrient composition of plants may be enhanced for extended periods.

The amounts of soil amendments applied to vegetation types must be computed for each site based on the nutrient and trace metal composition of amendments, the nutrient uptake potential of application sites, and the objective or desired result of the application. Berry[16] reported the heaviest rate of sludge applied to study sites in the southeastern U.S. was 68 Mg/ha. The primary interest for these applications was to use sewage sludge as a fertilizer as opposed to applying heavier applications which may be made when land disposal of sludge is the management objective. Berry[16] concluded that heavier amounts of the sludge used in the study could be applied to his sites with no negative impacts. However, one would not expect any increased growth rates of trees by applying amounts greater than approximately 34 Mg/ha. Berry[16] did report that the foliage of trees on sludge-amended plots was of greater nutritional quality with the greater loading rate.

III. EFFECTS OF SOIL AMENDMENTS ON WILDLIFE POPULATIONS

The effects of soil amendments on wildlife habitat can produce two types of responses in wildlife populations. Most herbivorous wildlife species will be attracted to areas that contain forages of higher nutritional quality than surrounding areas. Hanley[36] reported that deer and elk selected areas for foraging that provided high quality forage. Fertilized areas generally attract herbivorous wildlife due to the increased nutritional quality of forages following fertilization. Mule deer *(O. hemionus)* in New Mexico were found to increase utilization of an area fertilized with urea.[37] Bayoumi and Smith[22] found selective browsing by deer on fertilized range over unfertilized range. Thomas et al.[38] showed a sizeable grazing preference by deer for fertilized plots in the Black Hills. Woodyard[7] found significantly higher rates of browsing by white-tailed deer on sites treated with sewage sludge than untreated areas. Similarly, Campa et al.[30] found increased browsing by white-tailed deer and elk *(Cervus elaphus)* on sludge-amended sites, while Anderson[39] found that Columbian black-tailed deer *(O. hemionus columbianus)* preferred grasses in sludge-amended areas. These results reveal that a shift in habitat use by many herbivores can be achieved through the addition of soil amendments. Thus, application of soil amendments to a site should cause herbivorous species to be attracted to that site, and their foraging use of that area should increase while use of other areas will show a corresponding decline. This may increase the fitness of the individual animal by supplying it with increased food, or it may simply maintain the animal at its original fitness on a new foraging area. Ultimately, the only true gains to the population from soil amendments will occur if the increase in forage quantity or quality causes an increase in survival rates or reproductive rates that either increase the population size, or maintain the population size at higher levels or fitness than would have occurred without the soil amendment. In other words, it is possible for soil amendments to cause a shift in habitat use by a species in an area, without increasing the carrying capacity of the area for that species. This could occur in areas where a habitat component other than food is limiting the population, such as cover, chemical toxicities, water, or human related mortalities.

In some instances, shifting habitat use and maintaining existing population sizes are appropriate goals. If an impact from a development of some type is expected to cause a decrease in available habitat for a species, and the remaining acreage can be increased in habitat quality for the species, then a net loss in overall habitat may be avoided. Thus, soil amendments may be appropriate as a mitigation measure when a species is expected to decrease in abundance as a result of certain impacts. In these cases, the population size may be maintained by an increase in habitat quality for that species on a small acreage of habitat, replacing the additional habitat lost to development.

Few studies have actually documented changes in population dynamics as a response to soil amendments. Anderson[27] found higher productivity in Columbian black-tailed deer that fed on sludge-treated sites compared to deer that did not use treated sites. Small mammal populations have been found to respond favorably to sludge-applications, with specific responses noted for meadow voles *(Microtus pennsylvanicus)*, meadow jumping mice *(Zapus hudsonius)*, and eastern chipmunks *(Tamias striatus)*.[7] Haufler et al.[10] found that small mammal populations in four vegetation types treated with sewage sludge in Michigan had higher numbers of species, more individuals, and higher species diversity than populations in untreated areas.

A final consideration in any discussion of wildlife benefits associated with any habitat manipulations is that advantages provided to one species can cause reductions in habitat quality or populations of another species. Each wildlife species has evolved its own specialized niche, with a specific requirement for a combination of food, cover, and water components. Any manipulations that increase the supply of food or cover for one species, will often cause a decrease in the optimal combination of components for another species. For example, in sludge-treated aspen stands in Michigan, the small mammal community responded to the sludge with more species being present, more individuals, and greater species diversity.[10] However, the number of 13-lined ground squirrels *(Spermophilus tridecemlineatus)* in these stands declined due to the increase in ground cover caused by sludge treatment. Often shifts such as this appear to be of minor consequence, and the benefits obtained may far outweigh the negative changes. Nevertheless, manipulations must be examined from a community or ecosystem perspective prior to implementation to ensure that possible negative impacts are not incurred.

IV. ADVERSE EFFECTS OF SOIL AMENDMENTS ON WILDLIFE

While soil amendments have been found to have positive influences on wildlife habitats and populations, concerns have been raised about possible toxicants or hazards contained in some types of soil amendments, such as sewage sludges, and their possible effects on wildlife populations. Concerns have been expressed over heavy metals, especially lead and cadmium, and trace organics.

A. METALS

Sewage sludges contain much higher levels of many metals than standard fertilizers.[40] Some of these metals may pose toxicity problems to wildlife populations. Wildlife may ingest the metals contained in sludges by directly ingesting sludge materials applied to areas supporting wildlife populations, ingesting sludge materials adhering to plants in these areas, ingesting soils which contain sludges or sludge-borne metals applied to such areas, or by ingesting metals accumulated in plant tissues or prey species. No research, to date, is known to document direct wildlife consumption of sludge materials that have been land applied, let alone monitor metal transfer associated with any such consumption. A study on metal retention in tissues of cattle directly fed large quantities of sewage sludge found elevated levels of cadmium, lead, and copper in selected tissues.[41] The authors noted that this level of sludge consumption was extreme and would never occur under natural conditions, but that even at this level, no ill effects were observed in the cattle. Thus, it is conceivable that wildlife species could directly consume sewage sludge applied to their habitat. However, the relatively short duration of time that sludge would be available, and thus the relatively small amount that could be ingested, make any direct effects from metal consumption extremely unlikely. Direct consumption of sludge adhering to plants has also not been specifically investigated. As with direct consumption of sludge, the short duration of time and the relatively small amounts of sludge that could be consumed in this manner make direct effects from this source extremely unlikely.

Wildlife have been shown to consume considerable amounts of soil (geophagy) to obtain certain desired substances, especially salts.[42] Wildlife could ingest soils that have been amended with sewage sludges and could consume metals contained in the sludges. No record of this has been documented, and neither have any possible effects. It is unlikely, however, that enough soil would be directly consumed to constitute a hazard for metal consumption from sludges.

Clearly, the greatest potential source of metals contained in sludges to wildlife populations is through entry into wildlife food chains. Numerous studies have investigated the uptake of metals contained in sludges by plants. Various factors will influence the uptake of metals contained in sludges by plants, including soil acidity,[40,43–46] metal concentrations in sludges or sludge application rates,[43,45,47–49] and plant species.[40,50] Most metals contained in sludges are not easily transferred to plant tissues,[40,51] but significant increases in metal concentrations in plants grown on sludge-amended soils have been reported for cadmium,[43,45,47–50,52–53] zinc,[43,45,47–48,52–53] nickel,[43,45,52] and copper,[45,52–53] with lead showing both an increase[43] and a decrease.[52] Fresquez et al.[54] found no significant differences in levels of cadmium or lead in blue grama grown on sludge-amended semiarid grasslands. Similarly, Woodyard et al.[55] and Woodyard and Haufler[56] found no differences in accumulation of cadmium, copper, chromium, nickel, or zinc in forages growing in four different forest types amended with sewage sludge in Michigan as compared to vegetation from control sites.

These studies reveal that several metals can be transferred from sludge-amended soils to plants, and could then be ingested by herbivorous species of wildlife. Of the metals that can be transferred, cadmium is of greatest concern because of its potential toxicity at relatively low levels.[40] Zinc, copper, and nickel have never been found to be accumulated at levels considered to be toxicity threats to wildlife populations.

Several studies have reported higher levels of cadmium in livers and/or kidneys of herbivores foraging in sludge-amended areas. Anderson et al.[57] and Levine et al.[50] found higher levels of cadmium in meadow voles from sludge-amended areas compared to controls. Hegstrom and West[58] found higher cadmium levels in deer mice *(Peromyscus maniculatus)* living in sludge-amended areas. None of these studies suggested that the cadmium levels found represented a threat to the populations. Baxter et al.[59] investigated cadmium levels in cattle grazing on a sewage sludge disposal site compared to cattle on a control site, and found higher levels of cadmium in the kidneys of cattle from the sludged site. These levels were not above the range that could normally be expected from older range cattle.

In contrast, studies in Pennsylvania[53] found no accumulation of cadmium in cottontail rabbits *(Sylvilagus floridanus)* living on sludge-amended areas. In Michigan, no increases in cadmium levels were found in 13-lined ground squirrels, eastern chipmunks, woodland jumping mice *(Napaeozapus insignis),* or deer mice (*Peromyscus* spp.) living on sludge-amended sites.[55–56] Woodyard and Haufler[56] also investigated the possible transfer of cadmium and other metals through a wildlife food chain in a laboratory setting. They amended soils in growth chambers with fertilizer, sludge from Alpena, Michigan wastewater treatment facilities, and sludge from Detroit, Michigan wastewater treatment facilities. Rye-grass *(Lolium perenne)* grown on these three amended soils was harvested and combined with soybean and corn meal. This mixture was then fed to white-footed mice *(P. leucopus)* scheduled to be the food for red-tailed hawks *(Buteo jamaicensis)* and great-horned owls *(Bubo virginianus).* Three metals, cadmium, chromium, and zinc, had higher concentrations in ryegrass grown on the two sludge-amended soils than on the inorganic fertilizer-amended soil. However, after 60-day feeding trials on the different forages, no differences could be found in any of the tissues of the white-footed mice, therefore, the raptor tissues were not analyzed. Obviously, in this potential food chain, a transfer barrier for the metals existed between the plants and the herbivores.

Plants are not the only possible entry route of metals into wildlife food chains. Sludges may also be directly or indirectly ingested by certain detritivores, such as earthworms *(Lumbricus terrestris),* which may in turn be consumed by a number of wildlife species. Pietz et al.[60] found significantly higher concentrations of cadmium and copper in earthworms living in sludge-amended soils than in control sites. They cautioned that these elevated metal levels could pose a hazard to predators. To further examine this possibility, Woodyard and Haufler[56] tested transfer of cadmium, copper, chromium, nickel, and zinc in a soil-earthworm-American woodcock *(Philohela minor)* food chain. They amended separate soils with inorganic fertilizer, sludge obtained from Alpena, Michigan wastewater treatment facilities, and sludge obtained from Detroit, Michigan wastewater treatment facilities. Earthworms were raised in each type of amended soil and fed to woodcock to test for metal transfer. They found higher concentrations of cadmium, copper, chromium, and zinc in earthworms from sludge-amended soils when compared to earthworms raised in inorganic fertilizer-amended soils. However, levels of chromium, copper, and zinc were lower in the earthworms than in the soils. Cadmium concentrations in earthworms were approximately five-times greater than in the soils. Woodcock-fed earthworms raised in sludge-amended soils had significantly higher levels of cadmium than woodcock-fed earthworms from inorganic fertilizer-amended soils. Other metals did not differ among woodcock treatments. It was felt that woodcock, if maintained on a similar diet for 2 years, could develop toxicity problems from cadmium in the sludge. Woodyard and Haufler[56] recommended avoiding land application of sewage sludges containing substantial levels of cadmium, especially to sites with significant invertebrate detritivore populations such as those with high soil fertility and organic matter.

These results suggest that metals contained in sewage sludges are generally not considered to be a threat to wildlife populations. Seaker[61] felt that surface application of sewage sludges with low to medium levels of metals did not pose threats to the environment or wildlife food chains. An exception is cadmium, however, even this metal does not appear to be a problem to herbivores. Cadmium can enter detritivore food chains and may accumulate at higher trophic levels. Therefore, the use of sewage sludges should avoid substantial cadmium levels, and applications should avoid sites that typically contain high invertebrate detritivore populations.

B. TRACE ORGANICS

Few investigations have addressed the relationship between trace organic compounds contained in soil amendments, such as sewage sludges, and their effects on wildlife populations. Baxter et al.[59] examined cattle grazing on a sludge disposal site as compared to cattle on a control area for levels of 22 persistent organics. Only one substance had significantly different levels in fat, and this was lower from the sludge-amended site. Baxter et al.[52] also found no differences in 22 persistent organics in forage samples from the sludge disposal site and control areas. Bellin and O'Connor[62] investigated plant uptake of pentachlorophenol, an organic compound of environmental concern that was found in 156 of 223 sewage sludges tested in Michigan. These authors found that rapid degradation of this compound in the soil limited its availability to plants. Similarly, Jin and O'Connor[63] looked at toluene added to sludge-amended soils and found that gaseous transfer removed this compound within 10 days of application. Aranda et al.[64] found no uptake by plants of di-(2-ethyl hexyl) phthalate (DEHP), a priority organic pollutant frequently found in municipal sludges.

Elliott[40] felt that contamination of above-ground plant parts was unlikely from organic compounds found in sewage sludges, but that more research on levels in roots and tubers may be warranted.

Thus, no significant concerns with trace organics contained in sewage sludges and their potential effects on wildlife populations are presently known. It should be noted, however, that little research has been directed towards this subject, and new problems with trace organics in sludges could be discovered.

V. CONCLUSIONS AND RECOMMENDATIONS

Applying soil amendments to appropriate lands, prior to the growing season, may increase the quantity and quality of vegetation for wildlife use. Plant species which will benefit the most from additional nutrients are species which characteristically assimilate nutrients readily and/or those which have shallow root systems, such as herbaceous species. These species may benefit primarily because nutrient enhancements may be restricted to surface soil layers. The assimilation of nutrients, such as nitrogen and phosphorus, by plants will aid in maintaining wildlife forage species in a productive state of vigor which is attractive and can be used readily by herbivores. These increases in plant nutrients will usually increase their nutritional quality to wildlife and lead to their increased utilization by herbivores. Because soil amendments can increase production and nutritive qualities, however, plant maturation may be accelerated and therefore, the benefits of soil amendments short-lived as plants grow beyond the reach of wildlife species. Additional information regarding the duration of different types of soil amendment benefits to various vegetation types should be the focus of future research. Providing natural resource managers with this information will enable them to use soil amendments as a habitat management tool to maintain desirable forage species in a high nutritional status by treating various areas on a rotational basis.

Heavy metals and trace organic compounds are not felt to represent threats to wildlife from sludge application. Cadmium has been found to enter detritivore food chains, so sludge with high cadmium levels should be avoided, especially on sites with high numbers of detritivores such as earthworms.

REFERENCES

1. Barrett, M.W., Evaluation of fertilizer on pronghorn winter range in Alberta, *J. Range Manage.*, 32, 55, 1979.
2. Lemlich, S.K., and Ewel, K.C., Effects of wastewater disposal on growth rates of cypress trees, *J. Environ. Qual.*, 13, 602, 1984.
3. Haufler, J.B., and West, S.D., Wildlife responses to forest application of sewage sludge, in *The Forest Alternative for Treatment and Utilization of Municipal and Industrial Wastes*, Cole, D.W., Henry, C.L., and Nutter, W.L., Eds., University of Washington Press, Seattle, 1986, 110.

4. Grant, W.E., Dyer, N.I., and Swift, D.M., Response of a small mammal community to water and nitrogen treatments in a short grass prairie ecosystem, *J. Mammal.*, 58, 637, 1977.
5. Reed, F.C.P., Plant species number, biomass accumulation, and productivity of a differentially fertilized Michigan old-field, *Oceologia,* 30, 43, 1977.
6. Harcombe, P.A., The influence of fertilization on some aspects of succession in a humid tropical forest, *Ecology*, 58, 1375, 1977.
7. Woodyard, D.K., Response of wildlife to land application of sewage sludge, M.S. thesis, Michigan State University, East Lansing, 1982.
8. Thomas, A.H., First-year responses of wildlife and wildlife habitat to sewage sludge application in a northern hardwoods forest, M.S. thesis, Michigan State University, East Lansing, 1983.
9. Seon, E.M., Nutritional, wildlife, and vegetative community response to municipal sludge application of a jack pine/red pine forest, M.S. thesis, Michigan State University, East Lansing, 1984.
10. Haufler, et al., unpublished data, 1985.
11. Edmonds, R.L., and Cole, D.W., *Use of Dewatered Sludge as an Amendment for Forest Growth*, Vol. 2, Center for Ecosystem Studies, College of Forest Resources, University of Washington, Seattle, 1977.
12. Taber, R.D., and West, S.D., Biological processes in the control of risk tree species on rights-of-way in forested mountains: Pacific Northwest, in *Third Symposium on Environmental Concerns in Rights-of-Way Management*, Crabtree, A. Ed., Mississippi State University, Mississippi State, 1982, 410.
13. Sopper, W.E., and Kardos, L.T., Eds. *Recycling Treated Municipal Wastewater and Sludge through Forest and Cropland*, Pennsylvania State University Press, University Park, 479 pp., 1972.
14. Urie, D.H., Nutrient recycling under forests treated with sewage effluents and sludge, in *Utilization of Municipal Sewage Effluent and Sludge on Forests and Disturbed Land*, Sopper, W.E., and Kerr, S.N., Eds., Pennsylvania State University Press, University Park, 1979.
15. Anderson, T.J., and Barrett, G.W., Effects of dried sewage sludge on meadow vole *(Microtus pennsylvanicus)* populations in two grassland communities, *J. Appl. Ecol.*, 19, 759, 1982.
16. Berry, C.R., Use of municipal sewage sludge for improvement of forest sites in the southeast. U.S. For. Serv. Res. Pap. SE-266, 1987.
17. Fresquez, P.R., Francis, R.E., and Dennis, G.L., Sewage sludge effects on soil and plant quality in a degraded, semiarid grassland, *J. Environ. Qual.*, 19, 324, 1990.
18. Jokela, E.J., Smith, W.H., and Colbert, S.R., Growth and elemental content of slash pine 16 years after treated with garbage composted with sewage sludge, *J. Environ. Qual.*, 19, 146, 1990.
19. Safford, L.O., and Filip, S.M., Biomass and nutrient content of 4-year-old fertilized and unfertilized northern hardwood stands, *Can. J. Res.*, 4, 549, 1974.
20. Cole, D.W., Rinehart, M.L., Briggs, D.G., Henry, C.L., and Meciti, F., Response of Douglas-fir to sludge application: volume growth and specific gravity, in *TAPPI Research and Development Conference,* Appleton, Wis., Tech. Assoc. Park, Atlanta, 1984, 77.
21. Thomas, J.R., Cosper, H.R., and Bever, W., Effects of fertilizers on the growth of grass and its use by deer in the Black Hills of South Dakota, *Agron. J.,* 56, 223, 1964.

22. Bayoumi, M.A., and Smith, A.S., Response of big game winter range vegetation to fertilization, *J. Range Manage.*, 29, 44, 1976.
23. Hahn, E.P., Penkala, J.M., and Tourine, F., The use of sewage sludge as a fertilizer on outer-coastal plain soils, *Trans. N.E. Sec. Wildl. Soc.*, 1977, 33.
24. Dressler, R.L., and Wood, G.W., Deer habitat response to irrigation with municipal wastewater, *J. Wildl. Manage.*, 40, 639, 1976.
25. Blessin, C.W., and Garcia, W.J., Heavy metals in food chains by translocation to crops grown on sludge-treated strip mine land, in *Utilization of Municipal Sewage Effluent and Sludge on Forest and Disturbed Land*, Sopper, W.E. and Kerr, S.N., Eds., Pennsylvania State University Press, University Park, 1979, 471.
26. Campa, H., III, Nutritional responses of wildlife forages to municipal sludge application, M.S. thesis, Michigan State University, East Lansing, 1982.
27. Anderson, D.A., Reproductive success of Columbian black-tailed deer in a sewage-fertilized forest in western Washington, *J. Wildl. Manage.*, 47, 243, 1983.
28. Bailey, J.A., Effects of soil fertilization on the concentration of crude protein in witchhobble browse, *N.Y. Fish Game J.*, 15, 155, 1968.
29. Goetz, H., Effects of site and fertilization on protein content of native grasses, *J.Range Manage.*, 28, 380, 1975.
30. Campa, H., III, Woodyard, D.K., and Haufler, J.B., Deer and elk use of forages treated with municipal sewage sludge, in *The Forest Alternative for Treatment and Utilization of Municipal and Industrial Wastes*, Cole, D.W., Henry, C.L., and Nutter, W.L., Eds., University of Washington Press, Seattle, 1986, 188.
31. Murphy, D.A., and Coates, J.A., Effects of dietary protein on deer, *Trans. N. Am. Wildl. Conf.*, 31, 129, 1966.
32. Bledsoe, C.S., and Zasoski, R.J., Growth and nutrition of forest tree seedlings grown in sludge amended media, in *Use of Dewatered Sludge as an Amendment for Forest Growth: Management and Biological Assessments*, Edmonds, R.C., and Cole, D.W., Eds., College of Forest Resources, University of Washington, Seattle, 1979, 75.
33. Probasco, G.E., and Bjugstad, A.J., Influence of fertilizer, aspect, and harvest date on chemical constituents and *in vitro* digestibility of tall fescue, *J. Range Manage.*, 33, 244, 1980.
34. Kozlowski, T.T., and Keller, T., Food relations of woody plants, *Bot. Rev.*, 32, 293, 1968.
35. Cook, W.C., and Harris, L.E., The nutritive value of range forage as affected by vegetation type, site, and stage of maturity, *Utah Agric. Exp. Sta. Tech. Bull. 344*, 1950.
36. Hanley, T.A., Cervid activity patterns in relation to foraging constraints: Western Washington, *Northwest Sci.*, 56, 208, 1982.
37. Anderson, B.L., Pieper, R.D., and Howard, V.W., Jr., Growth response and deer utilization of fertilized browse, *J. Wildl. Manage.*, 38, 525, 1974.
38. Thomas, J.R., Cosper, H.R., and Bever, W., Effects of fertilizers on the growth of grass and its use by deer in the Black Hills of South Dakota, *Agron. J.*, 56, 223, 1964.
39. Anderson, D.A., Influence of sewage sludge fertilization on food habits of deer in western Washington, *J. Wildl. Manage.*, 49, 91, 1985.
40. Elliott, H.A., Land application of municipal sewage sludge, *J. Soil Water Cons.*, 41, 5, 1986.
41. Baxter, J.C., Barry, B., Johnson, D.E., and Kienholz, E.W., Heavy metal retention in cattle tissues from ingestion of sewage sludge, *J. Environ. Qual.*, 11, 616, 1982.

42. Jones, R.L., and Hanson, H.C., *Mineral Licks, Geophagy, and Biogeochemistry of North American Ungulates,* Iowa State University Press, Ames, 1985, 301pp.
43. Singh, B.R., and Narwal, R.P., Plant availability of heavy metals in a sludge-treated soil: II. Metal extractability compared with plant metal uptake, *J. Environ. Qual.*, 13, 344, 1984.
44. Pierzynski, G.M., and Jacobs, L.W., Extractability and plant availability of molybdenum from inorganic and sewage sludge sources, *J. Environ. Qual.*, 15, 323, 1986.
45. Heckman, J.R., Angle, J.S., and Chaney, R.L., Residual effects of sewage sludge on soybean: I. Accumulation of heavy metals, *J. Environ. Qual.*, 16, 113, 1987.
46. Jackson, A.P., and Alloway, B.J., The transfer of cadmium from sewage-sludge amended soils into the edible components of food crops, *Water, Air, Soil Pollut.,* 57, 873, 1991.
47. Chang, A.C., Page, A.L., Warneke, J.E., Resketo, M.R., and Jones, T.E., Accumulation of cadmium and zinc in barley grown on sludge-treated soils: a long term field study, *J. Environ. Qual.*, 12, 391, 1983.
48. Pietz, R.I., Peterson, J.R., Hinesly, T.D., Ziegler, E.L., Redborg, K.E., and Lue-hing, C., Sewage sludge application to calcareous strip-mine spoil: II. Effect on spoil and corn cadmium, copper, nickel, and zinc, *J. Environ. Qual.*, 12, 463, 1983.
49. Berry, C.R., Growth and heavy metal accumulation in pine seedlings grown with sewage sludge, *J. Environ. Qual.*, 14, 415, 1985.
50. Levine, M.B., Hall, A.T., Barrett, G.W., and Taylor, D.H., Heavy metal concentrations during ten years of sludge treatment to an old-field community, *J. Environ. Qual.*, 18, 411, 1989.
51. Elliott, H.A., Dempsey, B.A., and Maille, P.J., Content and fractionation of heavy metals in water treatment sludges, *J. Environ. Qual.*, 19, 330, 1990.
52. Baxter, J.C., Aguilar, M., and Brown, K., Heavy metals and persistent organics at a sewage sludge disposal site, *J. Environ. Qual.*, 12, 311, 1983.
53. Dressler, R.L., Storm, G.L., Tzilkowski, W.M., and Sopper, W.E., Heavy metals in cottontail rabbits on mined lands treated with sewage sludge, *J. Environ. Qual.*, 15, 278, 1986.
54. Fresquez, P.R., Aguilar, R., Francis, R.E., and Aldon, E.F., Heavy metal uptake by blue grama growing in a degraded semiarid soil amended with sewage sludge, *Water, Air, Soil Pollut.,* 57, 903, 1991.
55. Woodyard, D.K., Campa, H., III, and Haufler, J.B., The influence of forest application of sewage sludge on the concentration of metals in vegetation and small mammals, in *The Forest Alternative for Treatment and Utilization of Municipal and Industrial Wastes*, Cole, D.W., Henry, C.L., and Nutter, W.L., Eds., University of Washington Press, Seattle, 1986, 199.
56. Woodyard, D.K., and Haufler, J.B., *Risk Evaluation for Sludge-Borne Elements to Wildlife Food Chains*, Garland Publishing Co., New York, 1991, 188pp.
57. Anderson, T.J., Barrett, G.W., Clark, C.S., Elia, V.J., and Majeti, V.A., Metal concentrations in tissues of meadow voles from sewage sludge-treated fields, *J. Environ. Qual.*, 11, 272, 1982.
58. Hegstrom, L.J., and West, S.D., Heavy metal accumulation in small mammals following sewage sludge application to forests, *J. Environ. Qual.*, 18, 345, 1989.
59. Baxter, J.C., Johnson, D.E., and Kienholz, E.W., Heavy metals and persistent organics content in cattle exposed to sewage sludge, *J. Environ. Qual.*, 12, 316, 1983.

60. Pietz, R.I., Peterson, J.R., Prater, J.E., and Zenz, D.R., Metal concentrations in earthworms from sewage sludge-amended soils at a strip mine reclamation site, *J. Environ. Qual.*, 13, 651, 1984.
61. Seaker, E.M., Zinc, copper, cadmium, and lead in minespoil, water, and plants from reclaimed mine land amended with sewage sludge, *Water, Air, Soil Pollut.*, 57, 849, 1991.
62. Bellin, C.A., and O'Connor, G.A., Plant uptake of pentachlorophenol from sludge-amended soils, *J. Environ. Qual.*, 19, 598, 1990.
63. Jin, Y., and O'Connor, G.A., Behavior of toluene added to sludge-amended soils, *J. Environ. Qual.*, 19, 573, 1990.
64. Aranda, J.M., O'Connor, G.A., and Eiceman, G.A., Effects of sewage sludge on di-(2-ethylhexyl) phthalate uptake by plants, *J. Environ. Qual.*, 18, 45, 1989.

CHAPTER 4

Forestry

Charles L. Henry, Robert B. Harrison, and Dale W. Cole

TABLE OF CONTENTS

0-87371-860-7/95/$0.00+$.50
© 1995 by CRC Press, Inc.

I. INTRODUCTION

Municipalities and industries are faced with managing increasing quantities of waste products, both the solid waste and the residuals from primary and secondary treatment of their waste waters. Two events have recently occurred which increase the attractiveness of utilization of these waste products. The first is promulgation of 40 CFR 503[1] which has solidified U.S. Environmental Protection Agency's preference for beneficial use of biosolids (municipal sewage sludge) as a management alternative. The second is the goal to reduce the solid waste stream entering landfills, since siting and permitting landfills are becoming more difficult, expensive, and discouraged by recent legislation. Incineration is also under increased regulatory scrutiny as a high public health risk alternative and it is very difficult to site a facility.

Many research studies have investigated and established the benefits of use of organic residuals as soil amendments. Additionally, changing characteristics of organic residuals favor land application. Biosolids have been improved by many industrial pretreatment programs, and pulp and paper (P&P) sludges are being produced which have lower organic contaminants due to changes in P&P making processes. Hazardous waste collection programs have the potential to reduce the contaminants in solid waste materials. Physical changes have also made these materials more attractive for use, such as drier products or composts. Yet in many instances, landfilling continues to be the major management strategy. It is difficult to understand why land application has played such a minor role for residuals management considering the positive growth benefits that can be derived from land application in light of the increasing demand for fiber.

There are a number of reasons for considering forested sites as potential candidates for utilization of organic residuals:

1. Many forests are limited in productivity due to deficiencies in major nutrients that can be found in organic residuals (such as biosolids), especially nitrogen and phosphorus. A lack of adequate nutrition is the main factor limiting forest productivity, and consistent responses are seen with fertilization with nitrogen.[2]
2. Many forest soils are of marginal productivity due to poor soil textural characteristics. Highly productive land has often been converted years before to agricultural uses. The addition of organic matter from organic residuals can greatly improve both moisture and nutrient retention.
3. Since relatively small amounts of food are gathered from a forested site in comparison to an agricultural site, many of the public health concerns and land application regulations should not be as critical as those associated with agricultural sites.[3]
4. Forest soils theoretically have properties well suited to receiving organic residuals additions, including a forest floor with a great deal of organic carbon which can immobilize available nitrogen, a high infiltration rate which should minimize the potential for surface runoff, and a perennial root system which in some cases allow for year round uptake of available nutrients.

A. TYPES OF SOIL AMENDMENTS

A number of potential soil amendments exist which can be used in a forest environment. These include (1) municipal biosolids, (2) pulp and paper sludges, (3) incinerator ash, (4) composts derived from a variety of feed stocks, such as biosolids, yard waste, municipal solid waste (MSW), and co-composts, and (5) forest products waste. However, in comparison to agricultural applications of organic residuals, research on forest applications has been relatively minimal.

Land application of municipal biosolids has been successfully practiced for many years and has proven to be an effective fertilizer and soil conditioner when properly applied to certain types of forest lands.[4-8] P&P sludges also have been successfully used for soil conditioning and soil nutrition enhancing properties for tree growth.[9-10] A 20-year research program in biosolids applications to forest lands has been conducted by the University of Washington and funded by the Municipality of Metropolitan Seattle. Other research programs have been carried out by the University of British Columbia, University of Michigan, University of New Hampshire, and Pennsylvania State University. In addition, both New Zealand and Australia have initiated programs in forest applications of biosolids. Pulp and paper sludge applications have occurred operationally in the northeastern states and in Wisconsin. Research has also been conducted by the University of Washington and University of British Columbia.

Incinerator ash has been studied and applied to forest lands in the northeast, while applications of compost to forest stands have been studied by the University of Florida.

B. TYPES OF APPLICATIONS

These materials can be used in a variety of scenarios including: (1) application to recent clearcuts, (2) application over the canopy of a young plantation, and (3) application under the canopy of an older stand.

1. *Clearcuts*

Clearcuts offer the easiest, most economical sites to apply organic residuals. Since application will take place prior to tree planting, many agricultural application methods are appropriate. Vehicles delivering organic residuals can discharge directly on the land followed by spreading by a dozer and disking. Ease of delivery will depend upon the amount of site preparation (stump removal, residual debris burning, etc.), slopes, soil conditions, and weather. Other options available are temporary spray irrigation systems, injectors and splash plates for more liquid material, or manure spreaders for more solid material. Again, site characteristics and preparation are major factors influencing application technique.

While clearcuts are easier to apply to, they also may have major drawbacks depending upon application method and rate. A site surface applied with a relatively heavy rate of biosolids suddenly has an abundance of nutrients, and grass and weed growth can be vigorous. Plantation establishment becomes much harder due to vegetative competition, and leads to much higher mortality if grasses are not held in check by extensive plantation maintenance. Vigorous subordinate vegetation also offers excellent protection for rodents such as voles, which can proliferate and add to seedling survival problems from trunk girdling. Deer browse can also be a major concern before young trees are above browse height. If application of high nutrient organic residuals to a clearcut is planned, a program of periodic disking and herbiciding should be followed to control grasses and voles. Tree trunk protection devises are also available which provide a barrier against rodent girdling. Additionally, a fence or bud capping may be required to prevent excessive deer browse. One excellent clearcut application scenario is used in Christmas tree stands. Typically in this case a high level of maintenance is common and weed establishment and populations of rodents are minimized.

Conversely, an organic residual with a relatively high carbon to nitrogen (C:N) ratio can act as a mulch, reducing weed competition and decreasing moisture losses.[11]

2. *Young Plantations*

Applications of organic residuals to existing stands decrease problems with competition, rodents and deer. However, some application alternatives appropriate for clearcuts are not appropriate for existing plantations.

Semi-liquid materials are typically applied by a tanker/sprayer system, which can apply 18% biosolids over the canopy 50 m into the plantation. This method requires application trails at a maximum of 100-m intervals. Timing of applications is important with over the canopy applications. To aid in washing biosolids from the foliage and to keep biosolids off new foliage, spraying should take place during the rainy non-growing season. Biosolids sticking on new foliage could retard the current years growth.

Liquid materials have also been successfully applied via sprinkler irrigation system. Clogging of nozzles has been the major drawback to application by sprinkler irrigation. Manure spreaders are capable of applying materials which cannot be sprayed. Depending upon the range of biosolids trajectory, application trails may need to be at closer intervals than with other methods.

3. Older Stands

Applications to older stands have the advantage that organic residuals can be applied year-round because the foliage will not be affected, since application takes place under the canopy. Application methods are similar to those described for young plantations. Typically stands are not in rows, which may eliminate some of the alternatives available for plantations.

II. BENEFICIAL EFFECTS OF AMENDMENTS

Potential benefits from applications of organic residuals to forested ecosystems can be classified in three categories: (1) soil improvement, (2) increased timber production, and (3) secondary benefits from understory vegetation response.

A. SOIL IMPROVEMENT

The value of organic residuals is their ability to amend the soil by improving soil textural characteristics. The fine particles and organics can immediately and permanently enhance soil moisture and nutrient holding characteristics. Additionally, nutrients can be provided by some amendments (biosolids provide an excellent balance of virtually all nutrients) and some adjust the soil pH (such as an ash which is typically pH>9). Both short-term and long-term soil productivity can be improved. In the long-term many organic soil amendments provide continual slow release inputs of nutrients as the organics decompose. Applications of composted MSW to forest soils have been shown to improve the soil conditions for the developing stand in addition to restoring nutrients lost when trees are harvested.[12-13]

B. GROWTH RESPONSE

1. Biosolids

Numerous studies have been conducted on growth responses following biosolids applications. Increased tree growth have been reported in a number of symposium proceedings.[4-6,8] Growth response has been well documented in stands of Douglas fir *(Pseudotsuga menziesii)* in Washington.[14] Growth responses can range from 2 to

100% for existing stands, and over 1000% for trees planted in soils amended with heavy applications of biosolids. The magnitude of this response depends upon a number of site characteristics and stand ages. Some of the main site differences affecting the response include: (1) Site class. In both young Douglas fir plantations and older stands, greater responses have been found at lower productivity sites. (2) Thinned vs. unthinned stands treated with biosolids. There appears to be little difference in volume production for unthinned vs. thinned stands which have received biosolids applications. However, it is expected that growth is concentrated on higher value trees in thinned stands. (3) Response by species. Positive response to biosolids appears to exist for most species of trees, yet the magnitude of these responses range considerably.

Although the growth response of plantations established in biosolids-amended soils has been dramatic, survival is a major problem. It should be noted that some species suffered almost total mortality and would not be recommended for use under these circumstances. Additionally, the problem with plantation establishment has led to the recommendation that, where possible, biosolids be applied to existing stands rather than clearcuts. One exception to this recommendation is application to Christmas tree plantations where weed, small mammal, and deer control is usually under control.

It is very speculative at this time to estimate the monetary returns from accelerated tree growth for a number of reasons: (1) only a few stands of trees have been measured for growth response, (2) growth responses have only been followed for a short time, i.e., it is not known how long the effects of a biosolids application will continue, and (3) benefits have to be estimated many years into the future which depends upon a number of greatly fluctuating factors, such as discount rate and future value of wood. A conservative estimate of the value of biosolids could be on the value of the nitrogen fertilizer potential alone, which would be in the neighborhood of $30/dry tonne. Preliminary studies, however, have shown a greater response to biosolids than nitrogen fertilizer. Additionally, the effect appears to be much longer lasting, with some studies showing continued growth response eight years after application.

2. P&P Sludges

There typically exist two types of P&P sludges. The main constituent of primary P&P sludge is waste wood fiber; high in carbon and low in nutrients, thus potentially acting as a nutrient sink. In contrast, the microbial biomass of secondary sludge generally will release nutrients into soil as it decomposes.

Few studies have been conducted on growth response of tree species following P&P sludge additions. P&P sludges incorporated in as a soil amendment was investigated in nursery beds with three species of seedlings: Douglas-fir, western white pine *(Pinus monticola)*, and noble fir *(Abies procera)*.[15] The incorporation of secondary P&P sludge produced excellent growth responses; average height responses of 246, 107, and 207% of the controls were observed for Douglas-fir, western white pine and noble fir, respectively; and average diameter responses of 214, 116, and

167% of the controls, respectively. In contrast, with incorporated primary biosolids additions average height growth was significantly reduced; 68, 85, and 78% of the controls for Douglas-fir, western white pine and noble fir, respectively. Similarly, diameter growth was 66, 89, and 70% of controls, respectively. In a field study hybrid cottonwood (hybrid 11-11:*Populus deltoides x P. trichocarpa*) were surface-applied with P&P sludges.[11] Secondary P&P sludge addition resulted in average height and diameter increases for a total 3-year growth period of 256 and 281%, respectively, greater than the controls. Surface-applied primary sludge also resulted in increased average height and diameter responses of 94 and 92%, respectively, greater than untreated controls. The increase in hybrid cottonwood growth rates associated with primary biosolids application was suggested as being a mulching effect.

Douglas-fir seedlings were grown in a 4:1 primary/secondary P&P sludge mixture at rates of 50 to 450 mg ha^{-1}.[16] Visual evaluations suggested plots receiving additional nitrogen showed better response. Shields et al.[17] reported 41% greater growth of cottonwoods in soil amended with primary sludge compared to fertilized controls, and 68% greater growth when combined sludge was used (1:1 mixture). Secondary papermill sludge caused foliar nitrogen concentration and biomass to significantly increase in both over- and understory vegetation in a red pine plantation[10].

3. MSW Compost

A number of studies using MSW compost have been conducted in Florida. Growth response for young plantations amended with composted MSW has been well documented.[13,18-20] Application technique seems to affect response, however. In 2-year-old slash pine on sandy Florida soils, incorporation of 4.4 to 44 mg ha^{-1} showed significantly greater response than a similar surface application.[18]

C. SECONDARY BENEFITS

The third type of benefit from organic residuals additions is much more subjective. Although immediately after application the site is greatly altered in appearance, within 6 months understory growth is often much more vigorous than prior to application. Stimulation of understory herbaceous and shrub growth in a planted slash pine stand occurred during the first 5 years following addition of MSW and biosolids compost.[13] This is not only visually pleasing, but can be of commercial value to "brush pickers" harvesting ferns and other vegetation for floral arrangements. Increased understory is also typically higher in nutrients and can provide better habitat for wildlife. Researchers to-date have found no significant adverse effects on health of wildlife.[21] Although elevated levels of trace metals have been found in the organs of some of the small mammals that live in the biosolids-soil, there is no evidence of significant accumulation in the food chain. Positive effects have also been shown; increased vigor in animals that browse on the higher nutritional foliage. Increased populations of wildlife have also been found.

III. FOREST SOIL CHARACTERISTICS IMPORTANT TO FOREST ECOSYSTEMS IMPACTED BY SOIL AMENDMENTS

The desired objective of a land application program using organic residuals is to treat and utilize the waste in an environmentally safe and cost effective manner, including full utilization of the nutrients and soil conditioning properties. Research and demonstration sites have shown that this goal can be met on a variety of site and vegetation conditions. However, it is necessary to recognize all aspects of the environment potentially affected by a biosolids application, as discussed below.

A. CHEMICAL

Most of the ionic constituents in leachates from organic residuals are initially removed from solution through chemical reactions. Some of these reactions, such as ionic replacement on soil exchange sites, provide only temporary removal where as precipitation reactions can result in the permanent removal of the reacting ions from the soil drainage waters.

In general, anions will leach through the soil far more rapidly than cations due to the negative surface charge of most soil colloids. Two in particular, NO_3^- and Cl^-, will rapidly leach if present in sufficient concentrations. In contrast, K^+, Ca^{2+}, and Mg^{2+} will be removed by cation exchange reactions. There are important exceptions, however, to this general statement. Phosphorus, an anion and major constituent in biosolids is readily precipitated from the soil solution typically as a calcium or iron phosphate depending on the pH of the soil. The capacity of the soil to remove P is exceptionally high resulting in P not being a problem in any land application system.

Ammonium (NH_4^+) is removed from leachate by the exchange potential of the soil colloids, the same basic process discussed above for K^+, Ca^{2+} and Mg^{2+}. However, in the case of NH_4^+, this storage can be very temporary if nitrification occurs.

Almost all trace metals are strongly adsorbed on soil particles. This is especially true if the soil is neutral or basic. Trace metals can also be bound to organic chelating agents that convert them to anion complexes. In this form they readily move through the soil similar to the movement of other non-negative anions. Much research has been done on trace metal in agricultural applications. A specific review article was written by Logan and Chaney.[22] Harrison et al.[23] reported that trace metals are relatively immobile in soils in a forest site.

B. BIOLOGICAL

The "living" phase of the soil is also critical to the overall renovation process associated with land application. For example, it is responsible for the removal of pathogenic organisms and viruses as well as the uptake of nutrients, especially nitrogen, by the crop plants. The role of the "living" phase in the renovation process is perhaps even more important in forest systems than for agriculture crops.

Microorganisms associated with biosolids are initially filtered out and then replaced by the indigenous organisms of the soil. The survival time for most microorganisms following land application is typically very short but is dependent on a variety of soil and climatic conditions including temperature, moisture content, and pH. Bacterial pathogens will generally die off to negligible numbers within 2 to 3 months following application.[24-25] Viruses will commonly survive a maximum of 3 months, while protozoa will survive for only a few days.[26] In any case these microorganisms will not leach through the soil system and present a public health problem for the receiving ground waters.[25] They will remain in the surface soils for the duration of their survival period. Only if the site is subject to surface runoff will pathogens enter into receiving water bodies.

Forest soils, especially those with a well-developed forest floor layer, are excellent systems to renovate organic solutes, compounds, and solids. Typically, toxic organic compounds are not important constituents in most organic residuals including biosolids, P&P sludges and composts, especially those derived principally from domestic sources, and therefore not a problem in land application systems. Even when toxic organic compounds have been present, they have not presented a limiting parameter for the function of the land treatment system.[27] Organic compounds are readily sorbed to the organic surfaces of the soil system and thus have limited mobility through the soil profile. In that these organic compounds are typically biodegradable, they will not accumulate to any extent in the soil.

The living phase of the soil is also responsible for nutrient removal by plant uptake processes. In order for soil nutrient renovation, especially nitrogen renovation, to function over an extended period of time, the added nutrients must be removed by plant uptake. The rate of uptake that can be expected for forest species in land application systems was reported for each major region by McKim et al.[28] and Brockway et al.[29] In general, uptake and storage by forests can be as large as that of agriculture crops if the system is correctly managed and species capable of responding to nutrients are selected. Uptake and storage rates exceeding 300 kg/ha/year have been reported for 3 to 5 year-old poplar and 200 kg/ha/year for Douglas-fir ecosystems of the same age.[30] This utilization rate for nitrogen compares favorable with uptake for almost any agricultural crop.

C. PHYSICAL

The physical filtering process in a land application system is very important in assuring desired drainage water standards. Most importantly, it slows down the rate of percolation, increasing the time for chemical and biological reactions to occur. This filtering process also results in the removal of microorganisms and organic matter where they can be degraded over time. Non-biodegradable organic matter will be adsorbed to the soil surfaces and will not leach through the soil profile.

Much of the filtering process in forests takes place in the forest floor layer. Therefore, there is less of a tendency to clog the surface soil pores with organics (which decrease the infiltration capacity of the soil) associated with liquid organic residuals like biosolids.

IV. POTENTIAL ADVERSE EFFECTS ON SPECIFIC PROCESSES AND FUNCTIONS

A. GROUND WATER

The solutes from organic residuals that enter into the soil solution are retarded while passing through the soil system, that is, the movement of water is greater than the movement of the solutes. This delay is caused by a variety of chemical, physical, and biological processes. It is from a combination of such processes that allows treatment and beneficial use of the biosolids constituents to occur. However, not all solute constituents move through the soil at the same rate, nor is the amount of treatment provided the same. Potentially the most problematic are the anions, such as nitrate, which have a minimal delay and can rapidly move to the ground water. Consequently, it is important to minimize the presence of nitrate through the proper design and management at the land treatment system as will be discussed later.

B. SURFACE WATER

Proper design of the land application site will minimize the possibility of surface runoff. To minimize runoff it is essential to provide buffers from water bodies, and not exceed maximum slope recommendations, such as those suggested later. In general, forest soils have excellent infiltration rates because of the presence of a forest floor layer. In addition most forest soils have far greater porosities than agricultural soils. Both of these facts result in little, if any, runoff and make forests excellent candidates for land application systems.

C. AEROSOLS

The third way in which contaminants from organic residuals can leave a land application site is from aerosol dispersion during periods of applications. This movement is dependent on the design of the application system, and the rate of wind movement during an application period. It has been well documented that the rate of wind flow is greatly modified by the presence of a forest stand. Typically a 70 to 90% reduction in wind speed is found 50 m within a forest stand. Consequently, buffers to prevent the movement of aerosols from a forest land application system can be relatively narrow. Edmonds and Mayer[31] found some airborne bacteria 50 m from application of biosolids in a forest stand, but none at 100 m.

D. HUMAN AND ANIMAL HEALTH

Potential health risk from exposure of humans and animals to biosolids contaminants include (1) direct ingestion of biosolids, (2) intake through consumption of plants, and (3) intake through consumption of animals living on biosolids sites. Minimizing these risks is achieved through controlled loading limits of contaminants. Development of acceptable limits of contaminants for application of biosolids to forest systems has occurred and is reflected within EPA's 40 CFR 503.[1] Documentation of the risk assessment supporting these numeric limits is included in the Technical Support Documents of 40 CFR 503. These limits are probably also acceptable for application criteria for other organic residuals, as currently regulations for other organic residuals have not been developed.

V. MANAGEMENT PRACTICES TO MINIMIZE THE POTENTIAL FOR ADVERSE EFFECTS

A. APPLICATION RATE AND TIMING

Application rates for organic residuals will normally be based on one of two criteria: (1) nitrogen loadings for high nutrient waste materials (i.e., biosolids or secondary P&P sludges), or (2) organic matter contribution to the soil to amend the soil to a target percent organic matter. An amendment solely on the basis for increasing the organic matter is normally only a one-time application. In contrast, two application rate/timing philosophies exist for organic residuals with high nitrogen contents:

1. Annual applications designed to meet only the annual uptake requirements of the trees, considering volatilization and denitrification losses, and mineralization from current and prior years.
2. A heavier application rate 1 year followed by a number of years when no applications are made. This schedule depends on the soil storage to temporarily tie up excess nitrogen which will become available in later years.

Both philosophies have advantages. In the second alternative, application costs are lower due to entry into a site less frequently, and the public can use the site for recreation in the nonapplied years. Additionally, the excess available nitrogen present during the first year is no longer at elevated levels in following years, and leaching of nitrate in these years usually will not occur. If the organic residuals are at a low percent solids (<10%), annual applications may be preferred. With heavy applications at low percent solids, the total liquid quantity applied may exceed the site's capacity to treat it. Surface sealing may occur increasing the potential for runoff, and increasing anaerobic conditions which can cause odor problems or stress the plants.

Heavy applications should be made in a series of three or more partial applications depending upon the percent solids. This allows more even applications to be made and time for stabilization or drying of the liquid materials to occur; important for maintaining infiltration and controlling runoff. The "rest" between applications will range from 2 to 14 days depending on weather conditions.

B. IMPORTANT SITE SELECTION CRITERIA

Since the objectives of applying organic residuals to a forested site are to enhance tree growth, prevent environmental contamination and minimize operational costs, it is important that potential sites be evaluated with these objectives in mind. In particular consideration must be given to physical factors of a potential site, such as: (1) topography, (2) transportation and forest access, (3) soils and geology, (4) vegetation (stand and understory characteristics), (5) water resources, and (6) climate. A relative ranking of conditions within each of these factors is shown in Table 1. These numbers are suggested only for evaluating one site compared to another. In general, however, a ranking of "0" for any of the criteria means that the site should not be considered further for use as an application site. A ranking of 1 to 3 means that the site has some problems, and that there should be some very positive factors that offset these low rankings.

1. Topographic Factors

Probably the most important criterion which will quickly indicate the usability of a site is the topography. Important considerations are slopes and site continuity (land which is not broken up by waterways, cliffs, etc.). Both have major impacts on the usability of sites and development costs. Table 1 describes how sites may be evaluated according to differences in these two characteristics. Site continuity refers to the number of areas within a site which may require buffering, such as waterways, roads, steep slopes, etc. Depending upon the buffer distance, these discontinuities can rapidly decrease usable acreage and complicate applications. Another factor which may enhance operations is aspect of the site. The amount of solar radiation an area receives is affected by aspect; this radiation will aid stabilization of the biosolids, and also melt snow or thaw frozen ground.

2. Transportation and Forest Access Factors

Quite often hauling organic residuals to the site is the greatest expense in an application program. Therefore distance should be included as a major site selection criteria. In addition to the distance, the class of roads affect hauling costs. Also, routing of long-haul vehicles is an important factor, with consideration given to weight limits of roads and bridges, and travel through sensitive areas (population centers).

A major cost of site development is construction of appropriate access road and trail systems. On many forest sites a significant portion of these systems already

Table 1 Relative numeric ranking of forest sites for application of organic residuals

Factor	Relative numeric rank
Topography	
Slope	
<10%	10
10–20%	6
20–30%	3
over 30%	0
Site continuity (somewhat subjective)	
No draws, streams, etc., to buffer	10
1 or 2 requiring buffers	6
Numerous discontinuities	0–3
Transportation	
Distance	1–10
Condition of the roads	1–10
Travel through sensitive areas	1–10
Forest access system	
Percent of forest system in place	0–10
Ease of new construction	
Easy (good soils, little slope, young trees)	7–10
Difficult	1–5
Erosion hazard	
LIttle (good soils, little slope)	7–10
Great	1–5
Soil and geology	
Soil type	
Sandy gravel	10
Sandy	8
Well graded loam	5–6
Silty	3–6
Clayey	1–3
Organic	0
Depth of soil	
deeper than 3 m	10
1–3 m	8
0.3–1 m	4
<0.3 m	0
Geology (subjective, dependent upon aquifer)	
Sedimentary bedrock	6–9
Andesitic basalt	6–9
Basal tills	3–6
Lacustrine	1–3
Vegetation	
Tree species	
Hybrid cottonwood (highest N uptake rates)	10
Douglas-fir	9
Other conifers	7
Other mixed hardwoods	4
Stand age	
Hardwoods:	
<4 years	10
4 years or greater	8
Conifers:	
>30 years	7–10
>1 m high to 10 years	7–10
10–20 years	6–8
20–30 years	0–6
<1 m high	0–3

Table 1 (Continued) Relative numeric ranking of forest sites for application of organic residuals

Factor	Relative numeric rank
Stand condition	
Applying over the tree tops:	
Well stocked	10
Poorly stocked	2–6
Applying under the tree tops:	
Well thinned	10
Poorly thinned or dense	2–8
Understory	
Average 15 cm high or greater over 90% of site	10
Average 5 cm high over 70% of site	7
Partial vegetative cover	2–7
Water resources	
Ground water, depth to average seasonal water table	
Deeper than 5 m	10
2–5 m	7
0.6–2 m	4
<0.6 m	0
Ground water flow	
Away from usable aquifer	10
Significant contribution to usable aquifer	2–5
Distance to domestic use of aquifer	
>1500 m	10
300–1500 m	7
<300 m	3
Surface water, time to contribute to	
>1 h	7–10
Immediate	3
Surface water channels	
Easily defined and buffer	7–10
Difficult to define and buffer	1–5
Domestic use of surface water	
Not used	10
Some distant downstream single family uses	5
Used by municipality (depending upon distance from)	0–5
Climate (i.e., rainfall, temperature)	1–10

exist, especially those which are intensely managed and have had recent commercial thinning operations. Often, the investment made in construction of access systems will be at least partially recovered during final harvest.

3. Soil and Geologic Factors

Fortunately, some of the easiest sites to develop and operate on have soils which can biologically benefit the greatest from biosolids addition, such as the gravelly outwash soils. These soils not only are deficient in properties which can be supplied by many organic residuals (nutrients and soil tilth), but also usually are better drained. Trafficability, infiltration/ percolation, depth of soil, pH, and cation exchange capacity (CEC) all play important roles. Table 1 presents relative ratings of different soils in terms of texture and parent material. The importance of geologic factors lies in their relationship to ground and surface water hydrology in creating zones impermeable to downward water movement.

4. Vegetation Factors

Site vegetation is an important consideration in immediate acceptability of a site. Stand age and condition affect method of application and effectiveness in achieving even applications. Tree species affect the rate of application of high nutrient organic residuals. A well-established understory provides increased nutrient uptake, aids in stability of liquid materials and provides irregular surfaces which will minimize runoff potential.

5. Water Resource Factors

Application rates are designed to minimize impact on ground water. For most applications of high nutrient organic residuals, NO_3^- will be the only concern. Proper applications are designed so that excess N is not applied, thus significant leaching of NO_3^- is not expected. Because nutrient utilization takes place in the top layers of soil, distance to ground water table should be greater than 0.7 m throughout the year.

Nearness to waterways will eliminate portions of a site to protect against pollutants entering water bodies via runoff or biosolids erosion. If there are a number of drainage ways within or adjacent to a site, the site will obviously have less net usable area. An application area should not generally be a significant portion of a watershed used for drinking water supplies. Sites with standing surface waters and flooding potential should always be avoided. These factors are rated in Table 1.

6. Climatic Factors

Climatic factors to consider in site evaluation include temperature and precipitation; important in the number of days during the year when snow, freezing conditions or rain can be expected to interrupt operations. Elevation will affect the number of days a site may be unsuitable for applications due to snow or freezing.

C. ALLOWABLE CONTAMINANT LOADINGS

Controlled loading limits of contaminants and proper management practices keep the risks associated with application of organic residuals to forest lands exceptionally low as evidenced by the newly developed regulations for biosolids.[1] At this time it is reasonable to use these contaminant loadings as guidelines for other organic residuals until specific ones are developed. Table 2 presents the EPA's contaminant loadings for forest applications, which are the same as those for agriculture. EPA chose to eliminate regulation of trace synthetic organics because they were found only in very low concentrations in biosolids and it would require relatively high concentrations to cause an environmental or health concern. There is the possibility

Table 2 Trace Metal Loadings for Forest Applications of Biosolids[1]

Contaminant	Limit (kg/ha)
Arsenic	41
Cadmium	39
Chromium	3000
Copper	1500
Lead	300
Mercury	17
Molybdenum	18
Nickel	420
Selenium	100
Zinc	2800

of higher concentrations of trace synthetic organics in other organic residuals as compared to biosolids, and each new waste material should be evaluated. However, generally concentrations of consequence have not been found.

D. PATHOGEN REDUCTION

As with trace metals, little guidance is available for the majority of organic residuals in regards to pathogens, with the exception of biosolids. EPA has established two classifications for reduction of the total number of pathogen in biosolids. One form of evaluation of the classes is: (1) Class A will be met if the concentration of fecal coliform is less than 1000 per g dry solids or *salmonella* is below the detection limit, and (2) Class B biosolids will be met if the concentration of fecal coliform is less than 1,000,000 per g dry solids. A Class A biosolids has no public access restrictions, whereas a Class B biosolids requires public access restriction for 1 year.

E. NUTRIENT LOADINGS

As mentioned earlier, application rates for organic residuals will be based either on nutrients or organic matter. For high nutrient waste materials (i.e., biosolids or secondary P&P sludges), a balance must be achieved between application rates low enough that no environmental integrity is compromised from excess nutrients leaving the site and application rates that assure enough nutrients for a maximum growth response. At the same time rates should be responsive to site development and application costs. The three macronutrients most commonly considered in terms of need by plants are nitrogen, phosphorus, and potassium. Nitrogen has been traditionally considered the most important element in determining fertilizer or biosolids application rates because it is needed as a soil supplement in much greater amounts than the other two, and can, in the form of nitrate, be quite mobile in the soil.

Table 3 **Example Application Rate Assumptions and Calculations of Biosolids Based on Available Nitrogen for Two Different Types of Douglas-Fir Stands**

Assumptions	Young stand	Older stand
Initial NH_4^+-N, AN (g g^{-1} of dry solids)	0.01	0.01
Organic N, ON (g g^{-1} of dry solids)	0.04	0.04
Volatile loss, V (of initial NH_4^+)	0.25	0.10
N mineralized, M (of organic N)	0.25	0.25
Denitrification, D (of net available N)	0.25	0.10
Uptake by trees, TU (kg ha^{-1})	100	45
Uptake by understory, UU (kg ha^{-1})	90	25
Soil immobilization, SI (kg ha^{-1})	200	50
Target nitrate leaching (kg ha^{-1})	0	0
Equation: AR = (TU+UU+SI)/[1000*(AN*(1–V)+ON*M)(1–D)]		
Application rate, AR (Mg ha^{-1})	30	7
Total N to be applied (kg ha^{-1})	1500	350

1. Nitrogen Loading Rates

Excess N which exceeds the assimilative capacity of soil will result in nitrate leaching. A number of studies conducted at the University of Washington's Pack Forest confirmed this; heavy applications resulted in substantial increases of NO_3^- in the ground water.[32-35]

Biosolids or secondary P&P sludges may contain large amounts of nitrogen. Sommers[36] reported the average concentration of nitrogen as 4.2% in anaerobic biosolids. Anaerobic biosolids contain nitrogen existing in two main forms: organic N (typically about 80%) and mineral forms as ammonium (typically about 19%) and nitrate (<1%).[37] Nitrogen immediately available from biosolids addition (NH_4^+), and available shortly after application through mineralization can follow a number of pathways. The nitrogen can be (1) lost through ammonia volatilization; (2) taken up by the trees and understory; (3) temporarily held on soil exchange sites as NH_4^+; (4) stored long-term in the soil through microbial immobilization; or (5) microbially transformed into NO_3^-, which can either be lost to the system through leaching of the NO_3^-, or as N_2 or N_2O through denitrification. Table 3 presents two examples of assumptions and calculations that can be made for different types of Douglas-fir stands. These calculations, with a rather excessive range from 7 to 30 mg ha^{-1}, reflect the need to characterize the site appropriately and incorporate site assimilative capacities into rate calculations.

2. Phosphorus and Potassium

Phosphorus is generally found to be at concentrations in biosolids equal to about two-thirds of nitrogen concentrations, yet only needed by plants at rates equal to about 20 to 50% of that for nitrogen. Therefore it seldom limits plant growth when application rates are based on nitrogen needs. Precipitation and sorption of phosphorus in the soil are responsible for its limited availability, and this fixation in the soil restricts the

Table 4 Maximum Recommended Slopes for Applications of Biosolids to Forested Sites[38]

Dry season	
Good vegetative cover	30%
Poor vegetative cover	15%
Wet season	
Good vegetative cover	15%
Poor vegetative cover	8%

movement of it to ground water to very low levels. Furthermore, excessive phosphorus buildup in the soil poses no threat to forest plants or the environment.

In contrast to phosphorus, potassium is supplied at only about one-tenth the concentration of nitrogen from biosolids, yet is removed by plants at levels equaling nitrogen. Therefore, it is unlikely that a buildup of potassium can occur from a biosolids application. Potassium is generally found in comparatively high concentrations in soil and seldom is a limiting factor for growth.

F. SLOPE RESTRICTIONS

Application of liquid organic residuals to excessive slopes will enhance surficial flow (runoff) from an application site. Table 4 has been suggested for biosolids from field studies conducted in Washington State, and could readily be adapted to other organic waste materials.

G. BUFFER REQUIREMENTS

Buffers serve at least two purposes: (1) to provide a factor of safety against over oversprays or errors even when proper application and management techniques are used, and (2) to provide treatment and filtering of organic residuals and/or runoff from applied surfaces. The condition of the ground surface is critical. In other words, bare soil will provide virtually no filtering, a grassed surface will provide fair treatment, while a porous forest floor can provide excellent treatment and filtering. Differences should also be dependent upon the type of waterway. There are at least three reasons for this: (1) larger waterways generally are more difficult to cleanup once a mishap occurs (i.e., it is usually not possible to seal off a stream that requires biosolids cleaned up from it, like a ditch can be), (2) larger waterways have more potential for domestic use, and (3) larger streams often have a lesser defined bank and that is farther away from the normal waterline. Table 5 presents recommended buffers from waterways from biosolids sites.

VI. SUMMARY

Because of their high nutritional content and soil conditioning properties, organic residuals can serve as soil amendments for nutritionally deprived or organically poor

Table 5 Buffer Recommendations for Applications of Biosolids to Forested Sites[38]

	Type of water body			
Application method and buffer condition	**Large continually flowing**	**Small tributary**	**Ephemeral**	**Ditches**
Surface applied				
Undisturbed buffer	60 m	30 m	15 m	10 m
Disturbed buffer	60 m	60 m	30 m	15 m
Injected of incorporated	30 m	30 m	15 m	10 m

soils on forest sites. Studies conducted over the past 20 years on a number of different organic residuals at a number of different locations across the U.S. have largely confirmed the potential of these materials to increase the productivity of many forest lands. These studies clearly demonstrated that organic residuals, applied at environmentally acceptable rates, will result in growth responses for both young seedlings as well as established stands.

A silviculture program utilizing organic residuals must be carried out using environmentally responsible management techniques. A recent exposure risk assessment performed for the U.S. Environmental Protection Agency on nonagricultural land (including forests) has suggested that the contaminants in typical biosolids in the U.S. pose insignificant risk to humans and the environment when properly utilized. This is especially true since the quality of residuals has increased dramatically over the last 20 years. However, application rates should not exceed the capacity of the system to utilize and retain the nutrients applied, or losses through leaching can be expected. The most likely nutrient to be leached following application is N in the form of NO_3^-. For this reason, sites already high in N or those that receive high levels of N from atmospheric deposition should be avoided. When properly applied, organic residuals can provide an excellent alternative to chemical fertilizers as a means of enhancing forest production. Growth response can be greater and can last longer when compared to chemical fertilization. Careful consideration has to be made of the site conditions to be certain that environmental risks are minimal and losses through leaching and overland flow will not occur.

REFERENCES

1. U.S. Environmental Protection Agency, 40 CFR 503 Use and Disposal of Sewage Sludge, 1992.
2. Heilman, P.E., Minerals, chemical properties and fertility of forest soils, in *Forest Soils of the Douglas-fir Region,* Heilman, P.E., et al., Eds, Washington State Cooperative Extension Service, Pullman, 1981, 121.
3. Henry, C.L., Evaluation of comments on the proposed standards for management of sewage sludge: Non-agricultural land application, U.S. EPA-NNEMS Publication, 1989.

4. Sopper, W.E., and S.N. Kerr, *Utilization of Municipal Sewage Effluents and Sludge on Forest and Disturbed Land*, Pennsylvania State University Press, University Park, 1979.
5. Bledsoe, C.S., *Municipal Sludge Application to Pacific Northwest Forest Lands*, Inst. Forest Resources Contrib. No. 41, College of Forest Resources, University of Washington, Seattle, 1981.
6. Sopper, W.E., E.M. Seaker, and R.K. Bastian, *Land Reclamation and Biomass Production with Municipal Wastewater and Sludge*, Pennsylvania State University Press, University Park, 1982.
7. Henry, C.L. and D.W. Cole, *Use of Dewatered Sludge as an Amendment for Forest Growth. Vol. IV,* Inst. For. Resources, University of Washington, Seattle, 1983.
8. Cole, D.W., C.L. Henry and W. Nutter, *The Forest Alternative for Treatment and Utilization of Municipal and Industrial Wastewater and Sludge,* University of Washington Press, Seattle, 1986.
9. Henry, C.L., Nitrogen dynamics of pulp and paper sludge amendment to forest soils, Ph.D. dissertation, University of Washington, Seattle, 1989.
10. Brockway, D.G., Forest floor, soil, and vegetation responses to sludge fertilization in red and white pine plantations, *Soil Sci. Soc. Am. J.*, 47, 776, 1983.
11. Henry, C.L., Nitrogen dynamics of pulp and paper sludge to forest soils., *Wat. Sci. Tech.,* 24, 417, 1991.
12. Smith, W.H., and J.O. Evans, Special opportunity and problems in using forest soils for organic waste applications, in *Soils for Management of Organic Waste and Waste Waters*, 428, 1977.
13. Smith, W.H., D.M. Post, and F.W. Adrian, Waste management to maintain or enhance productivity, in *Proc. of Impact of Intensive Harvesting on Forest Nutrient Cycling,* State University of New York, College of Environmental Science and Forestry, Syracuse, 304, 1979.
14. Henry, C.L., D.W. Cole, T.E. Hinckley, and R.B. Harrison, The use of municipal and pulp and paper sludges to increase production in forestry, *J. Sustainable Forestry*, (in press).
15. Henry, C.L., Growth response, mortality and foliar nitrogen concentrations of four tree species treated with pulp and paper and municipal sludges, in *The Forest Alternative for Treatment and Utilization of Municipal and Industrial Wastewater and Sludge,* Cole, D., C. Henry, and W. Nutter, Eds, University Press, Seattle, 258, 1986.
16. Aspitarte, T.R., Agricultural utilization of wastewater treatment sludges, in *Proc. 1979 NCASI Central-Lakes States Regional Meeting*, NCASI Special Report No. 80-02, 1980.
17. Shields, W.J., M.D. Huddy, and S.G. Somers, Pulp mill sludge application to a cottonwood plantation, in *The Forest Alternative for Treatment and Utilization of Municipal and Industrial Wastewater and Sludge*, Cole, D., C. Henry, and W. Nutter, Eds., University Press, Seattle, 533, 1986.
18. Bengtson, G.W., and J.J. Cornette, Disposal of composted municipal waste in a plantation of young slash pine: Effects on soil and trees, *J. Environ. Qual.,* 2(4), 441, 1973.
19. Fiskell, J.G.A., W.L. Pritchett, M Maftoun, and W.H. Smith, Effect of garbage compost rate and placement on a slash pine forest and metal distribution in an acid soil, in *2nd Ann. Conf. of Applied Res. and Practice on Municipal Industrial Waste, Madison, WI*, 17-21 September, 302, 1979.

20. Jokela, E.J., W.H. Smith, and S.R. Colbert, Growth and elemental content of slash pine 16 years after treatment with garbage composted with sewage sludge, *J. Environ. Qual.*, 19, 146, 1990.
21. Henry, C. and R. Harrison, *Literature Reviews on Environmental Effects of Sludge Management: Trace Metals, Effects on Wildlife and Domestic Animals, Incinerator Emissions and Ash, Nitrogen, Pathogens, and Trace Synthetic Organics,* Regional Sludge Management Committee, Seattle, 1991.
22. Logan, T.J. and R. Chaney, Metals, in *Utilization of Municipal Wastewater and Sludge on Land,* Page, A.L., et al., Eds, University of California, Riverside, 235, 1983.
23. Dongsen, X., Harrison, R., and C. Henry, Long-term effects of heavy applications of biosolids to coarse-textured soils: 2) trace metals, draft manuscript, 1993.
24. Gerba, C.P., Pathogens, in *Utilization of Municipal Wastewater and Sludge on Land,* Page, A.L., et al., Eds, University of California, Riverside, 147, 1983.
25. Edmonds, R.L., Microbiological characteristics of dewatered biosolids following application to forest soils and clearcut areas, in *Utilization of Municipal Sewage Effluent and Biosolids on Forest and Disturbed Land,* Sopper, W.E., and S.N. Kerr, Eds, Pennsylvania State University Press, University Park, 1979.
26. Kowel, N.E., An overview of public health effects, in *Utilization of Municipal Wastewater and Sludge on Land,* Page, A.L., et al., Eds, University of California, Riverside, 329, 1983.
27. Overcash, M.R., Land treatment of municipal effluent and biosolids: specific organic compounds, in *Utilization of Municipal Wastewater and Sludge on Land,* Page, A.L., et al., Eds, University of California, Riverside, 480, 1983.
28. McKim, H.L., W.E. Sopper, D.W. Cole, W. Nutter, D.H. Urie, P. Schiess, S.N. Kerr, and H. Farquhar, *Wastewater Applications in Forest Ecosystems,* CRREL Report 119, U.S. Cold Regions Research and Engr. Lab., Hanover, NH, 1980.
29. Brockway, D.G., D.H. Urie, P.V. Nguyen, and J. B. Hart, Wastewater and sludge nutrient utilization in forest ecosystems, in *The Forest Alternative for Treatment and Utilization of Municipal and Industrial Wastewater and Sludge*, Cole, D., C. Henry and W. Nutter, Eds, University Press, Seattle, 533, 1986.
30. Schiess, P., and D.W. Cole, Renovation of wastewater by forest stands, in *Municipal Sludge Application to Pacific Northwest Forest Lands*, Bledsoe, C.S., Ed, Inst. For. Resources Contrib. No. 41, College of Forest Resources, University of Washington, Seattle, 131, 1981.
31. Edmonds, R.L. and K.P. Mayer, Aerial dispersion of coliform bacterial from sewage sludge spraying in a Douglas-fir forest, in *Use of Dewatered Sludge as an Amendment for Forest Growth Vol. IV*, Henry, C.L., and D.W Cole, Eds, Inst. For. Resources, University of Washington, Seattle, 1983, 104.
32. Riekirk, H., and D.W. Cole, Chemistry of soil and ground water solutions associated with sludge applications, in *Use of Dewatered Sludge as an Amendment for Forest Growth, Vol. I,* Edmonds, R.L., and D.W. Cole Eds, Center for Ecosystem Studies, Col. of For. Res., University of Washington, Seattle, 112, 1976.
33. Vogt, K.A., R.L. Edmonds, and D.J. Vogt, Regulation of nitrate levels in sludge, soil and ground water, in *Use of Dewatered Sludge as an Amendment for Forest Growth, Vol. III*, Edmonds, R.L., and D.W. Cole Eds, Inst. Forest Resources, University of Washington, 1980.

34. Henry, C.L., D.W. Cole, and R.B. Harrison, Nitrate leaching from fertilization of three Douglas-fir stands with municipal biosolids, draft manuscript, 1993.
35. Henry, C.L., R.C. King, and R.B. Harrison, Distribution of nitrate leaching from application of municipal biosolids to Douglas-fir, draft manuscript, 1993.
36. Sommers, L.E., Forms of sulfur in sewage sludge, *J. Environ. Qual.*, 6, 42, 1977.
37. U.S. Environmental Protection Agency, Process Design Manual: Land Application of Municipal Sludge, EPA-625/1-83-016, 1983.
38. Henry, C.L., Buffers and slopes recommendations (incorporated in), *Guidance for Writing Case-by-Case Permit Requirements for Municipal Sewage Sludge,* U.S. Environmental Protection Agency, draft, September 1988.

Chapter 5

AQUATIC SYSTEMS

K. E. Havens and A. D. Steinman

TABLE OF CONTENTS

0-87371-860-7/95/$0.00+$.50
© 1995 by CRC Press, Inc.

I. INTRODUCTION

A great quantity and variety of natural and synthetic materials are applied as amendments to agricultural soils. In general, the objectives are to enhance the quality of soils for crop production and to prevent the growth of nuisance organisms, including undesired plants, crop-destroying insects, and pathogenic fungi. Typical soil amendments include mineral fertilizers, synthetic pesticides (herbicides, insecticides, and fungicides), and waste materials from municipalities and other human activities.

Although soil amendments are targeted toward crop plants and their pests, it is often the case that they exit the treated locale and are transported into nearby aquatic ecosystems, including streams, rivers, lakes, and estuaries. At that point, the amendments may bring about undesired changes, both directly and indirectly, in the natural nontarget communities. Direct changes include increased biomass of certain components and decreases or extermination of others. Further, because the living organisms in aquatic ecosystems are linked in a complex trophic network or "food web", direct impacts of soil amendments on certain components of that web may have indirect impacts on others.

In this chapter, we review the impacts of soil amendments on aquatic ecosystems, focusing especially on the impacts of mineral fertilizers and pesticides. Both the direct physiological effects and the indirect ecological effects will be considered. We draw on examples from a variety of aquatic and semiaquatic habitats, but in particular the Lake Okeechobee and Florida Everglades ecosystems of south Florida, and the agricultural subbasins of Lake Erie, where extensive data exist regarding agricultural impacts.

II. THE NATURE AND FATE OF SOIL AMENDMENTS

A. NUTRIENTS

It has been estimated[1] that in the U.S. alone, over 45 million tons of mineral fertilizer are currently used each year. Primary plant nutrients account for nearly 20 million tons or 45% of this total. Nitrogen (N), potassium (K), and phosphorus (P) are the major constituents; total applications in 1992 alone were 11, 5, and 4 million tons, respectively. P and N are of particular interest to aquatic resource managers, because they are generally the limiting nutrients for aquatic plant and algal growth.[2–7] Hence, they have the greatest potential for aquatic community impacts. Generally, P is the limiting nutrient in temperate freshwaters, and N is limiting in tropical lakes and in marine ecosystems.

In agricultural watersheds, nutrient imports in fertilizer often exceed nutrient exports in crops. When this occurs, a balance of nutrients remains in the soil and/or is transported out of the system in surface and subsurface water flow. As an example, over 5400 metric tons of P were imported as mineral fertilizer into the agricultural basin of Lake Okeechobee, Florida during 1986.[8] The total yearly P export from the watershed, not only from agricultural crops, but also in milk, timber, and cattle, was 1400 metric tons or just 26% of the input from fertilizer alone. The result was a net accumulation of P in the watershed and an annual P load to the lake of 470 metric tons (Figure 1). As a consequence, Lake Okeechobee has experienced massive algal blooms and is now secondarily N-limited due to surplus P.[9,10]

On a larger scale, the total P and N exports from croplands into U.S. surface waters currently exceed 0.6 and 3.2 million metric tons per year (Table 1), and they account for 31 and 39% of all nonpoint and point sources of P and N, respectively.[11] Patrick[12] gave further details on the routes of nutrient transport, using as an example the highly agricultural Flint River watershed in Georgia. Approximately 95% of the P reaching surface waters was transported through direct surface runoff and most was associated with eroded sediments. In contrast, 80% of N was transported in subsurface flow, with surface runoff accounting for only 20%. Nearly all of the surficial N flux was as eroded sediments. These patterns were confirmed by recent research on Lake Erie tributaries draining agricultural watersheds.[13] Storm chemographs showed that P concentrations closely tracked suspended solids levels in runoff water (Figure 2); in both cases, there were rapid increases to maximal concentrations coincident with peak flow. In contrast, N concentrations increased gradually and reached maximal levels several days after peak flow.

The nutrient concentrations of surface waters draining agricultural areas are often extremely high. A well-studied example is runoff from the Everglades Agricultural Area (EAA) of Florida, U.S. into the marsh ecosystems of the Florida Everglades. Three tributaries draining the EAA sugarcane fields were found to account for 58% of the surface water inputs to the Everglades "Water Conservation Area" marshes; those same tributaries had total P concentrations averaging 190, 105, and 200 μg/l.[14] Phosphate concentrations in the marshes were highest at the northern ends near the EAA inflows, and declined exponentially with distance from the sources (Figure 3).[15]

On a larger scale, Omernik[16] examined relationships of nutrient (N and P) concentrations in U.S. surface waters to watershed land use. He found strong positive correlations between the nutrient concentrations and the percent of land in agriculture.

Crop type, geographic location, and fertilization practices also affect runoff water chemistry. A detailed review is beyond the scope of this chapter, but representative examples may be found in a report of experimental studies in the Florida EAA.[17] Baseline total P concentrations in runoff from different experimental crops averaged 43 μg/L for sugarcane, 69 μg/L for radish, 44 μg/L for cabbage, 49 μg/L for radish-rice rotation, and 78 μg/L for drained fallow fields. Total N averaged 8.5, 4.8, 14.4, 5.4, and 11.6 mg/L for the same crops. As an example of geographic variation, however, total P varied from 17 to 69 μg/L and total N varied from 7 to 10 mg/L for sugarcane crops at two different EAA sites. Fertilized fields had significantly higher total P and N concentrations in runoff than did unfertilized controls, and band fertilization gave lower nutrient exports than broadcast fertilization.

The use of fertilizers in agricultural crop production is a necessity, especially in regions where the natural soils do not contain sufficient quantities of essential plant nutrients. Because agricultural nutrient runoff can have dramatic impacts on aquatic

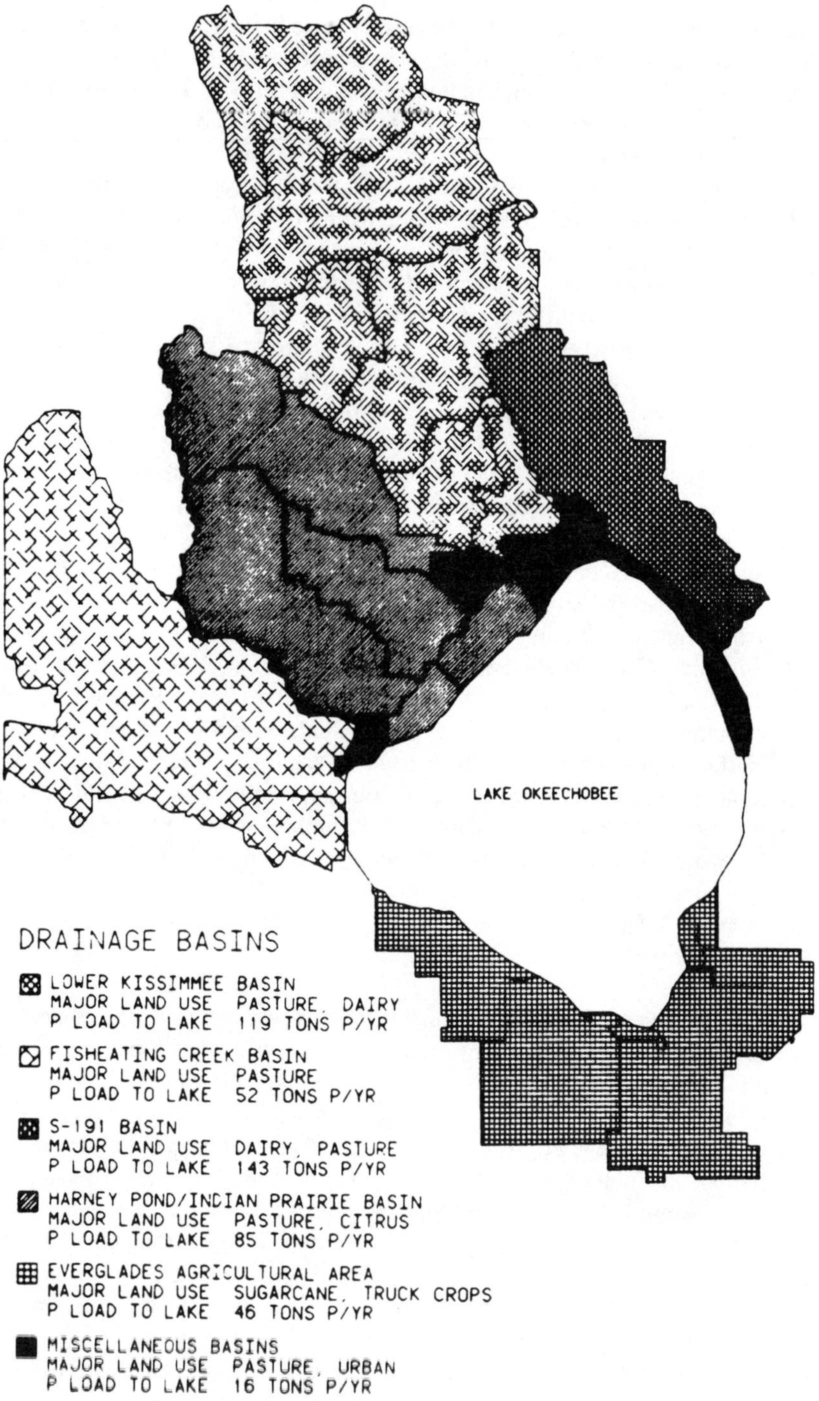
LAKE OKEECHOBEE
DRAINAGE BASINS
LOWER KISSIMMEE BASIN
MAJOR LAND USE PASTURE, DAIRY
P LOAD TO LAKE 119 TONS P/YR
FISHEATING CREEK BASIN
MAJOR LAND USE PASTURE
P LOAD TO LAKE 52 TONS P/YR
S-191 BASIN
MAJOR LAND USE DAIRY, PASTURE
P LOAD TO LAKE 143 TONS P/YR
HARNEY POND/INDIAN PRAIRIE BASIN
MAJOR LAND USE PASTURE, CITRUS
P LOAD TO LAKE 85 TONS P/YR
EVERGLADES AGRICULTURAL AREA
MAJOR LAND USE SUGARCANE, TRUCK CROPS
P LOAD TO LAKE 46 TONS P/YR
MISCELLANEOUS BASINS
MAJOR LAND USE PASTURE, URBAN
P LOAD TO LAKE 16 TONS P/YR

Figure 1 Total P loads to Lake Okeechobee, Florida from six agricultural subbasins in its watershed. (From Fluck, R. C., Fonyo, C., and Flaig, E., *Appl. Eng. Agric.*, 8, 813, 1992. With permission.)

Table 1 Water Pollutant Discharge Estimates from Point and Nonpoint Sources in the U.S. (in 10^3 Tons per Year)

Nonpoint sources	Phosphorus	Nitrogen
Cropland	615	3204
Pasture	95	292
Range	242	778
Forest	495	1035
Other rural lands	170	659
Streambanks	<1	<1
Gullies	<1	<1
Roads	<1	<1
Construction	<1	<1
Other	64	691
Point sources	330	1495
Total	2011	8154

From Gianessi, L. P., Peskin, H. M., Crosson, P., and Puffer, C., United States Department of Agriculture and United States Environmental Protection Agency Report, Washington, D.C., 1986. With permission.

ecosystems, it is important that farming practices be adopted which minimize nutrient export. This topic will be addressed in greater detail at the end of the chapter.

B. PESTICIDES

Pesticides are used extensively in agriculture, and they, too, are transported into surface waters. It was recently estimated that worldwide annual pesticide use (in 1990) exceeded 2.5 million tons.[18] In the U.S. alone over 0.5 million tons are used each year (based on 1989 data),[19] and 75% of that use is agricultural. Approximately 62% of the agricultural use is herbicidal, 22% is insecticidal, and 9% is fungicidal.[19]

In a comprehensive survey of pesticide use in the U.S.,[20] it was found that for corn, soybeans, cotton, and peanuts over 90% of fields were regularly treated with herbicides and approximately 30% were treated with insecticides. The most commonly used herbicides (based on 1987 U.S. EPA figures) are Alachlor and Atrazine. Alachlor is an organic acetanilide and Atrazine is an organic triazine. Both compounds target preemergent weeds and inhibit seed germination and root elongation by preventing cell division.[19] Worldwide, they also rank among the most widely used herbicidal chemicals. The most commonly used insecticides, both internationally and in the U.S., are carbamates, organophosphates, pyrethrins, and pyrethroids. These compounds are generally used to combat nematodes and insects, and they exert their effects on nervous system functioning.[19]

The overall pesticide application rate has declined somewhat since the early 1980s in the U.S., but on a global scale that decline has been offset by increased usage in other countries.[18] Further, long-lived organochlorine pesticides like DDT, now

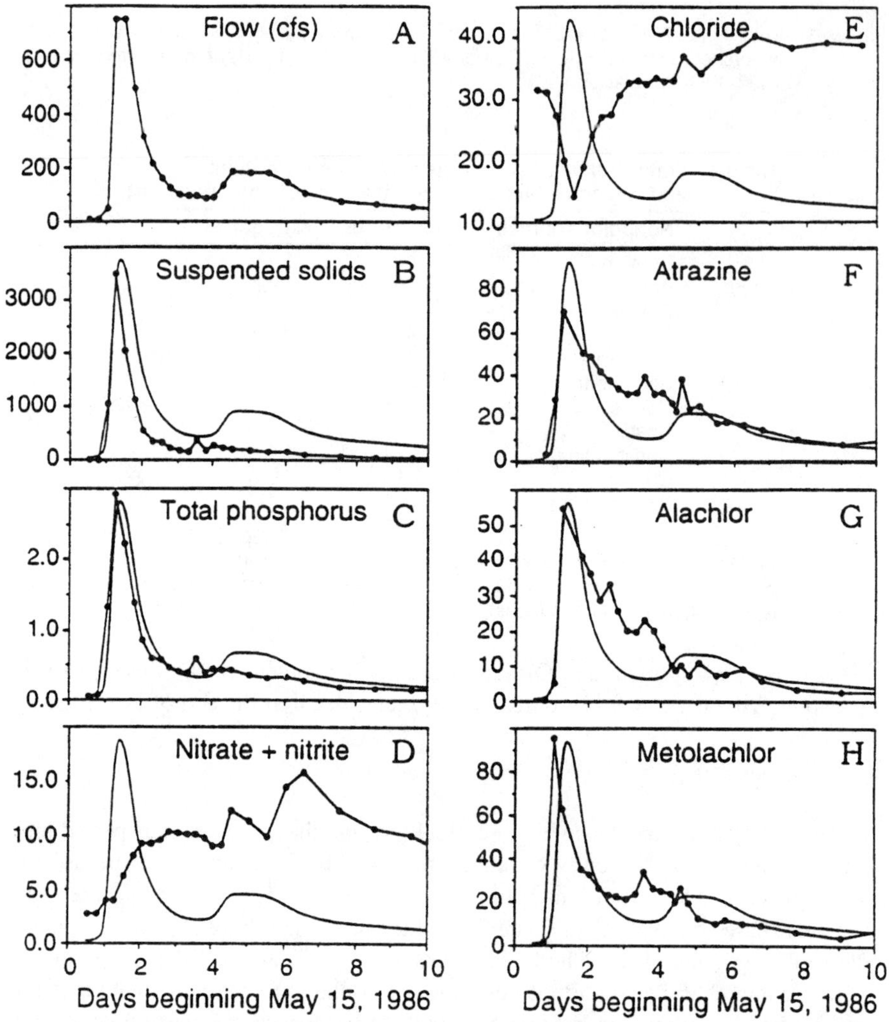

Figure 2 Storm hydrograph and chemographs of suspended solids, total P, nitrate, chloride, Atrazine, and Alachlor for a typical runoff event in the Honey Creek agricultural watershed, Ohio, U.S. (From Richards, R. P. and Baker, D. B., *Environ. Contamination Toxicol.,* 12, 13, 1993. With permission.)

banned in the U.S., continue to be used in developing countries.[19] Today's most commonly used pesticides (above), in contrast to DDT, are characterized by more rapid biodegradability and a limited capacity to bioaccumulate. However, they can be highly toxic to certain aquatic invertebrates.[20]

Transport of pesticides to the aquatic environment occurs primarily by volatilization, leaching, and surface erosion.[21] Pesticide runoff into surface waters varies with time after application, and upon entering the aquatic ecosystem, most pesticides tend

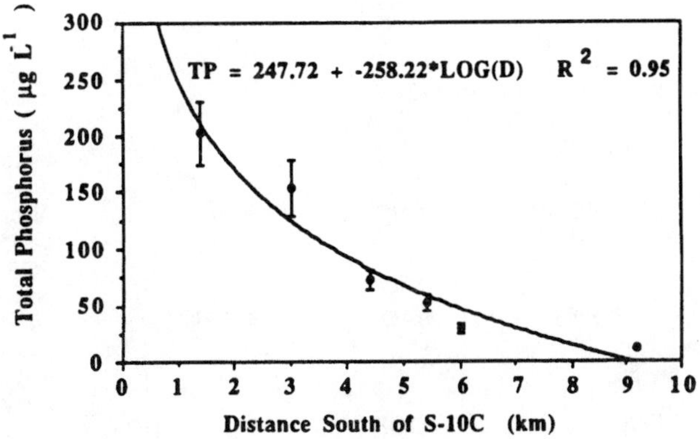

Figure 3 Surface water total P concentrations as a function of distance from an agricultural water inflow structure to Everglades Water Conservation Area 2A in south Florida. (From Koch, M. S. and Reddy, K. R., *Soil Sci. Soc. Am. J.*, 56, 1492, 1992. With permission.)

to become concentrated in the sediments, rather than the overlying water.[22] The pathway of pesticide transport from croplands to surface water differs from that of the major nutrient elements.[13] Soil chemographs (Figure 2) suggest that pesticides are continuously dissolved from the surface and near-surface soils during storm events, and that both overland and subsurface flow are important delivery routes to aquatic ecosystems.[13] Not surprisingly, there is often a distinct temporal relationship between peak period of agricultural activity and maximum flux rate of pesticide residues into the aquatic environment. As an example, pulses of several pesticides entering Lake Utah corresponded to the times of year when each pesticide was applied to watershed crops.[20] However, this relationship does not always exist; it has been reported by various authors that sediment transport and associated pesticides to agricultural watersheds in Ontario reach maxima between January and April. During that period of high pesticide runoff, agricultural activity is low, but soil erosion is high due to spring thaw and snowmelt.

Some pesticides are very persistent, while others have short half-lives and degrade within days or even hours.[23] For that reason, assessments of ecological impact should consider both the toxicity and persistence (as well as biomagnification) of agricultural pesticides.

It has been estimated that for every dollar spent on agricultural pesticides, over $4 of crop loss is prevented.[19] Therefore, the use of toxic pesticides will likely continue, despite potential adverse impacts on nearby aquatic ecosystems. Conservative application and careful choice of farming practices can limit pesticide runoff and minimize ecological impacts.

III. IMPACTS ON AQUATIC ECOSYSTEMS

A. AQUATIC ECOSYSTEM STRUCTURE AND FUNCTION

To fully understand the impacts of agricultural soil amendments, it is necessary to consider the ecosystem *in toto*, rather than as isolated parts. Because of the

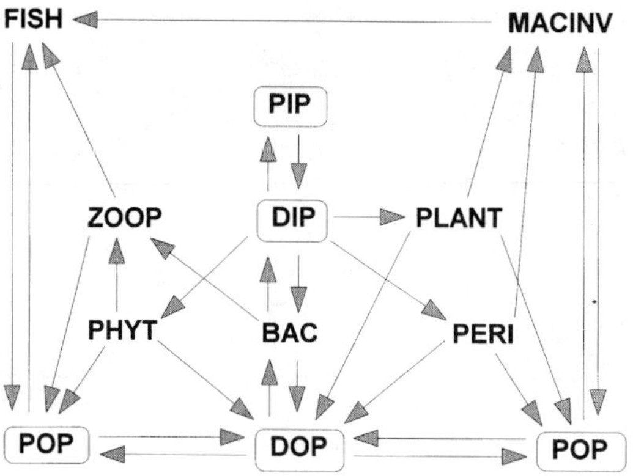

Figure 4 Major pathways of phosphorus cycling in the aquatic ecosystem. Abiotic pools are DIP (dissolved inorganic P), DOP (dissolved organic P), POP (particulate or detrital organic P), and PIP (particulate inorganic P, such as silt or clay particles). Biotic pools are fish, macroinvertebrates (MACINV), zooplankton (ZOOP) phytoplankton (PHYT), aquatic macrophytes (PLANT), periphyton (PERI), and bacteria (BAC).

interactive nature of the food web, pollutants may have both direct toxic effects on the organisms and indirect effects, mediated through altered predator-prey, host-parasite, and mutualistic interactions. Additional indirect effects, resulting from alterations of the physical environment such as reduced light penetration due to suspended soil particles from agricultural erosion, may affect the biotic components of the ecosystem.

A simplified diagram of an aquatic ecosystem is shown in Figure 4. The major pathways of P cycling indicate the complex linkages among components. As a hypothetical example of an indirect impact, consider some agricultural pesticide whose direct effect is the extinction of zooplankton from the system. Indirect effects that might follow include: a reduction in the growth rate of planktivorous fishes, particularly taxa and developmental stages which rely heavily upon zooplankton; and an increase in the biomass of phytoplankton and bacteria, which are normally grazed by the zooplankton. Shifts in the species composition of algae and bacteria might also occur, as species formerly suppressed by grazers outcompete others for essential nutrients.

Likewise, if an agricultural nutrient directly stimulates the growth of certain algal and bacterial taxa, the zooplankton which prey upon them might increase or decrease, and this effect might continue upwards through the food web. Such "cascading" food web effects have been documented in response to experimental perturbations of lake ecosystems,[24] although the strength of that cascade appears to be related to the trophic status of the water body.[25]

It is also necessary to consider spatial zonation in the aquatic ecosystem (Figure 5) when considering pollutant effects. In lakes, estuaries, and large rivers, extensive near-shore macrophyte communities may absorb the majority of nutrient and pesticide inputs from the adjacent watershed, and act as a natural buffer zone between the land and the open water. Wetzel[26] described wetland-littoral zones as "metabolic gates" which can control the inputs of carbon and inorganic nutrients to pelagic regions. Near-shore communities may act as nutrient sinks; they may introduce time lags into nutrient inputs (storing nutrients in plant tissue, and then releasing them later in the season with senescence) or they may transform nutrients from readily

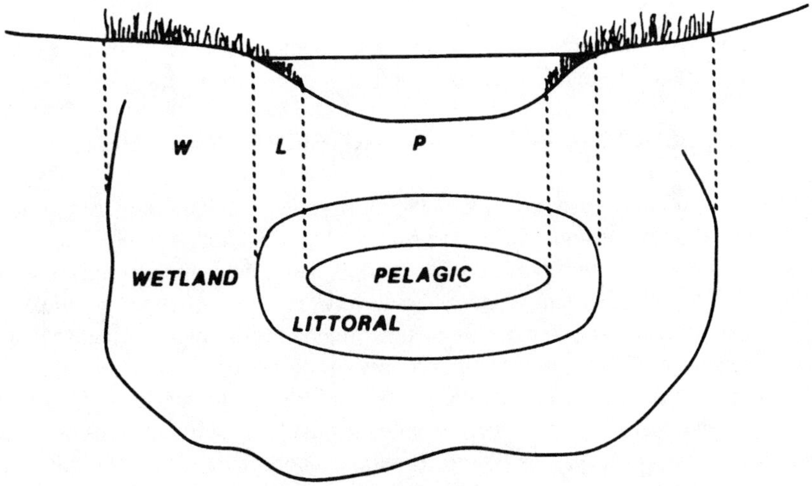

Figure 5 Spatial zonation in the aquatic ecosystem. Wetlands and littoral regions account for much of the total surface area in many aquatic ecosystems, and they may control overall ecosystem functioning. (From Wetzel, R. G., *Verh. Int. Vereinigung Theoretische Augewandte Limnol.,* 24, 6, 1989. With permission.)

available forms into less available ones. The latter function was demonstrated by Heath,[27] who studied P movement from an agricultural watershed into a shallow, macrophyte-covered wetland, and ultimately into Lake Erie. He found that nearly all of the P entering the system was in the form of soluble reactive P, while P exiting the system was particulate P, which was not readily available to the phytoplankton of Lake Erie.

In lotic ecosystems, riparian vegetation may play an analogous role as macrophytes in the littoral zone of lentic ecosystems. Research on the Rhode River watershed, located on the western shore of the Chesapeake Bay, revealed that croplands are the largest source of nutrient discharge per unit surface area. Consistent with the findings of Omernik,[16] nutrient discharge from these watersheds increased as the percentage of cropland increased in the area.[28] Riparian forests had a major influence on nutrient fluxes from croplands into stream channels in the Rhode River watershed; the forest caused a 84 to 87% decrease in the nitrate concentration of groundwater entering a first-order stream.[28] The efficiency of riparian forests in trapping nitrate and suspended sediments from overland runoff in the Rhode River watershed, as well as other coastal plain systems,[29] may be variable. Correll et al.[28] noted that the prevalence of aquicludes or confining layers near the soil surface restricts infiltration to deeper strata, thereby assuring extensive contact between shallow groundwater and the roots of riparian vegetation.

Because littoral and riparian zones act as buffers to nutrient and carbon inputs, cultural impacts on littoral and riparian zones can indirectly impact open water regions. The impacts might be especially pronounced for biota which spend part of their life cycle in each zone, for example, many economically important fishes. Ecologists have just recently recognized the potential importance of littoral-pelagic and riparian zone interactions to aquatic ecosystem function,[30,31] so the extent of indirect impacts of agricultural amendments is largely unknown.

B. THE ECOLOGICAL IMPACTS OF NUTRIENT ENRICHMENT

The general impact of nutrient enrichment from agricultural runoff is a hastening of the natural aging or eutrophication process. Rapid cultural eutrophication due to human influences is of worldwide concern to the scientific community and the general public users of lakes. Recently, there also have been concerns about the impacts of agricultural nutrients on wetland ecosystems. Mineral fertilizers, in particular P and to a lesser extent N, directly impact the primary producers of the aquatic community, enhancing their biomass and also the biomass of higher trophic levels. Qualitative changes often occur in the producer assemblage, involving shifts towards dominance by undesirable species such as bloom-forming cyanobacteria. Extensive autotrophic biomass ultimately can be detrimental to aquatic animals if oxygen depletion accompanies the decay of dead plant and algal materials. Winter "fish kills" due to oxygen starvation are a common concern in culturally eutrophic waters which experience winter ice cover.[32] Summer fish kills may also occur in hypereutrophic waters.

1. *Algae and Aquatic Plants*

The responses of planktonic algae to nutrient enrichment have been thoroughly studied, largely because algal blooms are one of the most visible signs of eutrophication. Limnological investigations in the 1970s and 1980s, involving multilake comparisons[33,34] and whole-lake fertilizations,[35,36] showed conclusively that algal biomass in freshwater temperate lakes is a direct function of P loading rates (Figure 6).

In a classic experiment at the Experimental Lakes Area (ELA), Canada, P, N, and carbon (C) were added to one half of a small lake; the other half, separated by a plastic curtain, received only N and C. Results showed that algal biomass was not affected in the N and C treated half. A massive surface bloom of cyanobacteria developed in the P, N, and C treated half, and surplus P drove the plankton toward N limitation. Because cyanobacteria are capable of fixing atmospheric N, they were able to take advantage of the high P levels and form blooms. The same events have occurred in Lake Okeechobee, Florida, due to excessive agricultural P inputs from the watershed.[9,10]

In a more recent review of nutrient addition experiments, Hecky and Kilham[3] found overwhelming support for the role of P as the limiting nutrient in temperate fresh waters. In tropical and marine systems, in contrast, N seems to most often limit algal production.[3,7]

Along with the increased algal biomass that occurs due to nutrient enrichment, there is generally an increase in the relative biomass of nuisance cyanobacteria and an increase in the average size of phytoplankton cells. This trend has been clearly demonstrated in whole-lake P fertilization experiments at the ELA[36,37] and in multilake survey studies.[38,39]

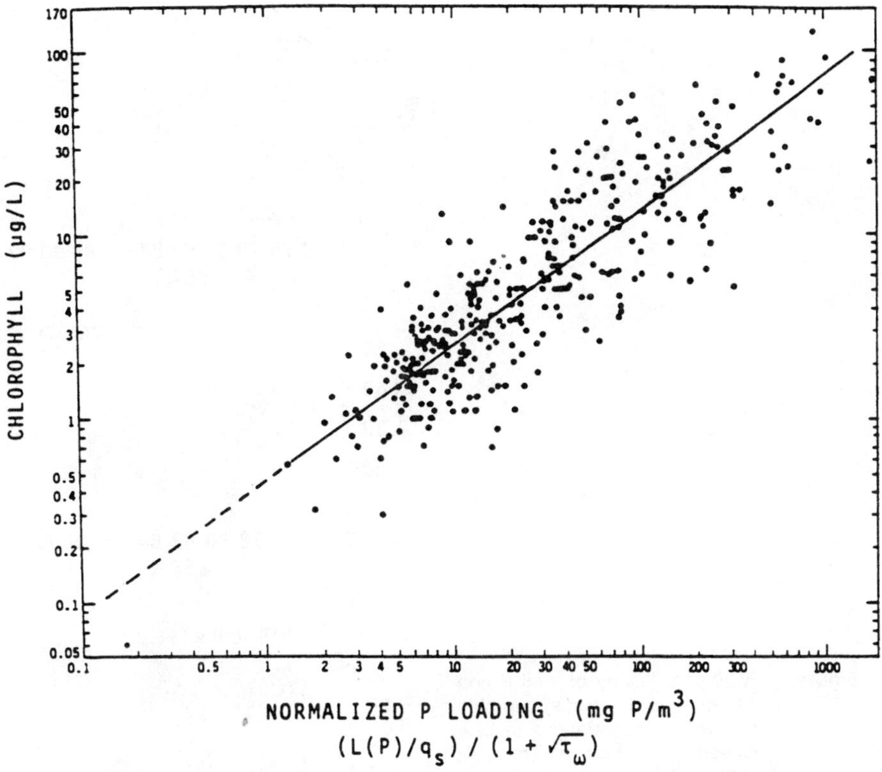

Figure 6 The empirical relationship between normalized P loading and total chlorophyll concentrations in North American lakes. (From Lee, G. F. and Jones, A., *Lake Line,* 12, 13, 1992. With permission.) The data originate from the OECD National Eutrophication Survey. Normalized loading is the rate of P loading relative to lake depth and flushing rate, i.e., lakes with shallower mean depths and/or slower flushing rates experience greater chlorophyll yields at given P loading rates.

Cyanobacteria can have deleterious impacts on the ecosystem when they become abundant. The common bloom-forming taxa *Anabaena, Aphanizomenon,* and *Microcystis* can produce endotoxins which are harmful or even fatal to aquatic animals.[40,41] Most planktonic cyanobacteria have the ability to float to the surface via intracellular gas vacuoles,[42] where they often form extensive floating mats or scums.[43] These impair the recreational value of the water body, and when they decay, severe oxygen depletion may occur. Finally, cyanobacteria are generally not exploited by zooplankton herbivores,[44–46] either because they are noxious or simply too large to ingest. Hence, when cyanobacteria become dominant, the dominant zooplankton tend to be taxa which can exploit alternative food resources, such as bacteria, and the phytoplankton → zooplankton → fish food chain is disrupted. The potential impacts of cyanobacterial blooms on higher trophic levels will be discussed below.

It has been suggested that cyanobacterial dominance in culturally eutrophic lakes and rivers is a result of both the high P concentrations and the low N:P ratio that

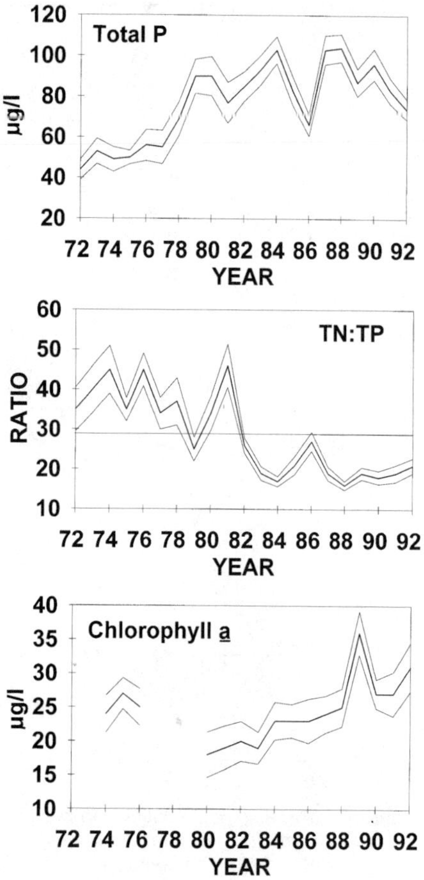

Figure 7 A 20-year history of total P concentrations, N:P ratios, and chlorophyll *a* concentrations in Lake Okeechobee, Florida. The solid lines represent yearly means, and the dashed lines are 95% confidence boundaries.

results from the P excess.[47,48] Many of the frequent bloom-forming species are "nitrogen fixers", and can obtain their required quota of N from the atmosphere.[48] Because the atmospheric N pool is unavailable to other phytoplankton, N-fixing cyanobacteria have a competitive advantage in aquatic ecosystems where N is naturally limiting, or where it becomes limiting due to excessive watershed P inputs. A well-studied example is Lake Okeechobee, Florida. As a result of P inputs from the agricultural subbasins north of the lake, total P concentrations in the lake water have increased from below 50 µg/L in the mid-1970s to near 100 µg/L in recent years (Figure 7). During that same period, the N:P ratio has declined, N has become secondarily limiting to algal production, and the frequency of cyanobacterial blooms has increased.[49–51] Massive blooms of N-fixing *Anabaena* and *Microcystis* in the late 1980s have covered up to 42% (700 km^2) of the lake surface and N-fixation now accounts for up to 30% of the total N inputs to the lake.[52] Agricultural BMPs (best management practices) are now being implemented to reduce P inputs to the lake, as mandated in the regional Surface Water Improvement and Management (SWIM) Plan.[53] One benefit of P reductions in Lake Okeechobee will be a higher N:P ratio, thereby making conditions in the lake less conducive for cyanobacterial blooms.

In estuarine ecosystems, changes in the ratio of N:P loading associated with nutrient runoff also have resulted in shifts in the phytoplankton community structure. In both the German Bight[54] and waters off Hong Kong,[55] the community has shifted from being diatom-dominated to flagellate-dominated. This phytoplankton shift may have consequences for trophic level interactions in these systems, with possible implications for regional fisheries as well.

Agricultural fertilizer runoff affects macrophyte-dominated communities. Research has shown that coastal wetlands are generally N limited, while inland marshes and fens are generally P and sometimes K limited.[5] Recent studies of the Florida Everglades showed high concentrations of soil N, high N:P ratios, and probable P limitation of macrophyte production.[15] Nutrient runoff from the EAA has impacted the natural vegetation of the Everglades marshes; the most alarming pattern (Figure 8) has been a replacement of the native Everglades sawgrass (*Cladium jamaicense*) community by pollution-tolerant cattails (*Typha domingensis*). Research by the South Florida Water Management District and University of Florida scientists[15] indicated that excessive P loadings, rather than N, were responsible for that trend. The excessive P inputs also have resulted in increased periphyton biomass, and the development of a species-poor community dominated by *Microcoleus* (a filamentous cyanobacterium) and a few pollution-tolerant diatoms.[14]

Increased nutrient influxes may also overwhelm the assimilative capacity of littoral macrophytes, potentially resulting in increased biomass of pelagic autotrophs (phytoplankton) and increased biomass of select species (cyanobacteria) that are best adapted to take advantage of the new nutrient regime.

In the Florida Everglades, nutrient-enriched sites have also experienced a suppression of normal litter decomposition, due to anaerobic conditions and a reduced number of decomposer microbes in the sediments.[14] Hence, the inputs of P from agriculture have affected the overall nutrient cycling in this marsh ecosystem. Further, the documented community changes have important implications for higher trophic levels, which depend upon the primary producers for food and habitat. These secondary impacts will be discussed below.

Efforts are now underway to construct artificial wetlands at the southern periphery of the EAA, in order to remove P from the agricultural water before it enters the Everglades marsh ecosystem.[14] The goal is to reduce the P concentrations of the runoff water to near natural levels, and the constructed wetlands are intended to serve as a natural pollution filter.

2. Zooplankton and Macroinvertebrates

Aquatic invertebrates also respond to nutrient enrichment, although the impacts in this case are secondary. Biomass generally increases, due to increased food levels for both the herbivorous and carnivorous taxa in the community. Nutrient enrichment leads to an increase in the total biomass of zooplankton, in particular the very small "microzooplankton" rotifers and ciliates.[56] Large cladocerans, especially members of the genus *Daphnia,* tend to be rare in hypereutrophic waters.[57,58] This shift in

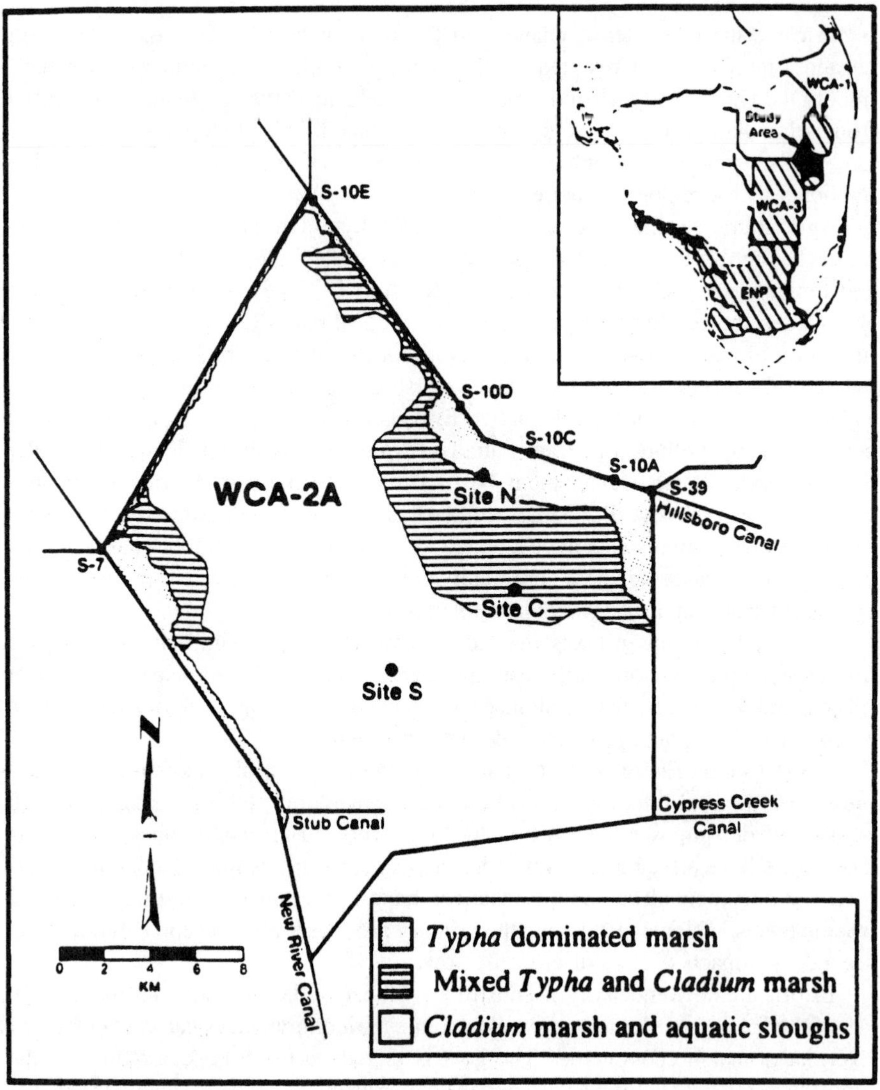

Figure 8 Invasion of *Typha* in regions near major P inputs (S-10 and S-7 structures) to the *Cladium*-dominated Everglades WCA-2A marsh ecosystem. The inset map shows the location in south Florida. (From Koch, M. S. and Reddy, K. R., *Soil Sci. Soc. Am. J.,* 56, 1492, 1992. With permission.)

community structure has important implications for the functioning of the community because large *Daphnia* are keystone taxa due to their intensive grazing and broad spectrum diet.[59–61]

Conditions in lakes which have been enriched by agricultural fertilizers are generally unfavorable for *Daphnia*. A high abundance of toxic cyanobacterial filaments is problematic because *Daphnia* are filter-feeders that invariably gather toxic filaments from the water. They must expend energy rejecting them[46] or else consume them. The latter may lead to mortality through direct toxicity.[40] Smaller zooplankton do not collect large filaments, but specialize on small phytoplankton and bacteria. They are not so adversely affected by toxic cyanobacteria and can persist in the community. Eutrophic lakes generally support high densities of planktivorous fishes (see below), and this, too, is unfavorable for large zooplankton like *Daphnia,* which are highly visible and therefore most susceptible to predation.[62,63] There is also evidence that large *Daphnia* are most sensitive to suspended silt particles in the water (they expend considerable energy eating them, but gain little or no nutritional value in return)[64] and to agricultural pesticides.[65]

The shift to microzooplankton dominance in culturally enriched lakes can indirectly affect higher trophic levels in the community. Because microzooplankton do not exploit large cyanobacterial filaments and colonies,[44] a considerable amount of primary production may go ungrazed. In fact, there is evidence that by consuming small phytoplankton and bacteria, and subsequently releasing nutrients into the water, microzooplankton may facilitate cyanobacterial growth. Havens[44] experimentally removed the macrozooplankton from a freshwater plankton assemblage; cyanobacterial biomass increased significantly during the week following the treatment, suggesting grazing control of cyanobacteria by the macrozooplankton. However, when both macro- and microzooplankton were experimentally removed, cyanobacterial biomass did not increase, suggesting that microzooplankton were needed to stimulate the cyanobacterial growth.

Another effect of microzooplankton dominance involves the microbial (bacteria and protozoan) component of the food web. Large *Daphnia* form an effective link between the microbial component and higher trophic levels (i.e., fish); microzooplankton do not.[59–61] Due to their small size, microzooplankton also are not exploited by many planktivorous fishes.[62,63] This may lead to a further reduction in the efficiency of energy flow through the food chain and, in extreme cases, a reduction in the biomass of taxa which depend upon zooplankton as a primary food resource.

Changes in the composition of littoral and benthic macroinvertebrate assemblages with eutrophication were documented long ago and serve as one of the earliest classification systems for a lake trophic state.[66] With respect to chironomids, lakes having *Tanytarsus*-dominated assemblages were classified as low-productivity or oligotrophic lakes, whereas lakes with *Chironomus*-dominated benthos were regarded as highly productive or eutrophic. In the most extreme cases, eutrophic lake benthos become largely dominated by oligochaete worms, which can tolerate prolonged periods of anoxia.

3. Fish, Water Fowl, and Other Aquatic Vertebrates

Fish occur at the top of the aquatic food web and their productivity is a function of energy flow from below. An indirect effect of increased nutrient loading is increased fish yield (Figure 9), and a moderate degree of cultural eutrophication might be perceived as beneficial by persons primarily interested in fish production. Carpenter and Lodge[67] developed a lake ecosystem model, which predicted that

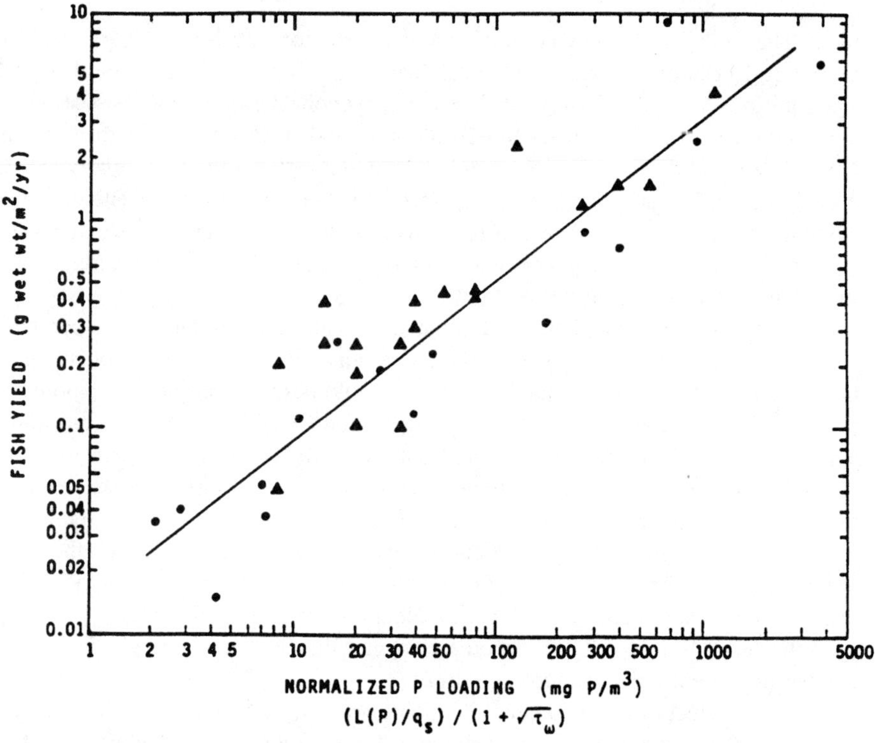

Figure 9 The empirical relationship between fish yield and normalized P loading in USA lakes. From Lee, G. F. and Jones, A., *Lake Line,* 12, 13, 1992. With permission.)

increased macrophyte biomass in enriched lakes would lead to an increase in the biomass of all autotrophs (including periphyton and phytoplankton), and subsequently an increase in the biomass of zooplankton, macroinvertebrates, and fish. Increased macrophyte biomass also benefits fish populations by providing greater cover for juvenile life stages.

At extreme levels of nutrient enrichment, qualitative changes occur in the fish assemblage, which also render the water body undesirable from a fisheries standpoint, despite high total fish biomass. Hypereutrophic lakes often support pollution-tolerant taxa such as carp, bullhead, and other rough fish generally deemed undesirable by the North American fishing public.[32] These taxa feed upon both plankton and benthic detritus, and their interaction with the sediments may further deteriorate water quality. In several experimental studies, it has been demonstrated that sediment feeding by benthivorous fish is a major contributor to water column P.[68–70] The fish act as "nutrient pumps", consuming benthic organisms and sediments, and then releasing nutrients into the water column in their feces. They also entrain sediments into the water by physical bioturbation. In an experimental study on a shallow freshwater estuary, Havens[68] enclosed fish (brown bullhead) in large mesocosms. In two mesocosms, the fish had access to the sediments; in two others, access was blocked by a coarse plastic net. Over the course of the 28-day experiment, maximal concentrations of total P and phytoplankton biomass were 150 to 200% higher where fish had access to the sediments.

Some highly turbid eutrophic lakes support large populations of stunted pan fish,[32] and this may be due to a lack of predation. Large sight-feeding predators, including bass and walleye, may decline in part because dense cyanobacterial blooms and silt turbidity make it impossible for them to find sufficient food. They also tend to be less tolerant of the anoxic conditions that often develop in highly eutrophic lakes than are rough fish like carp and bullhead.

Nutrient-induced changes in wetland vegetation also have been implicated in secondary impacts on aquatic vertebrates. In the Florida Everglades, bullfrog tadpoles feed primarily on the periphyton associated with sawgrass and other native vegetation. There is a significant relationship between the species composition of periphyton and the weight gain of tadpoles,[71] with less favorable growth where excessive nutrient levels cause periphyton dominance by cyanobacteria. Since the tadpoles are important dietary components of higher trophic levels in the marsh community, this impact has important ramifications for the entire food web. It also has been reported[72] that wading bird populations in the Everglades and in the littoral zone of Lake Okeechobee to the north are "delicately balanced with fish species that in turn depend on aquatic microorganisms". Hence, they too may succumb to the effects of nutrient enrichment.

4. *Nutrient Impacts at Population, Community, and Ecosystem Levels*

The effects of nutrient enrichment have been studied at various levels, from single populations to entire ecosystems. Nutrient bioassay experiments performed in the laboratory and field have provided valuable information regarding nutrient limitation, and algal compositional responses to increased nutrient availability.[2,3,9] Controlled experiments in microcosms and mesocosms have contributed information on responses at the planktonic assemblage level.[73,74] Ecosystem level responses, which include many complex indirect effects of nutrient enrichment, have largely been elucidated by experiments involving the fertilization of entire aquatic ecosystems. Much of that work has been done in the ELA of Canada, where lakes have been treated with various doses and combinations of C, N, and P.[4,35–37] Schindler[36] recently reviewed the results, and the reader is referred to that paper for a detailed description of ecosystem-level responses. In general, nutrient enrichment (with P, and sometimes N and P) produced the following results: (1) an increase in the amount of exported and unused primary production; (2) an increase in the average size of phytoplankton, with a shift to cyanobacteria dominance and large surface blooms; (3) a decrease in the average size of zooplankton, and a reduction in their ability to control the phytoplankton by grazing; and (4) "a general biotic impoverishment by extirpation of sensitive species and increased dominance by a few tolerant species". Overall, the nutrient enrichment caused major changes in the natural composition of the community and functioning of the lake ecosystem.

C. THE ECOLOGICAL IMPACTS OF PESTICIDE CONTAMINATION

The impacts of pesticide contamination on aquatic ecosystems also have been well studied, although not to the degree of eutrophication research. Pesticides impact organisms in the aquatic ecosystem which are trophically related to the terrestrial targets. As an example, pesticides which kill crop-destroying insects may also kill aquatic insects and insect larvae in wetlands, rivers, and lakes. This, in turn, may impact other organisms, such as the fish which utilize aquatic insects as a major food resource. Therefore, unlike nutrient runoff, which directly affects autotrophs and then indirectly affects organisms at higher trophic levels, pesticides have both direct and indirect effects at all trophic levels in the community. Further, because many of the common agricultural pesticides like Alachlor and Atrazine degrade quickly in the aquatic ecosystem,[23] their effects are "pulsed", rather than continuous as with nutrient enrichment. Communities tend to recover from pulsed stresses from the bottom up,[75] i.e., the microorganisms with the shortest generation times reappear first, followed by the zooplankton, and finally by the slowest-growing predators at the top of the food web. Unfortunately, top predators tend to be the most desired species from the standpoint of fisheries. In aquatic ecosystems experiencing frequent pulsed inputs of pesticides, there is the possibility that the community will be dominated by "pioneer" species at the base of the food web and also by pollution-tolerant rough fish like carp and bullhead at higher trophic levels. Pesticides with high solubilities in body fat also undergo "biomagnification", that is their concentration and potential toxic impacts increase with each successive trophic level in the food web.[19] Specific implications will be discussed below.

1. Algae and Aquatic Plants

Studies of pesticide impacts on aquatic autotrophs have been largely restricted to laboratory investigations, which remove the organisms from their natural environment, and thereby have the problem of underestimating or overestimating toxicity by preventing secondary detrimental or beneficial food web effects. Nevertheless, toxic effects have been demonstrated at pesticide levels below the maxima which often occur in aquatic ecosystems. Herbicide impacts have been most studied because they target weed plants in agricultural fields, and therefore they also impact autotrophs in the aquatic ecosystem. Krieger[76] discussed the impacts of the herbicide Atrazine on freshwater algae. At a dose of 1 μg/L, Atrazine inhibited the growth of important natural stream and lake taxa, including *Chlorella, Scenedesmus,* and *Nitzschia.* Aquatic macrophytes were similarly shown to be sensitive to Atrazine; 50% growth reductions were experienced by *Elodea* at 13 μg/L, *Vallisneria* at 1 μg/L, and *Lemna* at 50 μg/L. During storm runoff from agricultural watersheds, Atrazine concentrations can often exceed 70 μg/L.[13]

Agricultural insecticides have been shown to stimulate (indirectly) the growth of freshwater algae by temporarily removing their natural zooplankton predators from the community. Hughes et al.[77] showed that when small ponds were impacted by the insecticides Dursban, Abate, and Reldan (three commonly used organophosphates), an algal bloom occurred due to macrozooplankton extinctions. In a related study,[78] an algal bloom persisted for several weeks into zooplankton recovery because large algae became dominant and were unavailable to small microzooplankton, which were the first to recover. Most recently, Havens[79] showed experimentally that pesticide exposure caused a repartitioning of plankton biomass between zooplankton to phytoplankton in a freshwater plankton assemblage. Mesocosms containing natural lake plankton were treated with 10 different concentrations (0 to 200 μg/L) of the carbamate insecticide Carbaryl, and plankton biomasses were measured after 4 days exposure to the chemical. Zooplankton biomass declined from 500 to 200 μg dry weight/L over the chemical dose range; phytoplankton biomass increased from 200 to near 500 μg dry weight/L.

To some extent the ecological impacts of insecticides might be synergistic with those induced by nutrient inputs. Both nutrients and pesticides appear to stimulate the growth of large phytoplankton in the pelagic zone of lakes, and they both promote the dominance of small zooplankton and a generally inefficient food chain.

Herbicide impacts on aquatic vegetation can have ecosystem-level implications as well. Short-term responses to herbicides often are indirect and result from the decay of the plants: (1) declines in oxygen, pH, and alkalinity; (2) increases in free CO_2; (3) loss of habitat, cover, and attachment sites for epiphytes and epifauna; and (4) replacement of a herbivorous fauna with a detritivorous fauna.[67,80,81] Over a long time period, macrophyte reduction can lead to increases in sediment redox potential and declines in dissolved P and N, phytoplankton standing crop, and pelagic primary production.[67,82]

2. Zooplankton and Macroinvertebrates

Aquatic invertebrates are directly impacted by agricultural insecticides; some inhibit invertebrate nervous system functioning and others inhibit chitin production and exoskeleton molting.[19] In either case, the nontarget aquatic animals are detrimentally impacted in the same way as target crop pests.

The impacts of agricultural pesticides on zooplankton were recently reviewed[65] and only a brief account will be given here. The most commonly used carbamate and organophosphate insecticides cause massive extinctions of zooplankton at moderate and high doses,[65,83] and they bring about changes in species dominance at lower doses.[65,84,85] Microzooplankton are generally more tolerant of pesticide exposure than macrozooplankton,[65,85] perhaps reflecting their preadaptation to the natural toxins produced by co-occurring cyanobacteria.[65,86] Microzooplankton also have the shortest generation times, and therefore are often the first to recover following a pesticide pulse.[65,84]

Littoral and benthic macroinvertebrates are also impacted by insecticides. The important stream amphipod *Gammarus* is sensitive to organophosphorus compounds and so is the midge larva *Chironomus* and the common crayfish.[76] Sublethal exposures cause physiological stress and microhabitat abandonment; high level exposures cause massive mortalities. Species' life history patterns may be influenced by the frequency of the chemical pollution. Winged adults of aquatic insects represent a component of the population that is protected from a pesticide pulse which occurs during their life stage.[65] For species living in frequently impacted environments, it may be optimal to have asynchronous reproduction, so that a component of the population will always be in the air as adults.

Wallace[87] reviewed the recovery dynamics of macroinvertebrates following the introduction of pesticides into lotic ecosystems. He noted that recovery times varied from <1 month to >25 years and that 6 factors in particular were related to recovery time: (1) magnitude of the introduction and extent of continued use; (2) spatial scale of the disturbance; (3) persistence and toxicity of the pesticide; (4) vagility of the affected macroinvertebrates; (5) timing of the disturbance with respect to life history stages of the organisms; and (6) position within the watershed.

A study examining the long-term effects of insecticides and herbicides applied to fruit orchards revealed that over a period of 25 years, many invertebrates in ditches draining the orchards became totally resistant to the chemicals, whereas other taxa were completely eliminated.[88] Interestingly, some populations of chironomid midges actually increased in abundance, possibly because short generation times increased their chance of developing resistance or, alternatively, because the decrease in larvae of predatory beetles and odonates lowered predation pressure on the midges.

Herbicide-induced changes in the composition and biomass of primary producers might also impact the zooplankton and macroinvertebrate consumers. Research is underway to document these effects[89] which have not been quantified to date.

3. *Fish, Water Fowl, and Other Aquatic Vertebrates*

Pesticides impact vertebrate top predators in several ways: through direct environmental toxicity, through biomagnification in food chains, and by suppression of primary food resources. According to recent U.S. EPA data,[90] 6 to 14 million fish die per year as a result of pesticide effects, and it was noted that the figure is likely a gross underestimate. The American Fisheries Society[91] has estimated that the average economic value of a fish is $1.70; therefore, the pesticide-induced fish losses are roughly $10 to 23 million per year in the U.S. alone.

Given the economic value of fish, it is not surprising that fish toxicity is a research area that draws much attention. The subject has been a focus of numerous reviews,[21,92–95] and readers are referred to these reviews for detailed information. In general, it has been found that organochlorine compounds are more toxic than organophosphorus compounds, although some organophosphorus compounds (e.g., endrin and other cyclodienes) can be quite toxic to fish. The toxicity of herbicides to fish is low, but they can have indirect effects on fish by causing low dissolved oxygen concentrations associated with decaying aquatic vegetation.

Fish life stage influences susceptibility to pesticides; eggs and larvae with yolk sac are the least sensitive, whereas feeding larvae and juveniles are the most sensitive.[21] The greater toxicity of pesticides to juveniles than adults appears to be related to (1) higher rates of metabolism, and therefore greater food intake per unit of body weight in young fish; (2) greater intake of pesticide via gills because of the larger gill surface area relative to body mass in smaller fish; and (3) smaller lipid pools in the body, which are used to store toxicants.

Toxic pesticide effects on top predators of food webs, such as piscivorous fish, may result in indirect effects on competitors and prey organisms co-occurring in the impacted aquatic ecosystem. For example, a severalfold increase in density of aquatic insects occurred after a California mountain stream was treated with rotenone, which resulted in a massive fish kill.[96] Similarly, a 200-fold increase in chironomid midges occurred in a Wisconsin lake following the application of toxaphene, which eliminated the fish community.[97] Restocking of fish resulted in the decline of midge populations.

These indirect impacts might be of particular concern where the facilitated invertebrate species is also a human pest, such as the mosquito, which has an aquatic larval stage that is preyed upon by fish.

Wading birds are also adversely affected by pesticide contamination, both through direct exposure to the chemicals in their environment or by consuming contaminated food. Low-level exposures have been shown to cause reduced growth and reproductive output; higher doses cause mass mortalities.[98] Secondary impacts due to the decline of important food resources also have been observed.[99] Estimates of water fowl mortalities are sparse, but a few recent figures give an indication of the potential magnitude; Pimental et al.[18] give an excellent summary.

In terms of direct exposure, a study in the early 1980s[100] attributed the death of 1200 Canadian geese to spraying of methyl parathion (an organophosphorus compound) at a dose of 0.8 kg/ha on a wheat field. More recently,[101] it was documented that an application of Carbofuran (a carbamate insecticide) to alfalfa fields killed over 5000 ducks and geese. According to recent U.S. EPA data,[102] Carbofuran kills 1 to 2 million water fowl per year in the U.S. It also was estimated that from 0.25 to 9 birds per ha are killed in the U.S. from eating seeds treated with toxic insecticides.[103]

The application of herbicides in England has been indirectly linked with the demise of gray partridge populations.[99] When weed plants were eliminated from agricultural regions, so too were insects that occurred with them. Young partridges depended on those insects as a primary food source, and consequently local cases of starvation were documented. A similar fate has befallen the common pheasant in the U.S.[18]

The most publicized case of pesticide toxicity is that of DDT and predatory birds which feed on fishes in contaminated environments. The birds were top predators of the aquatic food chain and through biomagnification they experienced the greatest pesticide burdens. The result was widespread reproductive failure, often attributed to thinning of eggshells.[104] Although banned in the U.S., DDT use continues in many developing countries, including some important wintering areas for migratory birds.[104]

Pimental et al.[18] recently estimated the economic costs associated with pesticide poisonings of wild fowl. Assuming a value of $30 per bird, based on the economics of bird hunting and watching, yearly losses due to pesticides in the U.S. alone may exceed $2.1 billion.

Not all pesticides are directly toxic to fish and fowl, and not all pesticides biomagnify. However, secondary ecological impacts might still be a concern. In a recent study of the impacts of diflubenzuron, also known as dimilin, an inhibitor of chitin synthesis used to treat cotton boll weevil, mushroom flies, gypsy moths, and citrus pests, there was a high degree of tolerance in fish, amphibians, birds, and mollusks.[105] Further, the chemical did not bioaccumulate markedly in fish tissue, and it generally did not biomagnify in food chains. However, moderate exposures altered fish feeding habits and caused population declines in some aquatic invertebrates. The author[105] recommended that the chemical not be used within 5 km of coastal wetlands with economically important crustacean populations. He also suggested that frequent use of the chemical might suppress insect populations in both terrestrial and aquatic habitats adjacent to agricultural fields, and secondarily suppress birds which depend on those insects as a food resource.

4. Pesticide Impacts at Population, Community, and Ecosystem Levels

There is a growing appreciation that toxicity tests at the population level are not always useful for understanding community and ecosystem-level impacts.[106,107] In order to understand the ecosystem-level effects of a toxicant, the study must not isolate populations or subcomponents of the community from their natural environment, or else potentially important indirect impacts will be missed. Although ecosystem-level studies have been undertaken with the effects of nutrients,[35,36,108] little information is available on the ecosystem-level effects of pesticide contamination. One exception, a study by Gasith and Perry,[109] examined the effects of a single exposure of a fish pond to a relatively low level of the nonpersistent organophosphorus insecticide parathion. The authors found that shortly after its introduction, parathion accumulated several hundredfold in zooplankton and fish compared to the initial concentration in the water, but declined rapidly thereafter. The treatment had no significant impact on zooplankton community structure, but hemipteran beetles were temporarily eliminated and larval chironomid midges temporarily declined in abundance. The treatment adversely affected the growth and survival of silver carp and *Tilapia*. Overall, parathion affected certain biotic components, but had no detrimental effect on the community structure and function of the fish pond ecosystem. It is clear that other studies similar to this one are needed to address the ecosystem-level influence of pesticides. Future studies should examine the influence of different types of pesticides, different exposure periods of pesticides, and their influence on different types of ecosystems.

Given the difficulties and dubious environmental ethics of dosing natural aquatic ecosystems with pesticides, indirect impacts might be determined in part by studies utilizing small artificial ponds or *in situ* mesocosms. Such experiments allow for replication of treatments and controls, and if properly designed, the artificial systems can mimic many of the properties of the natural ecosystem. Most of the information regarding pesticide impacts on planktonic communities has come from such artificial system experiments. The results were reviewed by Havens and Hanazato,[65] and the major findings were described above. Important indirect effects were documented.

For example, a series of artificial pond experiments in Japan[110,111] demonstrated that zooplankton recovery after pesticide contamination is affected by the presence or absence of larval insect predators. In cases where the predators were abundant, they prevented recolonization by cladocerans, and rotifers dominated the community long after the pesticides degraded. When the predators were rare, cladocerans quickly recolonized the community, and outcompeted rotifers for resources. Previously described pond experiments[77,78] demonstrated that enhancements of algal production can secondarily occur when pesticides cause the extinction of grazing zooplankton.

There are limitations to the artificial system approach, however. The small ponds or mesocosms cannot mimic entirely the conditions of natural ecosystems, and they are not suitable for studying the impacts of chemicals on large vertebrate predators, such as piscivorous fish.

IV. AGRICULTURAL PRACTICES AND AQUATIC SYSTEM IMPACTS

Agricultural practices which prevent soil erosion may be generally effective in preventing excessive losses of sediment-bound P; carefully controlled use, not exceeding requirements, may be the most effective way to minimize N and pesticide losses, and thereby minimize impacts on aquatic ecosystems.

Baker[112] recently discussed the impacts of conservation tillage and no-till farming on nutrient and pesticide transport into Lake Erie from agricultural subbasins. He concluded that such management practices could substantially reduce the transport of soil, particulate P, and soil-bound pesticides like glyphosate and paraquat from agricultural fields into aquatic ecosystems. He cautioned, however, that runoff of soluble P might be enhanced due to surficial buildup of fertilizers and crop residue decay. Although soluble P generally accounted for less than 20% of the total inputs to surface waters (80% was particulate and associated with eroded soils), the soluble fraction is readily available to the aquatic biota.[113] Hence, on a gram for gram basis, soluble forms might have a greater impact on the ecosystem.

Baker[112] also noted that the effects of conservation tillage and no-till farming on N runoff are equivocal. The practices seemed to reduce loadings of particulate P, but, in some cases, enhanced the loadings of nitrate to surface waters. Nitrate is often the predominant form of N entering aquatic habitats. Thus, until further research is done, the best practice may be conservative use of N fertilizers.

Maas et al.[22] found that conservation tillage and no-till systems required greater pesticide doses, in particular herbicides, than conventional till systems. They also suggested that accumulated surface residue in no-till fields could intercept a portion of the applied pesticides and render it more susceptible to runoff losses. Maas et al. also discussed the potential ecological impacts of four other agricultural management practices: contour farming, cover crops, terraces, and sedimentation basins. In general, these practices reduce the export of sediments to surface waters, and thereby should reduce the exports of particulate P and soil-bound pesticides. However, little or no quantitative information is available.

Considering the great economic and societal values, the use of agricultural fertilizers and pesticides will continue, and perhaps increase along with increased resource demands by the growing human population. Because the runoff of agricultural chemicals can profoundly impact aquatic ecosystems, it is crucial that farming practices be adopted which will minimize that chemical flux. Such agricultural best management practices (BMPs) are ultimately a benefit to both the natural environment and the farmer, because the reduced fertilizer and chemical losses translate into dollars saved in reduced reapplications.

V. SUMMARY

It is clear that agricultural soil amendments contaminate aquatic ecosystems, and that they have deleterious impacts on the natural aquatic biota. Agricultural fertilizers, in particular P and N, stimulate phytoplankton, periphyton, and aquatic plant growth. Nutrient enrichment and alterations of natural nutrient ratios bring about changes in the taxonomic composition of the community, often involving shifts to dominance by undesirable forms like cyanobacteria in lakes and cattails in freshwater wetlands. These changes can, in turn, impact higher trophic levels. Cyanobacterial toxins can kill zooplankton and macroinvertebrates, inedible algal dominance can lead to food chain disruption, and decay of large macrophyte or cyanobacteria populations can cause oxygen depletion. Up to a point, nutrient enrichment may have the beneficial effect of stimulating fish productivity; however, even this use of the freshwater resource is threatened if nutrient inputs are extreme.

The impacts of agricultural pesticides, although not as well documented, are also profound. The various commonly used insecticides and herbicides kill aquatic plants, animals, periphyton, and phytoplankton. Pesticides entering the food web at lower trophic levels impact fish and birds at the top, both through alterations in the food base and through biomagnification. The impacts of specific pesticides appear to be highly varied, and only a small fraction of currently applied pesticides have been studied in detail. The extent of damage to the aquatic ecosystem from pesticide exposure depends on multiple factors, including toxicity, persistence, distance from the source, duration of application, soil-binding capacity, timing, and the structure of the impacted community.

Given the potential importance of indirect ecological effects, the impacts of agricultural soil amendments on aquatic ecosystems cannot be accurately assessed with single species laboratory toxicity tests. Some of the major findings regarding impacts have come from controlled experiments on natural ecosystems or artificial systems designed to mimic nature. Future research should continue in this area, in particular for pesticide effects, which are less understood.

To protect the natural aquatic resources, it is crucial that inputs of agricultural soil amendments to be minimized. A combined strategy of optimal agricultural practices and watershed management may be the best way to achieve that goal. In regard to watershed management, it must be recognized that riparian and wetland zones are natural buffers for the aquatic ecosystem. They have the capacity to absorb a considerable portion of the chemical inputs from the watershed, and concerted efforts should be made to protect these zones. However, the riparian and littoral communities may themselves succumb to the impacts of excess chemical runoff, and thereby lose their natural buffering capacity. It is, therefore, also crucial that agricultural practices, such as conservation tillage and integrated pest management, be implemented to minimize the export of soil amendments.

Natural aquatic ecosystems are important economically, in terms of human uses such as fishing, boating, recreation, and water supply. They are also habitats for economically valuable water fowl. Most importantly, though, natural aquatic ecosystems contain a significant fraction of the earth's total biodiversity, and they deserve protection in their own right.

ACKNOWLEDGMENTS

The authors are grateful to the Science Applications International Corporation, which provided financial support to A.D.S. during the writing of this chapter.

REFERENCES

1. Sine, C., *Farm Chemicals Handbook,* 78th ed., Meister, Willoughby, OH, 1992.
2. Elser, J. J., Marzolf, E. R., and Goldman, C. R., Phosphorus and nitrogen limitation of phytoplankton growth in the freshwaters of North America: a review and critique of experimental enrichments, *Can. J. Fisheries Aquatic Sci.,* 47, 1468, 1990.
3. Hecky, R. E. and Kilham, P., Nutrient limitation of phytoplankton in marine and freshwater ecosystems, *Limnol. Oceanogr.,* 33, 776, 1988.
4. Schindler, D. W., Detecting ecosystem responses to anthropogenic stress, *Can. J. Fisheries Aquatic Sci.,* 44 (Suppl. 1), 6, 1987.
5. Valiela, I. and Teal, J. M., Nutrient limitation in salt marsh vegetation, in *Ecology of Halophytes,* Reimold, R. J. and Queen, W. H., Eds., Academic Press, New York.
6. Goodman, G. T. and Perkins, D. F., The role of mineral nutrients in *Eirophorum* communities. IV. Potassium supply as a limiting factor in an *E. vaginatum* community, *J. Ecol.,* 56, 685, 1968.
7. Henry, R., Hino, K., Tundisi, J. G., and Riberio, J. S. B., Responses of phytoplankton in Lake Jararetinga to enrichment with nitrogen and phosphorus in concentrations similar to those of the River Solimnes (Amazon, Brazil), *Arch. Hydrobiol.,* 103, 453, 1985.
8. Fluck, R. C., Fonyo, C., and Flaig, E., Land use based phosphorus balances for Lake Okeechobee, Florida drainage basins, *Appl. Eng. Agric.,* 8, 813, 1992.
9. Aldridge, F. J., Phlips, E. J., and Schelske, C. L., The use of nutrient enrichment bioassays to test for spatial and temporal distribution of limiting factors affecting phytoplankton dynamics in Lake Okeechobee, Florida, *Arch. Hydrobiol.,* in press.
10. Havens, K. E., Historical trends in nutrient limitation in a shallow subtropical lake, Lake Okeechobee, Florida, USA, *Environmental Pollution,* in review.

11. Gianessi, L. P., Peskin, H. M., Crosson, P., and Puffer, C., Nonpoint source pollution: a cropland controls the answer?, United States Department of Agriculture and United States Environmental Protection Agency Report, Washington, D.C., 1986.
12. Patrick, R., *Surface Water Quality. Have the Laws Been Successful?*, Princeton University Press, Princeton, NJ, 1992.
13. Richards, R. P. and Baker, D. B., Pesticide concentration patterns in agricultural drainage networks in the Lake Erie basin, *Environ. Contamination Toxicol.,* 12, 13, 1993.
14. South Florida Water Management District, Surface Water Improvement and Management Plan for the Everglades, Volume III, Technical Report, South Florida Water Management District, West Palm Beach, FL, 1990.
15. Koch, M. S. and Reddy, K. R., Distribution of soil and plant nutrients along a trophic gradient in the Florida Everglades, *Soil Sci. Soc. Am. J.,* 56, 1492, 1992.
16. Omernik, J. M., Nonpoint source, stream level relationships: a nationwide study, United States EPA Report, Corvallis, OR, 1977.
17. Izuno, F. T. and Bottcher, A. B., The Effects of On-Farm Agricultural Practices in the Organic Soils of the EAA on Nitrogen and Phosphorus Transport, Final Report, South Florida Water Management District, West Palm Beach, FL, 1991.
18. Pimental, D., Acquay, H., Biltonen, M., Rice, P., Silva, M., Nelson, J., Lipner, V., Giordano, S., Horowitz, A., and D'Amore, M., Environmental and economic costs of pesticide use, *BioScience,* 42, 750, 1992
19. Ware, G. W., *The Pesticide Book,* 3rd ed., Thomson, 1989.
20. Muirhead-Thomson, R. C., *Pesticide Impact on Stream Fauna with Special Reference to Macroinvertebrates,* Cambridge University Press, London, 1987.
21. Murty, A. S., *Toxicity of Pesticides to Fish,* Vol. 1, CRC Press, Boca Raton, FL, 1986.
22. Maas, R. P., Dressing, S. A., Spooner, J., Smolen, M. D., and Humenik, F. J., Best management practices for agriculture nonpoint source control. IV. Pesticides, United States Department of Commerce Report, 1984.
23. Caro, J. H., Pesticides in agricultural runoff, in *Control of Water Pollution from Cropland, Volume 2 — An Overview,* Stewart, B. A., Woolhiser, D. A., Wischmeier, W. H., Caro, J. H., and Frere, M. H., Eds., United States Department of Agriculture and United States Environmental Protection Agency, Washington, D. C., 1976.
24. Carpenter, S. R., Kitchell, J. F., Hodgson, J. R., Cochran, P. A., Elser, J. J., Elser, M. M., Lodge, D. M., Kretchmer, D., He, X., and von Ende, C. N., Regulation of lake primary productivity by food-web structure, *Ecology,* 68, 1863, 1987.
25. McQueen, D. J., Post, J. R., and Mills, E. L., Trophic relationships in freshwater pelagic ecosystems, *Can. J. Fisheries Aquatic Sci.,* 43, 1571, 1986.
26. Wetzel, R. G., Land-water interfaces: metabolic and limnological regulators, *Verh. Int. Vereinigung Theoretische Angewandte Limnol.,* 24, 6, 1989.
27. Heath, R. T., Phosphorus dynamics in Old Woman Creek National Estuarine Sanctuary: a preliminary investigation, Technical Memorandum 11, National Oceanic and Atmospheric Administration, Washington, D.C., 1987.
28. Correll, D. L., Jordan, T. E., and Weller, D. E., Nutrient flux in a landscape: effects of coastal land use and terrestrial community mosaic on nutrient transport to coastal waters, *Estuaries,* 15, 431, 1992.
29. Lowrance, R. R., Todd, R. L., Fail, J., Hendrickson, O., Leonard, R., and Asmussen, L. E., Riparian forests as nutrient filters in agricultural watersheds, *BioScience,* 34, 374, 1984.

30. Lodge, D. M., Barko, J. W., Strayer, D., Melack, J. M., Mittelbach, G. G., Howarth, R. W., Menge, B., and Titus, J. E., Spatial heterogeneity and habitat interactions in lake communities, in *Complex Interactions in Lake Communities,* Carpenter, S. R., Ed., Springer-Verlag, New York, 1988.
31. Gregory, S. V., Swanson, F. J., McKee, W. A., and Cummins, K. W., 1991, An ecosystem perspective of riparian zones, *BioScience,* 41, 540, 1991.
32. Lee, G. F. and Jones, A., Effects of eutrophication on fisheries, *Lake Line,* 12, 13, 1992.
33. Dillon, P. J. and Rigler, F. H., The phosphorus-chlorophyll relationship in lakes, *Limnol. Oceanogr.,* 19, 767, 1974.
34. Jones, R. A. and Lee, G. F., Recent advances in assessing the impact of phosphorus loads on eutrophication-related water quality, *J. Water Res.,* 16, 503, 1982.
35. Schindler, D. W., Whole-lake eutrophication experiments with phosphorus, nitrogen and carbon, *Verh. Int. Vereinigung Theoretische Angewandte Limnol.,* 19, 3221, 1975.
36. Schindler, D. W., Experimental perturbations of whole lakes as tests of hypotheses concerning ecosystem structure and function, *Oikos,* 57, 25, 1990.
37. Findlay, D. L. and Kasian, S. E. M., Phytoplankton community responses to nutrient addition in Lake 226, Experimental Lake Areas, northwestern Ontario, *Can. J. Fisheries Aquatic Sci.,* 44, 35, 1987.
38. Agusti, S., Duarte, C. M., Canfield, D. E., Jr., Biomass partitioning in Florida phytoplankton communities, *J. Plankton Res.,* 13, 239, 1991.
39. Canfield, D. E., Jr., Philips, E., and Duarte, C. M., Factors influencing the abundance of blue-green algae in Florida lakes, *Can. J. Fisheries Aquatic Sci.,* 46, 1232, 1989.
40. Benndorf, J. and Henning, M., *Daphnia* and toxic blooms of *Microcystis aeruginosa* in Bautzen reservoir, *Int. Rev. Hydrobiol.,* 74, 233, 1989.
41. Rothhaupt, K. O., The influence of toxic and filamentous blue-green algae on feeding and population growth of the rotifer *Brachionus rubens, Int. Rev. Hydrobiol.,* 76, 67, 1991.
42. Reynolds, C. S., Growth, gas vacuolation and buoyancy in a natural population of a blue-green alga, *Freshwater Biol.,* 2, 87, 1972.
43. Jones, B., Lake Okeechobee eutrophication research and management, *Aquatics,* 9, 21, 1987.
44. Havens, K. E., An experimental analysis of macrozooplankton, microzooplankton and phytoplankton interactions in a temperate eutrophic lake, *Arch. Hydrobiol.,* 127, 9, 1993.
45. Lampert, W., Inhibitory and toxic effects of blue green algae on *Daphnia, Int. Rev. Hydrobiol.,* 66, 285, 1981.
46. Gliwicz, Z. M., Why do cladocearns fail to control algal blooms?, *Hydrobiologia,* 200, 83, 1990.
47. Smith, V. H., Low nitrogen to phosphorus ratios favor dominance by blue-green algae in lake phytoplankton, *Science,* 221, 669, 1983.
48. Paerl, H. W., Nuisance phytoplankton blooms in coastal, estuarine, and inland waters, *Limnol. Oceanogr.,* 33, 823, 1988.
49. Bierman, V. J. and James, R. T., A preliminary modeling analysis of phosphorus and phytoplankton dynamics in Lake Okeechobee, Florida: diagnostic and sensitivity analyses, *Water Res.,* in review.
50. Havens, K. E., Phosphorus reduction goals and limiting nutrient status in Lake Okeechobee, Florida, USA, unpublished.

51. Havens, K. E., Hanlon, C., and James, R. T., Historical Lake Okeechobee trends in the ecosystem. V. Algal blooms, *Arch. Hydrobiol.,* in press.
52. Phlips, E. J. and Ihnat, J., Planktonic nitrogen fixation in a shallow subtropical lake, Lake Okeechobee, Florida, USA, *Arch. Hydrobiol.*, in press.
53. South Florida Water Management District, Lake Okeechobee SWIM Plan — Planning Document, S.F.W.M.D., West Palm Beach, FL, 1993.
54. Radach, G., Berg, J., and Hagmeier, G., Long term changes of the annual cycles of meteorological, hydrographic, nutrient, and phytoplankton time series at Helgoland and at LV ELBE 1 in the German Bight, *Continental Shelf Res.,* 19, 305, 1990.
55. Holmes, P. R. and Lam, C. W. Y., Red tides in Hong Kong waters — a response to a growing problem, *Asian Mar. Biol.,* 2, 1, 1985.
56. Bays, J. S. and Crisman, T. L., Zooplankton and trophic state relationships in Florida lakes, *Can. J. Fisheries Aquatic Sci.,* 40, 1813, 1983.
57. Havens, K. E., Zooplankton dynamics in a freshwater estuary, *Arch. Hydrobiol.,* 123, 69, 1991.
58. Crisman, T. L., Phlips, E. J., and Beaver, J. R., Zooplankton seasonality trends in Lake Okeechobee, Florida, *Arch. Hydrobiol.,* in press.
59. Porter, K. G., Paerl, H., Hodson, R., Pace, M., Priscu, J., Riemann, B., Scavia, D., and Stockner, J., Microbial interactions in lake food webs, in *Complex Interactions in Lake Communities,* Carpenter, S. R., Ed., Springer-Verlag, New York, 1988.
60. Burns, C. W., The relationship between body size of filter-feeding Cladocera and the maximum size of particle ingested, *Limnol. Oceanogr.,* 13, 675, 1968.
61. Riemann, B., Potential influence of fish predation and zooplankton grazing on natural populations of freshwater bacteria, *Appl. Environ. Microbiol.,* 50, 187, 1985.
62. Brooks, J. L. and Dodson, S. I., Predation, body size, and composition of plankton, *Science,* 150, 28, 1965.
63. O'Brien, W. J., The predator-prey interaction of planktivorous fish and zooplankton, *Am. Sci.,* 67, 572, 1979.
64. Kirk, K. L. and Gilbert, J. J., Suspended clay and the population dynamics of planktonic rotifers and cladocerans, *Ecology,* 71, 1741, 1990.
65. Havens, K. E. and Hanazato, T., Zooplankton community responses to chemical stressors: a comparison of results from acidification and pesticide contamination research, *Environ. Pollut.,* 82, 277, 1993.
66. Thienemann, A., Die binnengewasser mitteleuropas: eine limnologische einfuhrung, *Die Binnengewasser,* 1, 1, 1925.
67. Carpenter, S. R. and Lodge, D. M., Effects of submersed macrophytes on ecosystem processes, *Aquat. Bot.,* 26, 341, 1986.
68. Havens, K. E., Responses to experimental fish manipulations in a shallow, hypereutrophic lake: the relative importance of benthic nutrient recycling and trophic cascade, *Hydrobiologia,* 254, 73, 1993.
69. Straskraba, M., The effect of fish on the number of invertebrates in ponds and streams, *Mitt. Int. Verh. Limnol.,* 13, 106, 1965.
70. Sondergaard, M., Jeppesen, E., Mortensen, E., Dall, E., Kristensen, P., and Sortkjaer, O., Phytoplankton biomass reduction after planktivorous fish reduction in a shallow, eutrophic lake: a combined effect of reduced internal P loading and increased zooplankton grazing, *Hydrobiologia,* 200, 229, 1990.

71. Browder, J. A., Perspective on the Ecological Causes and Effects of the Variable Algal Composition of Southern Everglades Periphyton, South Florida Water Management District, West Palm Beach, FL.
72. Ogden, J. C., Freshwater marshlands and wading birds in south Florida, in *Rare and Endangered Biota of Florida, Volume 2, Birds,* Kale, H., Ed., University of Florida Press, Gainesville, FL, 1978.
73. O'Brien, W. J., Hershey, A. E., Hobbie, J. E., Hullar, M. A., Kipphut, G. W., Miller, M. C., Moller, B., and Vestal, J. R., Control mechanisms of arctic lake ecosystems: a limnocorral experiment, *Hydrobiologia,* 240, 143, 1992.
74. Mazumder, A., Taylor, W. D., McQueen, D. J., and Lean, D. R. S., Effects of fertilization and planktivorous fish on epilimnetic phosphorus and phosphorus sedimentation in large enclosures, *Can. J. Fisheries Aquatic Sci.,* 46, 1735, 1989.
75. Havens, K. E., Experimental perturbation of a freshwater plankton community, a test of hypotheses regarding the effects of stress, *Oikos,* 68, 1, 1993.
76. Krieger, K. A., Agricultural herbicides and insecticides and their effects on aquatic biota, Water Quality Laboratory, Heidelberg College, OH, 1987.
77. Hughes, D. N., Boyer, M. G., Papst, M. H., and Fowle, C. D., Persistence of three organophosphorus insecticides in artificial ponds and some biological implications, *Arch. Environ. Contamination Toxicol.,* 9, 269, 1980.
78. Papst, M. H. and Boyer, M. G., Effects of two organophosphorus insecticides on the chlorophyll *a* and pheopigment concentrations of standing ponds, *Hydrobiologia,* 69, 245, 1980.
79. Havens, K. E., An experimental comparison of the effects of two chemical stressors on a freshwater zooplankton assemblage, *Environ. Pollut.,* 84, in press.
80. Brooker, M. P. and Edwards, R. W., Aquatic herbicides and the control of water weeds, *Water Res.,* 9, 1, 1975.
81. Hurlbert, S. H., Secondary effects of pesticides on aquatic ecosystems, *Residue Rev.,* 57, 81, 1975.
82. Boyle, T. P., Responses of lentic aquatic ecosystems to suppression of rooted macrophytes, in *Aquatic Plants, Lake Management and Ecosystem Consequences of Lake Harvesting,* Breck, J. E., Pretnki, R. T., and Loucks, O. L., Eds., Institute of Environmental Studies, Madison, WI, 1979.
83. Hanazato, T. and Yasuno, M., Effects of a carbamate insecticide, carbaryl, on the summer phyto- and zooplankton communities in ponds, *Environ. Pollut.,* 48, 145, 1987.
84. Helgen, J. C., Larson, N. J., and Anderson, R. L., Responses of zooplankton and *Chaoborus* to temephos in a natural pond and in the laboratory, *Arch. Environ. Contamination Toxicol.,* 17, 459, 1988.
85. Hanazato, T., Effects of repeated application of carbaryl on zooplankton communities in experimental ponds with or without the predator *Chaoborus, Environ. Pollut.,* 73, 309, 1991.
86. Gliwicz, Z. M. and Sieniawska, A., Filtering activity of *Daphnia* in low concentrations of a pesticide, *Limnol. Oceanogr.,* 31, 1132, 1986.
87. Wallace, J. B., Recovery of lotic macroinvertebrate communities from disturbance, *Environ. Manage.,* 14, 605, 1990.
88. Heckman, C. W., Long-term effects of intensive pesticide applications on the aquatic community in orchard ditches near Hamburg, Germany. *Arch. Environ. Contamination Toxicol.,* 10, 393, 1981.

89. Hanazato, T., personal communication.
90. United States Environmental Protection Agency, Fish kills caused by pollution, 1977–1987, USEPA, Washington, D.C., 1990.
91. Monetary values of freshwater fish and fish-kill counting guidelines, Spec. Pub., #13, American Fisheries Society, Bethesda, MD, 1982.
92. Alabaster, J. S., Survival of fish in 164 herbicides, insecticides, fungicides, wetting agents and miscellaneous substances, *Int. Pest Control,* 11, 29, 1969.
93. Johnson, D. W., Pesticide residues in fish, in *Environmental Pollution by Pesticides,* Edwards, C. A., Ed., Plenum Press, New York, 1973.
94. Holden, A. V., Effects of pesticides on fish, in *Environmental Pollution by Pesticides,* Edwards, C. A., Ed., Plenum Press, New York, 1973.
95. Johnson, W. W. and Finley, M. T., Handbook of Acute Toxicity of Chemicals to Fish and Aquatic Invertebrates, Resource Publ. 137, U.S. Fish and Wildlife Service, Washington, D.C., 1980.
96. Cook, S. F., Jr. and Moore, R. L., The effects of a rotenone treatment on the insect fauna of a California stream, *Trans. Am. Fish. Soc.,* 98, 539, 1969.
97. Hilsenhoff, W. L., The effect of toxaphene on the benthos in a thermally-stratified lake, *Trans. Am. Fish. Soc.,* 94, 210, 1965.
98. McEwen, F. L. and Stephenson, G. R., *The Use and Significance of Pesticides in the Environment,* John Wiley & Sons, New York, 1979.
99. Potts, G. R., *The Partridge: Pesticides, Predation and Conservation,* Collins, London, 1986.
100. White, D. H., Mitchell, C. A., Wynn, L. D., Flickinger, E. L., and Kolbe, E. J., Organophosphate insecticide poisoning of Canada geese in the Texas panhandle, *J. Field Ornithol.,* 53, 22, 1982.
101. Flickinger, E. L., Juenger, G., Roffe, T. J., Smith, M. R., and Irwin, R. J., Poisoning of Canada geese in Texas by parathion sprayed for control of Russian wheat aphid, *J. Wildlife Dis.,* 27, 265, 1991.
102. United States Environmental Protection Agency, Carbofuran: a special review technical support document, USEPA, Washington, D.C., 1989.
103. Mineau, P., Avian mortality in agroecosystems. I. The case against granule insecticides in Canada, in *Field Methods for the Study of Environmental Effects of Pesticides,* Greaves, M. P., Smith, B. D., and Greig-Smith, P. W., Eds., Thornton Heath, London, 1988.
104. Stickel, W. H., Stickel, L. F., Dyrland, R. A., and Hughes, D. L., DDE in birds: lethal residues and loss rates, *Arch. Environ. Contamination Toxicol.,* 13, 1, 1984.
105. Eisler, R., Diflubenzron hazards to fish, wildlife, and invertebrates: a synoptic review, U.S. Environmental Protection Agency, Washington, D.C., 1992.
106. Kimball, K. D. and Levin, S. A., Limitations of laboratory bioassays: the need for ecosystem-level testing, *BioScience,* 35, 165, 1985.
107. Steinman, A. D., Mulholland, P. J., Palumbo, A. V., DeAngelis, D. L., and Flum, T. E., Lotic ecosystem response to a chlorine disturbance, *Ecol. Appl.,* 2, 341, 1992.
108. Peterson, B. J., Deegan, L., Helfrich, J., Hobbie, J. E., Hullar, M., Moller, B., Ford, T. E., Hershey, A. E., Hiltner, A., Kipphut, G., Lock, M. A., Fiebig, D. M., McKinley, V., Miller, M. C., Vestal, J. R., Ventullo, R., and Volk, G., Biological responses of a tundra river to fertilization, *Ecology,* 74, 653, 1993.
109. Gasith, A. and Perry, A. S., Fate of parathion in a fish pond ecosystem and its impact on food-chain organisms, in *Agricultural Residue-Biota Interactions in Soil and Aquatic Ecosystems, Panel Proceeding Series,* International Atomic Energy Agency, Vienna, Austria, 1980.

110. Hanazato, T. and Yasuno, M., Effects of a carbamate insecticide, carbaryl, on the summer phyto- and zooplankton communities in ponds, *Environ. Pollut.*, 48, 145, 1987.
111. Hanazato, T. and Yasuno, M., Influence of *Chaoborus* density on the effects of an insecticide on zooplankton communities in ponds, *Hydrobiologia,* 194, 183, 1990.
112. Baker, D. B., Regional water quality impacts of extensive row-crop agriculture: a Lake Erie Basin case study, *J. Soil Water Conserv.,* 125, 1985.
113. Krogstad, T. and Lovstad, O., Available soil phosphorus for planktonic blue-green algae in eutrophic lake water samples, *Arch. Hydrobiol.,* 122, 117, 1991.

CHAPTER 6

Humans

Burton C. Kross, Michael L. Olson, Amadu Ayebo, and J. Kent Johnson

TABLE OF CONTENTS

0-87371-860-7/95/$0.00+$.50
© 1995 by CRC Press, Inc.

I. INTRODUCTION: THE PUBLIC HEALTH CONTEXT

In evaluating the influence of soil amendments on the human biotic system, it is necessary to consider the context in which these amendments have been, are being, and will be applied. One of the essential characteristics of the 20th century has been the tremendous increase in the production and use of chemicals for industrial (including production agriculture), commercial, and consumer purposes. This development of chemical science has played an essential role in our contemporary world and brought enormous economic and social benefits.[1] Because the concept of waste is time period and/or culture bound, industrial societies were slow to adequately conceive that increasing chemical waste was an intrinsic part of the increase in chemical production and use after World War II.[2] It is now clear that part of our modern heritage is the increasing volume of waste, including hazardous chemicals, created by all industrial societies.[3]

An exponential increase has occurred in the development, production, and use of synthetic chemicals since World War II. By the 1970s, when the public health risk from environmental pollutants was becoming a public and regulatory concern, production of these chemicals had risen from 100 to 1000 million pounds before World War II to 1000 billion pounds.[4] Chemical production capacity has increased worldwide by 350-fold since midcentury.[5] Between 1000 to 1500 new chemical compounds are now produced each year. These new compounds are being added to the approximately 70,000 pure chemicals and the 2 million mixtures, formulations, and blends already in commercial use.[5,6] These compounds are manufactured and sold by some 115,000 industries and firms.[7]

By the mid-1960s, pollution from chemical wastes was evident in the air and water environmental compartments. The Air Quality Act of 1967 and the Water Pollution Control Act of 1972 restricted the indiscriminate release of hazardous pollutants into these receiving media. During the 1980s many of the land-based chemical waste sites resulting from chemical production and use also became a matter of regulatory concern and were declared hazardous.[3]

The actual hazard from long-term environmental exposures that this widespread production and use of chemicals now presents human populations is not known with any degree of certainty. The research evidence appears to indicate that the concern is well founded. At present, researchers have limited epidemiologic data for determining the prevalence or incidence of potential environmental diseases among exposed human populations. An adequate toxicological profile on these chemical agents is also lacking. A study by the National Academy of Sciences in the early 1980s indicated that no more than 20% of commercial chemicals have had adequate premarket toxicologic evaluation.[8]

II. AGRICULTURAL SOIL AMENDMENTS

This review will be concerned with two classes of soil amendments: (1) municipal sewage sludges and (2) commercial inorganic fertilizers, containing nitrogen. Municipal sludges have been a growing resource for agriculture since the early 1970s. The application of municipal sludges to agricultural cropland has also been a valuable and necessary waste disposal option for urban areas.

Nitrogen-based fertilizers have been an indispensable resource in the development of modern production agriculture after World War II. These benefits have, unfortunately, come at a cost that is yet to be determined. Nitrate levels in shallow groundwater in several regions of the U.S. and Europe have led to concern about contamination of water sources, particularly in shallow rural wells.

The use of both these classes of soil amendments illustrates the need for rational management practices based on present knowledge and research. Such rational practices will be necessary to minimize the risk to human health from sludge and fertilizer application to cropland. Modern industrial societies have little choice but to disperse some of their municipal waste on agricultural land. Nitrate fertilizer application will also need to be guided by sound management and application practices.

III. COMMERCIAL INORGANIC FERTILIZERS: BENEFITS AND COSTS

The commercial inorganic fertilizers (nitrate, phosphate, and potassium) have played a fundamental role in the development of modern agriculture and its ability to feed more people with less human labor on the farm. The largest increase has been in the use of nitrogen fertilizers, with a nearly 20-fold increase in application rates from 1945 to 1983.[9] Since they are a concentrated source of nutrients that can be easily transported, stored, and applied, the fertility of most soils in the U.S. and other industrialized countries is partially restored and maintained by applying these commercial fertilizers to cropland. This increased use of commercial fertilizers has helped to dramatically increase the per acre yields of agronomic and horticultural crops. The use of inorganic commercial fertilizers throughout the world increased ninefold and average per capita use increased fivefold between 1950 and 1985. By 1985 the additional food commercial fertilizers helped produce fed about one out of every three people in the world.[10]

Groundwater contamination and eutrophication of surface waters have been part of the costs of this widespread use. Nitrogen fertilizers in particular have been overapplied in the past. From 1965 to 1982, when fertilizer use in general began to level off, average nitrogen fertilizer rates increased from 65 to 135 lb./acre across the Corn Belt.[11] Estimates of crop absorption of applied nitrogen range from approximately 25 to 70%.[11,12] Unused nitrogen can be immobilized, denitrified, washed into streams and lakes, or leached from the soil into groundwater and the subsoil. The public health concern is that the nitrates in commercial fertilizers can leach into groundwater supplies, where excessive levels of nitrate ions can make drinking water toxic, most particularly for infants.[13] A detailed review of nitrate and nitrite toxicity is presented later in the chapter.

IV. MUNICIPAL SEWAGE SLUDGES: BENEFITS AND COSTS

By the early 1970s, serious research was beginning on the feasibility of applying municipal sewage sludge on agricultural cropland. The agricultural use of sludge includes applications to land used for a wide variety of crops including grains, animal feeds, food, and nonfood chain crops. The objective of this practice is to improve the soil-conditioning properties and nutrient status, and thus increase crop production.[14] The public health concern is that this practice may involve incorporation of toxic substances from the municipal wastewater stream in the human food chain.[14]

Before the passage of the Federal Water Pollution Control Act (FWPCA) in 1972, municipal sludge management was considered primarily a problem of dewatering and physical removal of sewage plant residues with little concern for the broader implications of disposal options. With the enactment of FWPCA, planning for sludge disposal became a required part of general wastewater treatment facility planning. Since this time, the land application of sewage sludges as soil amendments has supplied a disposal route of growing importance for urban waste products.

U.S. municipalities produced an estimated 6.8 dry metric tons of sewage sludge in 1983.[14,15] That figure is expected to grow to 12 million dry metric tons by the year 2000. [14] Various land disposal routes accounted for about 42% of sludge disposal in the early 1980s.[16] Since water and air have been severely restricted as waste disposal media, land application of this industrial and domestic waste is expected to grow as well.

Municipal sludge is a by-product of waste treatment systems and thus sludge represents an intentional concentration of inorganic and reduced organic materials that have deliberately been prevented from entering rivers, lakes, oceans, or other receiving media. The potentially toxic substances are removed because past experience and environmental research evidence have shown them to be pollutants of concern to the environment and biota, including humans. Since the hazardous constituents of the sludge will depend on the industrial, commercial, and residential inputs from each locality to the wastewater stream, each sludge must be individually characterized as to these pollutants.

There are three categories of toxic agents of potential concern with municipal sludge application on agricultural cropland: (1) metals, (2) toxic organic chemicals, and (3) pathogens. This review will cover the important metals and several of the toxic organic chemicals that are of public health concern and potentially found in sludge. Pathogens will not be reviewed here, as a number of recent reviews are available.[17-20]

V. ADVERSE CHARACTERISTICS OF SLUDGE: TOXICITY AND EXPOSURE/PATHWAY

To understand the potential risk to humans from these pollutants, it is necessary to consider the potential exposure pathways and the nature of the intrinsic toxicity of the agents. Both an exposure/pathway and intrinsic toxicity of an agent must be present before a hazard exists. The risk is then the likelihood that an adverse effect will occur. The critical factor in determining risk from chemical exposure is the likelihood that a given level of chemical exposure is sufficient to express its intrinsic toxicity, i.e., its ability to cause an adverse effect(s) on living organisms.[21] Investigating more completely the nature of intrinsic toxicity at low-level doses found from multiple environmental exposures is one of the most important concerns of environmental health research. More research information is particularly needed on the concept of a "critical effect" in the potential toxic process of environmentally caused (or influenced) chronic diseases.

Toxicity is usually defined most simply as the intrinsic ability of a substance to harm living things.[22] This is more precisely defined as causing adverse effects on the biochemical, physiological, or behavioral processes of living organisms. This harm or toxic effect is considered to be damage to an organism measured in terms of loss, reduction, or change of function/structure. It may be reported as a symptom or detected as a sign.[22,23] The chemical disposition of a substance in the intact body is an important factor in the concept of toxicity, since the toxic agent or its metabolic products after entering the body must then reach the target organ(s) at a sufficient concentration and for a sufficient length of time to produce a biological response for an adverse effect to occur.[23] Hence, a fundamental maxim of traditional toxicology, particularly for acute effects, is that "dosage makes the poison". The threshold level is considered the lowest dosage at which a chemical has a detectable effect. As the Office of Technology Assessment has put it, "In most cases, the toxic effects causing greatest concern are those that are most severe, occur at lowest exposures, and persist after exposure ceases."[23]

The threshold level effect is based on the concept of the biological organism as a homeostatic system. Up to a certain point, the living system can absorb, distribute, metabolize, and excrete many xenobiotic agents. Only when these detoxifying mechansims are overwhelmed does the toxic process begin. The metabolic process can also bioactivate and thus produce a more toxic metabolite than its parent compound. Many toxic effects occur through this metabolic activation mechanism. This homeostatic system paradigm of physiology and traditional toxicology is now being influenced by biomolecular findings from research in molecular toxicology and biology.

Hattis has pointed out that molecular toxicology (as well as molecular epidemiology) must consider a taxonomy of toxicity that includes two more conceptual categories beyond the traditional homeostatic system-influenced toxicity. In addition to the traditional acute and chronic toxicities, Hattis defines two more categories of effects resulting from insidious processes that are irreversible or poorly reversible at low doses or early stages of causation. These two categories are molecular biological (stochastic process) effects and chronic cumulative effects.[24] Molecular biological effects "occur as a result of one or a small number of irreversible changes in information coded in DNA, e.g., mutagenesis, most carcinogens, and some teratogenesis".[24] Chronic cumulative effects "occur as a result of a chronic accumulation of many small-scale damage events, e.g., emphysema, noise-induced hearing loss, atherosclerosis, and probably hypertension".[24] It is clear that environmental research (and biomedical research in general) is moving towards a molecular biological definition of disease. Both the stochastic and the deterministic paradigms will most likely have important roles to play in adequately characterizing disease at the biomolecular level.

Oral exposure through the human food chain is the potential route of most concern from sludge-amended soils. Researchers have identified a number of protective mechanisms in the sludge-amended soil-plant-animal-human exposure pathways that limit the bioavailability, and thus potential exposure, to human populations.

By the early 1980s, sludge researchers had recognized that the potential risk from the relatively high levels of metals and toxic organics found in the municipal sludges of the 1970s would be reduced by the nature of municipal sludges, sludge-soil relationships, soil-plant barriers, plant-animal barriers, and animal homeostatic mechanisms.[25] As Chaney and Hansen remarked, "Neither the benefits nor the hazards associated with the agricultural uses of even unsuitable sewage sludges conform to direct, much less linear, dosage-response relationships".[25] The specific nature, or characteristic, of sludge-amended soils usually reduces the bioavailability of a toxic agent to a human population, thus reducing the potential risk from toxic agents that might be hazardous from another exposure matrix.

Thus, to assess the risk from potential environmental exposure from sludge application to cropland requires knowledge of the potential for transfer of each toxic agent from the sludge or sludge-soil mixture to crops and to animals, including humans, that ingest sludge, sludge-soil mixture, or crops grown on the sludge-amended soil.[26] Without such a pathway analysis, it is not possible to adequately characterize the potential for sludge-amended soils to serve as a matrix for exposure to toxic environmental contaminants.

Current regulatory and scientific concerns associated with environmental toxicants focus largely on human exposure(s) to complex mixtures from multiple sources and via various media, particularly from low-level environmental exposures. Sludge-amended soils are only one possible matrix for exposure to toxic agents and must be evaluated within this larger environmental health context for the development of rational sludge application practices that will protect public health.

Trace elements illustrate the complexity of environmental contamination and transport processes and the complexity of adequately characterizing their potential adverse effects to humans upon uptake. It has been known for many years that essential, or nutrient, elements and nonessential elements interact within the human body. A fundamental concern of toxic metal research is how these interactions can become mechanisms of toxic action at sufficient doses. The margin of safety between deficiency and excess of an essential metal and its potential role in toxic actions is also of current scientific and regulatory concern.

A. TRACE ELEMENTS/METALS

Trace elements occur in natural systems in small amounts and when present in excessive concentrations they are toxic to living organisms. Certain of the trace elements are known as heavy metals, usually described as those metals with densities >5.0.[27] Because metals do not break down in the environment, but tend to remain in the ecosystem for a long time and are very difficult to remove once contaminated, a major objective of sludge research has been to determine the hazard, and thus risk, from trace metals in sludge-amended soil.

Most research on toxic metals has concentrated on the essential metals, such as iron, copper, zinc, and selenium, and on a limited number of nonessential metals, primarily cadmium, lead, mercury, and arsenic. These four elements are recognized toxic hazards and have all caused adverse health outcomes as a result of environmental exposures. A number of other metals present in sludges are of concern for toxicity to humans, e.g., aluminum, cobalt, nickel, molybdenum, and selenium.[28]

At least 12 metals are known to be essential for the maintenance of life.[28] Just as the interaction of toxic agents within the soil-plant-animal food chain must be characterized as fully as possible for understanding the potential for human exposure, so, too, must the interaction of metals within the human biologic system be characterized to understand the potential for adverse effects from metals. A particular concern is that certain metals, such as cadmium, can replace (by mimicking) an essential element in the microenvironment of the body and thus potentially disrupt the metal homeostasis of the biologic system.

Industrial societies make significant use of about 45 of the 80 metals and their compounds that are of concern to metal toxicology. Approximately 20 of these are known to produce specific effects in humans.[28,29] The major source of exposure to metals for the general public is through food, drinking water, and beverages. The primary concern for potential exposure to a human population from sludge-amended soils is through the food chain. Groundwater contamination from sludge-amended soils is not considered likely with sound sludge application practices.

Municipal wastewater sludges contain various amounts of aluminum, arsenic, boron, cadmium, cobalt, copper, lead, manganese, mercury, molybdenum, nickel, and zinc. Each municipal sludge will have a toxic metal profile based on these sources. Sludge from residential areas can contain greater concentrations of copper (>500 mg/kg) and zinc (>1000 mg/kg) than industrial source sludges, most likely from metal pipes and tanks during conveyance, storage, or treatments.[27]

Heavy metals constitute only a fraction of the sludge solids from municipal wastewater, usually <1% dry weight, but the metal contents of soils could potentially be raised through long-term application of sludges.[30] Sound application practices should keep this theoretical possibility of elevated metal concentrations in agricultural soils from occurring. A recent study of metal concentrations in U.S. agricultural cropland soils by Holmgren et al. indicates little evidence of significant accumulation of cadmium or lead in these soils, although the study found that copper and zinc have accumulated from normal agricultural practices.[31]

The role of trace metals in limiting the feasibility of sludge application on cropland has been an objective of sludge research since the early 1970s. In a 1972 article, Dean et al. listed cadmium, chromium, cobalt, copper, iron, lead, manganese, mercury, nickel, and zinc as the metals of most immediate concern to human health.[32] In 1976 the Council for Agricultural Science and Technology classified cadmium, copper, molybdenum, nickel, and zinc as potential hazards to human health from the application of sludges to cropland.[27] In 1982, Naylor and Loehr listed cadmium, chromium, copper, nickel, lead, and zinc as metals of concern found in municipal sludges.[33] The U.S. Environmental Protection Agency (U.S. EPA) has recommended cumulative limits for cadmium, copper, lead, nickel, and zinc as the metals of major concern.[14] Of these five, cadmium, lead, and nickel are of most concern for their potential toxicity to humans, with cadmium being of most concern from sludge-amended soils.

The concentrations of trace elements in many municipal sludges, particularly mercury, decreased during the decade of the 1980s as a result of industrial point-source control, and this trend is expected to continue.[34,35] The most cost-effective method for reducing trace element contamination in sludges is through removal at the industrial source.

B. SOIL-PLANT BARRIER

Since the start of sludge research there has been concern that sludge application to cropland would cause trace element phytotoxicities and accumulation of undesirably high levels of trace elements in the food chain, particularly with respect to cadmium. The context for this concern was two cadmium poisoning incidents from environmental exposures in Japan during the late 1960s. Chronic environmental oral exposure to high levels of cadmium in contaminated rice and water resulted in two separate outbreaks of "Itai-Itai (ouch ouch) Byo", a painful bone disease, in Japan.[36] See Chaney for the reasons why these cadmium contaminations are not relevant in other circumstances for determining potential health risks from cadmium in sludge-amended soil.[37]

The concept of the "soil-plant barrier", a term introduced by R. L. Chaney, uses a pathway analysis to examine what is known about the absorption of trace elements by plants from sludge-amended soil and how this is affected by trace element contents in sludge, by chemical properties of sludge, by chemical reactions in soil, and by plant factors as an important component in a "holistic" environmental toxicology.[35,37]

Researchers have identified two major (potential) flaws in past experimental methods of researching toxic metal uptake by plants. The first flaw is called the "salt vs. sludge" error. This error occurs when soluble metals are added directly to soil under experimental conditions. Such salts are nearly always taken up by plants in greater amounts and show more toxicity than when the metal is applied in equal amounts in sludge. The second flaw is the "greenhouse vs. field" error. This error results in higher contents of metals in crops grown in the greenhouse compared with those crops grown in the field for the same soil.[25,38]

The potentially toxic trace elements in the aerial portion of a plant can be limited by three protective mechanisms of the soil-plant system: (1) elements that are insoluble in soil and not accumulated in the plant (e.g., chromium, lead, mercury, and tin); (2) elements absorbed into the root, but that are insoluble in vascular fluids, having limited translocation to the shoot (e.g., aluminum, iron, and occasionally, lead and mercury); and (3) elements that when applied in excess cause phytotoxicity; thus, plants are not consumed by man or domestic animals (e.g., arsenic, cobalt, copper, nickel, and zinc).[27]

Cadmium, molybdenum, and selenium are among the metals that do not fall into one of these three categories. All three can cause toxicity in humans, although molybdenum and selenium are less of a toxic hazard and are usually present in low concentration in sludges. Thus, these elements do not generally limit application rates.[27] Molybdenum and selenium are of concern for animals, since livestock have been injured by soils with excessive concentrations of these two metals. Cadmium is generally considered the metal of most concern in the application of sludge to cropland. Most countries that practice sludge application have regulated application rates based on cadmium soil concentrations.

Chang et al. in 1987 reviewed newly available long-term field data on elemental uptake by plants in terms of their implications on land application of sludges. Their conclusions indicate that with rational management sludge-amended soils would not serve as a high-risk matrix for exposure to metals, including cadmium.[39]

C. METAL TOXICITY: OVERVIEW

Metals can induce a wide range of toxicity at sufficient doses from the very toxic lead and mercury to the relatively nontoxic metals. Although metals also have a variety of toxic properties, they share certain common biochemical features that are postulated to form the basis of at least some of their adverse effects. It is thought that a major mechanism of toxicity of certain heavy metals is the alteration and impairment of the structure and function of enzymes in the human body. Many metals have an affinity for sulfur and bind with the sulfhydryl groups in enzymes, thus inhibiting enzyme activity. Certain metals may also interact with phosphate biocompounds, protein amino groups, and carboxylic acid. Metabolism of the metals essential to the body is normally adjusted to a homeostasis to maintain body stores and concentrations at proper levels in biologic fluids. Nonessential heavy metals, such as cadmium and lead, have no effective metabolic pathway for detoxification and excretion. Unlike many synthetic organic chemicals, the heavy metal ions are resistant to biological oxidation processes. Certain nonessential metals, e.g., cadmium, can mimic the pathways of essential metals such as zinc. A number of specific transport and depository proteins have been identified, and the accumulation of certain of these metals in organ tissues such as the kidney and nervous system has been associated with toxicity in human and animal studies. Metals, such as cadmium and lead, may also bind to cell membranes, thus disrupting the transport processes through cell walls.

Mechanisms of metal toxicity have been analyzed into three categories: (1) blocking of the essential biological functional groups of biomolecules, (2) displacing the essential metal ion in biomolecules, and (3) modifying the active conformation of biomolecules.[40] This interference at the biomolecular level is thought to lead to physiological disruptions and potentially to pathological states in the organ system.

In contrast to the lipophilic toxic organic compounds, generally thought to pass into cells by passive diffusion (lead also can be so absorbed), metals are thought to be taken up principally by nonspecific binding to the plasmal membrane followed by endocytosis. Metal-bound metallothioein, a cytosolic, sulfhydryl-rich metal-binding protein, is taken up by pinocytosis in the renal proximal tubules. This appears to be a nonspecific process for general protein uptake whose probable function is amino acid recovery.[41]

Cadmium, copper, mercury, and zinc all bind to metallothionein on absorption and thus reduce the levels of free metals in the cell. Thus, induction of metallothionein markedly decreases the acute toxicity of cadmium and other metals in the liver.[29] Although it is known that metals such as cadmium and zinc induce metallothionein synthesis even at low levels, the precise mechanisms by which induction occurs remain to be completely established. In addition to metallothionein, a number of other metal-binding proteins exist, e.g., albumin plays an important role in the plasma binding of a wide variety of xenobiotics and may also serve to reduce the bioavailability of well-bound metals to the tissues.

Toxic metal research has investigated a number of potential effects to the organ systems of animals and humans. Aluminum, cadmium, lead, manganese, and mercury have all demonstrated adverse effects on the nervous system.[42] While these metals are all recognized as neurotoxicants at sufficient dosages, the current research concern is with the possibility that these metals may be associated with subtle nervous system dysfunction, particularly of the developing nervous system. Thus, it is necessary to differentiate between the neurotoxicity of metals on mature vs. the developing nervous system.[43]

Arsenic, beryllium, cadmium, chromium, lead, mercury, selenium, and zinc are the metals reported to produce immune dysfunction in animals or humans.[44-46] Since the immune system defends the biological organism against infectious agents and spontaneously arising neoplasms, it is not surprising that one of the most consistent effects of metal exposure in animals is a decreased resistance to infectious organisms.

The immunotoxic effects investigated in animal bioassays have been supported by some epidemiologic studies that associate certain environmental exposures to alteration, through immunosuppression or immunopotentiation, in the immune response.[46]

Cadmium, copper, lead, molybdenum, manganese, selenium, and zinc are among the metals known to be teratogenic in animal experiments.[47] Cadmium and lead have shown evidence of affecting the human reproductive organs and, in particular, lead, has shown evidence of inducing longer-term developmental effects. As with the other noncarcinogenic toxic adverse outcomes, the threshold level effect of classic toxicologic research is generally believed to exist for reproductive and developmental toxicants. The focus of current environmental health research is to determine more adequately than in the past the exposure/dose level at which these thresholds exist for particular toxic metals within the human body.

Arsenic, beryllium, cadmium, chromium, lead, nickel, and zinc are among the metals that have been investigated for carcinogenicity from environmental and occupational exposures and in animal experiments. Theoretically, the carcinogenic process could be initiated by a single molecule interacting with DNA. The U.S. EPA and many researchers accept this no-threshold level when attempting to determine risks from potential environmental carcinogens. Other researchers hold that for many potential carcinogens a threshold level most likely exists. Many environmental chemicals must be metabolized to become reactive. This conversion to the metabolite and the likelihood that it will interact with DNA may often be dose-dependent. Thus, this process would, theoretically, be nonlinear with a threshold level below which the carcinogenic process would not be initiated.

D. CADMIUM

The industrial production and use of cadmium during this century has grown from a few tons per year at the beginning of the century to over 18,000 tons by the last quarter of the century. In 1989 alone, four U.S. companies produced an estimated 3.9 million pounds of cadmium.[7] In the same year estimated U.S. cadmium imports were close to 6 million pounds. Cadmium is now widespread in the environment with about 2/3 of the cadmium produced dissipated into the environment.[7] Oral exposure to cadmium via food, drinking water, and beverage contamination constitutes the major environmental route of entry for the nonsmoking general population, with food being the primary route. The cadmium content of different foodstuffs ranges from 0.001 to 1.3 ppm, and the average intake from food and water ranges from 10 to 30 μg daily.[48]

Cadmium is considered the toxic metal of most concern in the application of municipal sludge on agricultural cropland because cadmium is more readily taken up by plants than other metals, such as lead, and is not sufficiently phytotoxic to protect the food chain. Cadmium is thought unlikely to affect crop growth until tissue concentrations exceed 10 mg/kg dry matter, but concentrations of much less than this could lead to potentially harmful human dietary intake of cadmium.[49] In addition to sludge application, another agricultural practice that can lead to cadmium exposure from soil is the use of cadmium-containing phosphate fertilizers.

The U.S. EPA's recently completed National Sewage Sludge Survey, a study of 209 wastewater plants randomly selected from all regions, detected cadmium 194 times in the treatment plants. The mean concentration was 65.5 mg/kg with a range of 0.7 to 8220 mg/kg.[50]

1. Toxicity

The absorption of cadmium depends on the route of exposure. The primary concern from sludge-amended soils is the soil-plant-animal-human ingestion pathway. Cadmium is a cumulative toxicant; thus, subtle increases in the diet sustained over long periods could potentially pose a problem of chronic toxicity to humans. It is estimated that only about 5% of the cadmium that is ingested is absorbed, although various conditions, including dietary status and iron deficiency anemia, can elevate this proportion.[48] Due to the poor level of cadmium absorption from the gastrointestinal tract, 95% of ingested cadmium is slowly excreted in the feces . Most humans slowly accumulate cadmium in the body until approximately age 50, after which levels diminish. The half-life of cadmium in the body is approximately 20 to 30 years.

Once absorbed, cadmium is bound to red blood cells and serum albumin and rapidly taken up by the liver and kidney, tissues with a high capacity for metallothionein systhesis. Generally, over 50% of the body burden of cadmium will be found in these organs. In the U.S. the mean level of cadmium in the kidney cortex is 20 to 35 μg/g with about 0.6% of the population exceeding 100 μg/g.[51] The metal-binding protein metallothionein, which is synthesized in response to cadmium exposure, will effectively bind cadmium and thus render it toxicologically inert when in the cell.[29] This concept that cadmium is detoxified by long-term storage rather than biotransformation or elimination has recently been challenged since cadmium-complexed metallothionein is reabsorbed from the renal tubules. At sufficient dosages kidney lesions predominate after long-term exposure. The primary site of action in the kidney is the proximal tubules. It is thought that there is a critical concentration of cadmium within the kidney (estimated at 200 to 300 μg/g), and that once this is exceeded, cadmium-induced nephropathy will occur as a result of the proximal tubules' inability to reabsorb small-molecule proteins.[51] Cadmium is stored in association with metallothionein for long periods, and one hypothesis is that extracellular cadmium in association with metallothionein may be the actual toxic agent to the kidneys.[52] Cadmium metallothionein, administered parenterally, is highly toxic to the kidney and will acutely induce a proximal tubular necrosis characteristic of long-term exposure to ionic forms of cadmium. Goyer favors the hypothesis that the cadmium that is not bound to metallothionein within the cells induces the nephrotoxicity. This occurs when the level of cadmium exceeds that of metallothionein available for binding, leading to renal cell injury with the associated proteinuria and calcuria.[29]

Studies by Fowler and co-workers indicate that the nonmetallothionein-bound cadmium ions are temporally associated with cytotoxicity through interfering with the normal process of lysosomal biogenesis.[53] Cadmium metallothionein induced the characteristic pattern of vesiculation and an increased number of electron-dense lysosomes. This was temporally associated with decreases in lysosomal protease activity, low molecular weight proteinuria, and calcuria.[53] Fowler considers the results of his recent *in vitro* studies to investigate the mechanisms of cadmium-induced renal cell injury to indicate that induction of stress proteins followed by cellular vesiculation occurs well before measurable alterations in intracellular cadmium ions, which occur only when the cells begin to die. Fowler suggests that the observed toxic effects are not secondary to altered calcium-induced mechanisms but rather a function of cadmium ions binding to effector molecules early in the toxic process. Calmodulin is considered the primary protein for this role.[53] Activation of calmodulin by cadmium ions not bound to metallothionein could damage the cytoskeleton by interfering with the process of lysosomal biogenesis.[53] Recent epidemiological evidence suggests that over a period of 50 years an oral intake of 140 to 260 μg of cadmium per day or a cumulative intake of 2000 mg induces an increased prevalence of low molecular weight proteinuria.[54]

In a recent *in vivo* study of human proximal tubule cells, Hazen-Martin et al. investigated the hypothesis that many of the overt toxic effects of cadmium ions *in vivo* appear to be a direct consequence of disrupted junctions between cells in various endothelial and epithelial surfaces.[55] This proposal is supported by findings of junctional disruption by cadmium ions *in vitro*. Tight junction structure demonstrated no significant alterations due to cadmium ion exposure. Cadmium ion exposure did alter the characteristics of the apical cell membrane as evidenced by the significant increase in apical intramembrane particles.[55]

Aminoaciduria is another result of damage to tubular cells that normally reabsorb the amino acids filtered through the glomeruli. Other related effects are glycosuria and decreased tubular reabsorption of phosphate.[29] Animal studies also show that exposure to cadmium impairs pancreatic secretory activity and has injurious effects on the cellular structure as well as on the functional capacity of the pancreatic tissue.[36] During prolonged exposure, accumulation of cadmium in the pancreas continues even after it has ceased in the liver and kidneys.

Another major hypothesis for the toxic actions of cadmium concerns cadmium's ability to interfere with the biologic pathways of several essential elements within the human body. Cadmium will often follow the biologic pathways of zinc metabolism, being taken up by cells through mechanisms normally devoted to zinc uptake. At the molecular level it is thought that many of the toxic effects of cadmium are due to its replacement of zinc in biologic systems. It is postulated that cadmium exerts its toxicity, at least in part, as antinutritional metabolites for essential elements such as zinc and copper, inducing toxic syndromes similar to nutritional deficiency of these elements.[42] Zinc deficiency may modify cadmium distribution and potentially enhance cadmium toxicity. Waalkes found that feeding rats a zinc-deficient diet increased the accumulation of cadmium in the liver, kidney, and testes. Reductions in renal and testicular zinc concentrations were induced by zinc deficiency in association with cadmium exposure. Metallothionein induction was also reduced in the kidney.[56]

Cadmium will also follow the biologic pathways of calcium. Both suppression of calcium uptake by cadmium or an elevated uptake of cadmium due to low dietary calcium can induce cellular dysfunction.[42] According to Chang, this interaction between cadmium and calcium should be considered an important factor when evaluating potential cadmium exposure and cadmium toxicity.[42]

The principal long-term effects of low-level exposure to cadmium are chronic obstructive pulmonary disease and emphysema (both from occupational inhalation exposures), and chronic renal tubular disease. Chronic renal dysfunction is the outcome of concern from environmental exposure and has been found to be irreversible and slowly progressive, even after exposure was reduced.[57]

Chronic environmental oral exposure to high levels of cadmium has resulted in Itai-Itai (ouch ouch) disease in Japan, with a higher incidence among postmenopausal women. The fully developed syndrome is characterized by severe pain in the bones and pathological fractures, waddling gait, aminoaciduria, osteomalacia, osteoporosis, proteinuria, glucosuria, and anemia.[36] Cadmium may also play a role in hypertension, which may result from sodium retention, vasoconstriction, and hyperreninemia, but there is considerable debate over what that role may be.[58] Chronic cadmium exposure has induced hypertension in rodent experiments.[48] Cadmium has also been shown to affect the immune system in animals.[44,59]

Teratogenic effects of cadmium have not been observed in humans. Cadmium has been shown to be a potent teratogen in animal models, particularly through the injection route. Its teratogenic effects can mimic those produced by zinc deficiency.[56]

2. *Carcinogenicity*

The National Toxicology Program (NTP) has classified cadmium among substances or groups of substances that may reasonably be anticipated to be carcinogens. This classfication is used for those substances for which there is limited evidence of carcinogenicity from studies in humans, i.e., the causal interpretation is credible, but that alternative explanations, such as chance, bias, or confounding, could not be adequately excluded, or for which there is sufficient evidence of carcinogenicity from studies in experimental animals.[7]

The International Agency for Research on Cancer (IARC) has classified cadmium and certain cadmium compounds as Group 2A (agents that are probably carcinogenic in humans).[60] The evidence for carcinogenicity to humans is limited. Epidemiological evidence is confined to occupational exposures of cadmium oxide, resulting primarily in respiratory, prostatic, and genito-urinary cancers. There have been several studies that have not detected a correlation between excess cancer mortality and exposure to cadmium.[61] However, Waalkes et al. have recently reported findings that support the prostate, through injection, as a site of carcingenesis in laboratory rats.[62,63] The lung has also been confirmed as a site of carcinogenic induction in rats. Evidence for carcinogenicity to animals was determined by the IARC to be sufficient.[60] The U.S. EPA has concluded that there is as yet no evidence of cadmium oncogenicity in chronic oral animal studies.[61]

The relevance of the then available studies for evidence regarding the potential carcinogenicity of cadmium to humans from environmental exposures was considered largely conjectural by Ryan et al. in the early 1980s.[51] Nordberg considers the experimental and epidemiologic research to provide increasing evidence that cadmium is carcinogenic. Nordberg also considers cadmium's carcinogenicity to be its critical effect at low doses, which is postulated to be stochastic in nature.[54]

E. LEAD

Lead is the most widely dispersed of all the toxic metals by human activity. Nearly all lead in the environment is inorganic and is found in the air, in dust and soil, in food, and in water. Ingestion of lead through food represents the majority of daily intake for most individuals. The natural lead levels in soil generally range from 2 to 200 ppm and average about 10 to 15 ppm.[64] Lead concentrations in sludges are usually higher than soil concentrations and repeated applications can cause its accumulation. In most cases, the root system of most plants will provide an adequate barrier against the translocation of lead to the plant .

Lead is also found in all biologic systems and is thus an environmental pollutant of great concern at the present time. Because lead is toxic to most living organisms at high exposures, the major issue regarding lead is the dose at which it becomes toxic from environmental exposures. The major risk is toxicity to the nervous system and hemapoietic system, with an extensive literature on heme synthesis. The most susceptible populations are children and infants in the neonatal period and the unborn fetus.[65,66] The aged could also be at increased risk, particularly from neurotoxicity.

Adults absorb 5 to 30% of ingested lead, while young children absorb about 50% of ingested lead. Upon absorption the highest concentrations of lead are found in the kidney, from which it is excreted and redistributed to other tissues, primarily bone. Up to 75% of absorbed lead can be excreted in the urine. Total body lead in humans is essentially distributed in three compartments: blood, soft tissue, and bone. Ninety percent of the body burden is found in bone. The biological half-life of lead in the blood and soft tissue is 35 to 40 days and in bone about 10 to 20 years.

The National Sewage Sludge Survey conducted by the U.S. EPA detected lead 213 times in the wastewater plants sampled, with the mean concentration 195.2 mg/kg and range from 9.4 to 1670 mg/kg.[50]

1. *Toxicity*

Absorbed lead is toxic to a variety of enzyme systems, having, like other metals, a particular affinity for the sulfhydryl groups that are present in many enzymes. Similar to cadmium, lead may be particularly toxic to enzyme systems that are zinc dependent. It is also thought that lead may replace calcium at the biomolecular level, thus altering cell functions such as ion transport, energy production, and the function of heme-containing enzymes. Its overt effects at high doses include hematological, renal, neurological, and reproductive impairments. At high levels of exposure, proximal tubular dysfunction and mitochondrial degeneration have been reported in humans and animals.[67]

The heme synthesis pathway has had the most systematic study. Lead is known to inhibit the porphobilinogen synthetase, coproporphyrinogen oxidase, and heme synthetase in the heme biosynthetic pathway. This effect takes place on bone marrow erythroblasts, which contain mitochondria. Because lead tends to accumulate in mitochondria and thus disrupt mitochondrial function and structure, the inhibition of cellular respiration is a major toxic effect of lead at sufficient doses. There is also evidence showing that with any incremental rise in lead, the dose of lead delivered to other parts of the body rises at a disproportionately greater rate. This is attributed to the fact that blood is a capacity-limited compartment for lead. Thus, an increase of lead uptake may result in even greater increases in the concentration of lead at other target sites for toxicity.[42]

Lead interacts with several essential metals. In 1970 Mahaffey-Six and Goyer demonstrated that a calcium-deficient diet will potentiate tissue lead deposit, as well as biochemical and pathological changes associated with lead toxicity.[68] Many investigators have demonstrated that calcium can influence lead metabolism and vice versa, as well as lead's interaction with other essential nutrients such as copper, zinc, and phosphate.[69] The interaction between lead and calcium is central to many of the biochemical mechanisms postulated to induce lead toxicity. It is thought that lead may block the ability of calcium to reach an active regulatory site in some instances. Upon entering the cell, lead may also mimic the action of calcium as regulator of cell function.[70]

Although the toxicity of lead was recognized centuries ago, concern was restricted to overt symptoms such as colic, anemia, renal disease, or encephalopathy.[67] Interest was limited to occupational exposures and there was a lack of awareness of specific biochemical or metabolic effects at the subclinical level. Identification of subclinical effects has been possible during the past two decades because of the development of more sensitive measures to detect cognitive and behavioral changes that are not apparent clinically, because of methods to measure the reduced activity of heme enzymes, and because of a better understanding of how lead can disrupt other biomolecular functions.

Lead crosses the blood-brain barrier and produces a diffuse toxic encephalopathy. It is thought that lead may cross the blood-brain barrier via the calcium transport channels since lead has a similar ionic radius and charge to the calcium ion.[42] Chang considers the disturbance of calcium metabolism as having the most direct relationship to cell injuries, both endothelial and neuronal, and neurodysfunctions induced by lead.[42] Goldstein has recently investigated blood-barrier cells in tissue culture and found that endothelial cells were very resistant to the overt toxicity of lead.[70] In contrast, astrocytes in culture developed cytotoxic changes at relatively low concentrations. Goldstein suggests that primary injury to astrocytes and subsequent loss of their ability to maintain expression of the blood-brain barrier in the endothelial cells may be the primary mechanism underlying the toxic action.[70]

Recent research has demonstrated that lead in substituting for calcium binds to and affects the brain protein kinase C at picomolar levels, thus effecting second messenger metabolism.[69,70] The hypothesis is that this could interfere with normal brain development by interfering with the cellular response to a given stimulus and thus lead to subsequent disorder in neuronal connective function. This may account for the learning and behavioral disorders found in lead-exposed individuals, particularly children.[70]

Lead is a cumulative toxicant and can affect multiple systems within the body. In addition to hematopoiesis and heme synthesis, lead can adversely affect the central and peripheral nervous system, the kidney, the gastrointestinal tract, reproduction in both sexes, and the developmental process. Lead causes a neuropathy in the peripheral nervous system of adults that primarily affects the motor nerves. Lead targets motor axons and produces axonal degeneration and segmental demyelination.[42] A number of studies have suggested that low-level lead exposure affects behavior and learning in children.[65] There is also growing evidence that low birth weight and deficits in postnatal behavioral development are more likely to occur with low levels of maternal exposure to lead.[71]

Current research implicates lead as a contributing etiologic factor in a number of common diseases affecting large portions of the population such as subtle cognitive and neurological deficits, hypertension, immunotoxicity, congenital malformations, and deficits in growth and development.[45,69,72] For each of these disorders there may be multiple etiologic factors. Other potential subtle health effects include the influence of small amounts of lead on cell proliferation and lead as a cofactor in carcinogenesis.

2. Carcinogenicity

The IARC has classified inorganic lead and certain lead compounds Group 2B (agents that are possibly carcinogenic in humans).[60] The epidemiologic evidence for respiratory, digestive, and stomach cancer among occupationally exposed workers was determined to be inadequate. The evidence for carcinogenicity to animals was judged sufficient. Renal tumors and gliomas have been induced by oral administration in rat studies.[60] Almost all the lead compounds tested for carcinogenicity in animals are soluble salts.

F. NICKEL

Nickel is widely distributed in the natural environment and is present in the soil in varying amounts and concentrations. The concentration of nickel in plants is closely related to the concentration in the soil used in cultivation.[28] Nickel concentrations in agricultural soils around the world range from 3 to 1000 mg/kg, with the wide range dependent on the mineral content of the top soil.[41] Because nickel is phytotoxic to plants, it should pose little risk to human health from sludge-amended soil. Studies of sludge-treated soils maintained at pH >6.0 have rarely shown phytotoxicity.[39] The highest level of exposure to nickel for the general population comes from food intake, since nickel occurs in most food items. Nickel is found in fruits, vegetables, grains, seafood, and mother's and cow's milk. Very little nickel ingested in food is absorbed. The U.S. EPA estimates that the total dietary intake ranges from 107 to 900 μg/day with average values of 160 to 500 μg/day.[73]

The National Sewage Sludge Survey conducted by the EPA detected nickel 201 times in the 209 wastewater plants sampled, with the mean concentration 77.0 mg/kg and range from 2.0 to 976.0 mg/kg.[50]

1. Toxicity

The route of exposure and chemical form are the primary factors in determining the toxicity of nickel. Inhalation during work is the primary toxic route associated with nickel exposure. After ingestion, approximately 10% of nickel appears to be absorbed in animal experiments.[41] Even less appears to be absorbed from food, remaining unabsorbed in the gastrointestinal tract until excreted in the feces. Serum albumin is the main carrier protein for nickel in humans and animals. There was no uptake of nickel in rats chronically exposed to levels of 5 ppm in drinking water over the lifetime of the animals.[73] Tissue distribution in animals after oral exposure to nickel is concentration dependent. Calves fed supplemental nickel in the diet at levels of 62.5, 250, or 1000 ppm showed somewhat elevated levels in the pancreas, testes, and bone at 250 ppm; pronounced increases were seen in these tissues at 1000 ppm.[74] Transplacental transfer of nickel to the fetus takes place in both animals and humans.[73] The main excretory route of absorbed nickel in human and animals appears to be urine. Unabsorbed dietary nickel is excreted in the feces.[73]

Lipid peroxidation with the generation of free radicals has been postulated as a possible mechanism for hepatocellular effects in acute nickel toxicity. The ability of nickel to interact with essential metal ions and thus disrupt normal physiological functions of metal homeostasis is considered the most likely hypothesis for nickel's toxicity at the biomolecular level.[40]

Experimental animal studies have examined systemic effects on cardiovascular, renal, hematological, hyperglycemia, immunological, and reproductive/developmental endpoints.[73,74] The functional effects of nickel on macrophages and natural killer cells in the immune system have been suggested as one possibility for the altered host resistance to viral, bacterial, and tumor challenges in laboratory animals.[75] It is one of the most common causative agents of dermal hypersensitivity reactions among the general public. The reactions usually follow contact with nickel-containing metal objects. No clinical or epidemiologic studies on the toxicity of nickel after oral exposure appear to be available in the literature.[74]

2. Carcinogenicity

Nickel is a human carcinogen among occupational workers. Nasal cancer from inhalation exposure appears to be the predominant type of neoplasm. Epidemiologic studies suggest nickel may also induce cancer of the lung, larynx, stomach, and possibly also the kidney.[60] Experimental evidence supports respiratory cancer as a potential outcome from inhalation.[40]

Nickel is a potent animal carcinogen.[40] Nickel's ability to bind with DNA and other nucleic acids and thus induce alterations in enzyme activities and DNA repair mechanisms/structure suggests its genotoxic nature. The carcinogenic nickel compounds are water insoluble, and thus can enter cells through phagocytosis. Sunderman has investigated the molecular mechanisms of nickel genotoxicity using as an experimental model the teratogenetic effects produced in South African frogs.[76] Sunderman found that the nickel ion binding protein pNiXa plays a key role in nickel ion teratogenesis. Sunderman considers the histidyl residues of specific nuclear proteins that bind to DNA or regulate cell division and gene transcription to be the most likely critical molecular targets. This formation of complexes with nickel ions could generate oxygen free radicals that can cause genetic damage.[76] Costa and coworkers have found that nickel sulfide produces selective damage in the heterochromatic long arm of the X chromosome in Chinese hamster ovary cells.[77] They postulate that since bivalent nickel ions have a high affinity for the amino acids and proteins that are highly concentrated in these regions, which contain relatively few genes, it is this interaction with proteins that produces the damage. Also, when nickel ions bind to proteins they can be oxidized and DNA damage can be induced.[77]

The IARC classifies nickel and certain nickel compounds as Group 1 (agents that are carcinogenic in humans).[60] Epidemiologic studies of occupational exposures were determined to provide sufficient evidence for carcinogenicity to humans. The evidence for carcinogenicity to animals was found to be sufficient. Metallic nickel is classified as Group 2B, a possible human carcinogen.

G. ZINC

Zinc is the cofactor in many metalloenzymes and is therefore an essential nutritional element. It may regulate the activity of protein kinase C, which is a critical enzyme induced by neuronal membrane receptors.[78] It is found normally in all tissues and tissue fluids. Muscle and bone contain the highest amounts of zinc. Deficiency of zinc can induce severe health consequences with a variety of effects on the nervous system, hematopoietic system, cell-mediated immune functions, skin, liver, eye, and testis.[41,78] Zinc deficiency has also been associated with growth retardation. Recent research has implicated stress, trauma, and inflammation as factors in interfering with zinc metabolism and internal redistribution.[79] Zinc is readily excreted from the gastrointestinal tract (primarily through the urinary tract) and excessive intake by the oral route is thus unlikely to induce toxic effects.[29] Prasad states that elemental zinc is virtually nontoxic if ingested orally in amounts up to 45 mg per day by an adult, causing no adverse effects on serum lipids or cell-mediated immune functions.[78]

Excessive exposure to zinc is relatively uncommon and requires heavy exposure. Zinc does not accumulate with continued exposure, but body content is modulated by homeostatic mechanisms that act principally on absorption and liver levels.[80]

About 20 to 30% of ingested zinc is absorbed. The mechanism is thought to be homeostatically controlled and is most likely a carrier-mediated process.[29] Within the mucosal cell, zinc induces metallothionein synthesis and, when saturated, may depress zinc absorption. In the blood, about two-thirds of the zinc is bound to albumin. About 2 g of zinc is filtered by the kidneys each day, and about 300 to 600 μg/day is excreted by adults.[29] Zinc concentrations in tissues can vary widely.[29] There is little information available to correlate serum and bone zinc with human health effects. Studies that examined the long-term ingestion of zinc sulfate tablets to treat chronic leg ulcers generally indicated that serum zinc levels of 90 to 192 μg zinc per 100 ml and 95 to 157 zinc μg per 100 ml were not associated with toxic effects.[80] In the liver, as well as other tissues, zinc is bound to metallothionein. The greatest concentration of zinc in the body is in the prostate.[29]

The National Sewage Sludge Survey conducted by the U.S. EPA detected zinc 239 times in the wastewater plants sampled, with the mean concentration 1692.8 mg/kg and range from 37.8 to 6800 mg/kg.[50] Zinc and copper were the only metals to be detected in greater mean concentrations when compared with an earlier 40-city study of municipal sewage plants.[50]

1. Toxicity

Zinc toxicity from excessive ingestion is uncommon, but gastrointestinal distress and diarrhea have been reported following high dose ingestion. Evidence of hematologic, hepatic, or renal toxicity has not been observed in individuals ingesting as much as 12 g of elemental zinc over a 2-day period.[29] A primary concern over excess zinc intake is the potential induction of copper deficiency syndrome. It is thought that induction of metallothionein synthesis in the intestinal mucosa is one mechanism for zinc's disruption of copper metabolism.[81] Metallothionein binds copper in the intestinal mucosa cells and thus lowers the amount of copper entering the body. Copper deficiency causes changes in cholesterol metabolism. In humans, 150 mg of zinc daily administered to young men caused a significant decrease in high-density lipoprotein cholesterol.[81] Decrease in the activity of copper-zinc superoxide dismutase in erythrocytes is another effect of copper deficiency. Based on available human studies, Sandstead considers 45 mg and above of absorbed zinc to be the amount of zinc that may inhibit copper retention.[81] Beyond copper deficiency, no disorders are known to be associated with excessive exposure to zinc in humans.[78]

Animal studies have investigated reproductive/developmental effects via oral exposure. These studies suggest that zinc is fetotoxic and may be teratogenic by the parenteral route.[80] Chandra reported immunological effects in humans following ingestion of zinc in food (153 ppm).[82] Two case studies in humans suggested neurological effects such as headaches and lethargy after ingestion.[80] Metal fume fever is the most common effect from occupational inhalation exposure to zinc.

2. Carcinogenicity

Testicular tumors have been produced by direct injection in the testes of chickens and rats. Michalowsky's injection of zinc chloride into chicken testes was the first experimental evidence of the carcinogenic action of metal compounds.[41] This effect is likely related to the concentration of zinc normally in the gonads and may be hormonally dependent.[83] The evidence that zinc compounds are carcinogenic by any other exposure route than direct injection is of limited interpretive value because of study weaknesses.[80] Prasad states that there is no evidence that zinc is genotoxic or carcinogenic.[78] The U.S. EPA designates zinc as a category III substance, which is considered a noncarcinogen.

H. SELENIUM

Concentration of selenium in soils varies from 0.03 ppm to more than 30 ppm in regions with seleniferous soils, such as the upper Great Plains in the U.S. and Enshi County, Hubei Province in the People's Republic of China. Concentrations of selenium in groundwater are usually quite low, although regions with seleniferous soils often have elevated selenium levels in water.[84] Irrigation drainage water can contain extremely high concentrations of selenium.

The bioavailability as well as the toxic potential for selenium and selenium compounds are related to chemical form and, most importantly, to solubility. Selenium occurs in nature and biologic systems as selenate, selenite, elemental selenium, and selenide. Deficiency of selenium leads to dysfunctions characterized by clinical "deficiency" syndromes.[85] Deficiency can also lead to a cardiomyopathy in mammals, including humans.[29] Selenium in foodstuffs provides a daily source of selenium. Seafoods, meat, milk products, and grains provide the largest amounts in the diet.

Elemental selenium is probably not absorbed from the gastrointestinal tract. Selenium is transferred through the placenta to the fetus, and it also appears in human milk.[86] Levels in milk are dependent on dietary intake. Selenium compounds may be biotransformed in the body by incorporation into amino acids or proteins or by methylation.[87] Selenium is essential for the activity of glutathione peroxidase, which functions to remove peroxides from cells. The biological role of selenium in mammals including humans is attributed to the presence of selenocysteine at each of four catalytic sites of the enzyme glutathione peroxidase.[87] The excretion pattern of a single exposure to selenite appears to have a rapid initial phase with as much as 15 to 40% of the absorbed dose excreted in the urine the first week. The remainder of the dose is excreted slowly with a half-life of 103 days. The major excreted form of selenium is the hepatic metabolite, trimethylselenonium.[88] Within certain physiologic limits, the body appears to have a homeostatic mechanism for retaining trace amounts of selenium and excreting the excess material.[29] Selenium toxicity appears when the intake exceeds the excretory capacity.

The National Sewage Sludge Survey conducted by the U.S. EPA detected selenium 163 times in the 209 wastewater plants sampled, with the mean concentration 6.2 mg/kg and range from 0.5 to 70 mg/kg.[50]

1. *Toxicity*

The potential toxicity of selenium was first suspected over 50 years ago, and, through the years, syndromes of toxicity have been described in animals, and humans living in seleniferous areas where the soil content is relatively rich in selenium, contributing to relatively high selenium in vegetation. Skin lesions, nail changes, liver toxicity, and increased prothrombin time are early critical effects of selenium toxicity in humans. This has been attributed to subclinical liver toxicity with depressed protein synthesis.[89]

Chronic selenium poisoning from environmental exposure has been described in North American and Chinese populations. During the 1930s, the health status of individuals residing in regions with seleniferous soils was examined. Common signs and symptoms found in individuals with the highest levels of selenium in urine (>200 μg per liter) were icteroid discoloration of skin, changes in fingernails, gastrointestinal disorders, and discoloration and decay of teeth.[84] These adverse health effects from chronic ingestion of selenium are not fully understood. Chronic selenium intoxication of horses, pigs, and cows also occurred in the areas. Headache and nausea have also been observed from high exposures.[88] Recently, Longnecker et al. investigated adults living in seleniferous areas and found that chronic daily selenium intakes as great as 724 μg manifested no adverse effects.[90] Selenium excess and deficiency have been associated with impairment of immunocompetence in animal studies.[91]

Selenium is an essential nutritional element and plays an important role in the glutathione regulation and thus has an antioxidation function in biological systems. Although selenium is toxic at higher dose levels (approximately three to five times higher via ingestion) than at dietary levels, selenium has been shown to reduce the toxicities of various heavy metals, particularly cadmium.[29,64] The basic mechanism of selenium's protection against cadmium toxicity is believed to be the ability of selenium (as selenide) to form large, high molecular weight selenoproteins. Selenoproteins are thiol rich and have an affinity towards metals such as cadmium; thus, these proteins can divert the binding of cadmium in tissues and plasma from otherwise sensitive sites of cells and tissues.[42] Prevention of testicular injury is the most significant interaction of selenium with cadmium, but reduction of cadmium-induced mortality, teratogenicity, and renal toxicity by selenium has also been reported.[42] Macaques fed selenomethionine, the most teratogenic selenium compound in avian species, showed no evidence of teratogenecity. Teratogenicity has not been demonstrated in humans.[89]

2. Carcinogenicity

Selenium has been characterized as a carcinogen or a tumor promoter and as an anticarcinogen. A number of animal studies indicate that various selenium compounds are carcinogens. The potential of selenium as a tumor promoter is demonstrated by the interaction of dietary selenium level with dietary fat intake in the development of pancreatic cancer in mice.[92] Selenium also inhibits development of chemically and virally induced tumors.[93] Although genotoxic effects have be observed in several *in vitro* systems, genotoxicity has not been demonstrated in humans.[89]

In human populations the cancer risk remains unclear with some epidemiological studies, suggesting a link between selenium deficiency and cancers at several sites.[94,95] Ecologic studies of selenium contents of foodstuffs and prevalence of certain tumors suggested an inverse (protective) relation between these variables.[96] Prospective studies of cancer risk and selenium status have not been consistent in their results.[97-99] The U.S. EPA designates selenium as a category III substance, which is considered a noncarcinogen.

I. COPPER

Copper is widely distributed in nature and is an essential element in mammalian systems. Copper is a cofactor for important oxidative enzymes such as catalase, peroxidase, and cytochrome oxides.[100] The principle route of exposure is through ingestion, but inhalation of copper dusts and fumes occurs in industrial settings, with metal fume fever the most common adverse effect. Adults ingest 1 to 5 mg/day, approximately half of which is absorbed.[101] Gastrointestinal absorption of copper is normally regulated by body stores. It is transported in serum bound to albumin to the liver and is stored mainly in the liver, heart, brain, kidney, and muscles. In the liver it is transferred to ceruloplasmin. Ceruloplasmin is the primary transport carrier for incorporating copper into the copper-dependent enzymes. Copper is also essential for hemoglobin formation. The normal serum level of copper is 1 μg/ml.[101] Phytotoxicity from copper accumulation has rarely been reported in sludge-treated soils maintained at pH > 6.0.[39]

The National Sewage Sludge Survey conducted by the U.S. EPA detected copper 239 times in the wastewater plants sampled, with the mean concentration 665.3 mg/kg and range from 6.8 to 3120 mg/kg.[50]

1. Toxicity

As with many essential elements metallic copper is not very toxic and is not considered an important source of acute posioning in most circumstances. Because many copper compounds do not dissolve readily, their toxicity is relatively low. The main adverse health effects observed result from inhalation of dusts and fumes in the workplace, such as the influenza-like syndrome, "metal fume fever". Excessive storage of copper in the liver, kidney, and brain as well as several other organs occurs in Wilson's disease. Wilson's disease is an inherited error of metabolism and it is thought that the excess storage in the liver may be caused by impairment of hepatic excretion into bile.[41] Animal experiments using excessive oral doses of copper have supported these depository sites. Copper levels in the rat brain tripled within a year when copper was added to their drinking water.[101] Inhibition of sulfhydryl groups that protect the cell from oxidative stress, particularly in glutathione glucose-6-phosphate dehydroginase and glutathione reductase, is thought to be the mechanism for copper toxicity, both acute and chronic.[100] Copper ions can also bind to cell membranes, disrupting transport processes through the cell wall.

Copper has an antagonistic action with zinc and cadmium. In a condition of low or marginal copper intake, exposure to cadmium could induce the onset of copper deficiency syndrome, which represents one of the earliest clinical manifestations of cadmium toxicity.[42] Clinical signs of copper deficiency include anemia, gastrointestinal disturbances, depressed growth dystrophy of bone, impaired reproduction, and heart failure.[102] Copper deficiency is teratogenetic in ungulates.[100] Excess molybdenum can also interact with copper producing a copper deficiency. Copper is also associated with the absorption and metabolism of iron, being essential for the incorporation of iron into hemoglobin.[103]

Since copper and zinc are two of the most important mineral elements for the metabolic and functional integrity of the nervous system, depletion of copper can induce adverse effects on the nervous system.[42,101] It has been reported that animals fed on cadmium-contaminated grass displayed signs of copper and zinc deficiencies, and their offspring had high incidences of demyelinating disease.[42]

Infants and children may be susceptible to the effects of copper as evidenced by the incidence of childhood cirrhosis and the reports of copper intoxication in young children in India caused by drinking milk boiled and stored in brass vessels.[104-106] It is thought that infants and children have increased susceptibility to copper toxicity because of the normally high hepatic copper levels in early life, and homeostatic mechanisms are not fully developed at birth.[29] In the U.S. there are case reports of severe liver disorders resulting from ingestion of 10 mg copper per 10-kg child per day from contaminated milk.[29]

2. *Carcinogenicity*

The U.S. EPA designates copper as a category III substance, which is considered a noncarcinogen. There appears to be no evidence that copper is mutagenic.[103] See Owen, Jr. for reference to early studies investigating the association between elevated human ingestion of copper and cancer.[102]

J. MOLYBDENUM

Hazardous exposure to molybdenum and related compounds usually occurs by the inhalation of dusts in occupational settings. Molybdenum is found widely in soils, normally with ranges between 0.1 to 10 mg/kg. Concentrations of molybdenum in plants vary, with higher concentrations found in leafy vegetables and legumes than in edible roots. Its concentration in plants, depending on soil characteristics, reflects that in soil.[100] Gastrointestinal absorption is about 50% of an ingested amount and depends on the water solubility of the compound involved.[107] Daily intake of molybdenum has been estimated at 100 to 500 μg, with an estimated average of 350 μg.[28,103]

Molybdenum is contained principally in liver, kidney, bone, fat, and blood. The major portion contained in the liver is contained in a nonprotein cofactor bound to the mitochondrial membrane. It may be transferred to an apoenzyme, transforming it into an active enzyme.[29] Molybdenum is rapidly excreted (greater than 50%) renally, mainly as molybdate. It may also be excreted through the bile when excess molybdenum is present.

Molybdenum is an essential metal as a cofactor for the enzymes xanthine oxidase, aldehyde oxidase, and sulfite oxidase.[103] In plants, it is necessary for fixing of atmospheric nitrogen by bacteria at the start of protein synthesis. Because of these factors, it is ubiquitous in food. Molybdenum is also added in trace amounts to fertilizers to stimulate plant growth. The concentration of molybdenum in urban air is minimal, but it is present in more than one third of freshwater supplies, and in certain areas the concentration may be near 1 μg/L. Excess exposure can result in toxicity to animals via ingestion and to humans through inhalation at work.

The National Sewage Sludge Survey conducted by the EPA detected molybdenum 148 times in the wastewater plants sampled, with the mean concentration 13.1 mg/kg and range from 2.0 to 67.9 mg/kg.[50]

1. Toxicity

Molybdenum has a complex chemistry and the biologic differences of molybdenum's valence forms are not well known. The soluble hexavalent compounds are well absorbed from the gastrointestinal tract into the liver. Molybdenum is a component of xanthine oxidase, which has a role in purine metabolism, and has been shown to be a component of aldehyde oxidase and sulfite oxidase as well. Increased molybdenum intake in experimental animals has been shown to increase tissue levels of xanthine oxidase. Increased xanthine oxidase activity has been postulated as the cause of a gout-like disease observed in humans living in a high molybdenum area of Armenia.[103] Molybdenum interacts metabolically with copper and iron and is present in inverse relation to copper in animals.[100] High doses of molybdenum disrupt copper metabolism.

Research is extremely limited on chronic toxicity from molybdenum and its compounds. Loss of appetite, listlessness, diarrhea, and reduced growth rate have been observed in chronic toxicity. Anemia is also common. Animals develop deformities of joints and long bones and mandibular exostoses.[103] Kidney disease and liver dysfunction are reported in animals after oral administration. There appear to be no reports of these two outcomes in humans.[107]

Pastures containing 20 to 100 ppm molybdenum may produce a disease known as "teart" in cattle and sheep.[29] It is characterized by anemia, poor growth rate, and diarrhea. It can also produce impaired reproduction.[103] Central nervous system degeneration has been observed in sheep.[100] The mechanism of action is thought to be related in part to lower copper levels.[100] Interactions of molybdenum with other metals with respect to toxicity in cattle and sheep have also been reported.[29]

2. Carcinogenicity

The U.S. EPA designates molybdenum as a category III substance, which is considered a noncarcinogen.

K. ARSENIC

Although arsenic is ubiquitous in the environment, its levels in water and ambient air are generally low. The major source of human exposure is food, which typically contains somewhat less than 1 mg/kg. In certain parts of Taiwan and South America, the water may contain hundreds of milligrams per liter. In these regions inhabitants may suffer from dermal hyperkeratosis and hyperpigmentation.[108] A more serious outcome was gangrene of the lower extremities, called the black-foot disease, resulting from peripheral endarteritis.[109] Cancer of the skin has also been studied in those areas, but a number of potential confounding variables make interpretation uncertain.

Industrially produced arsenic, usually in the trivalent form (arsenite), is more toxic than elemental arsenic and its pentavalent forms (e.g., arsenate) found in nature. Organic arsenic, although present in many foods, is also less toxic than the inorganic trivalent species.[110]

Once absorbed, arsenic is distributed predominantly into muscle, liver, skin, and hair. Arsenic has a biological half-life of 30 to 60 h. About 60 to 70% of an oral dose of inorganic arsenic is excreted in the urine in 4 days. Inorganic arsenic is methylated rapidly and excreted as mono- and dimethylated products.[111]

The National Sewage Sludge Survey conducted by the U.S. EPA detected arsenic 194 times in the wastewater plants sampled, with the mean concentration 12.4 mg/kg and range from 0.3 to 315.6 mg/kg.[50]

1. Toxicity

Arsenic toxicity varies with the oxidation state and the solubility of the species. Mitochondrion is a primary target site for inorganic arsenicals.[53] Arsenic inhibits the activity of many enzymes by reacting with sulfhydryl groups and interfering with phosphate metabolism. Such reactions are considered responsible for much of the toxic action of arsenicals. This is another example of the mimicry of essential metabolites by metals. Arsenic when in the biological system usually exists as arsenate. Arsenate is an oxyanion with a structure similar to that of the phosphate oxyanion. The interaction of arsenic with phosphate replaces phosphate with arsenic in the process of adenosine triphosphate (ATP) synthesis, which is essential for energy in metabolism. This uncoupling of ATP synthesis is posited as the basis of arsenic toxicty.[42]

Arsenate is transported by the kidney tubules where a small fraction is reduced to trivalent arsenic, which is more acutely toxic.[53] Recent studies have shown induction of several stress proteins in the kidney of animals exposed to arsenic. Fowler concludes that this indicates that the genetic mechanisms in the nuclei are also being affected, but that it is not clear whether this is a primary response to arsenic entering the nucleus or secondary to other aspects of the cell injury process such as decreased energy production.[53] Arsenic toxicity from chronic exposure may produce gastrointestinal pain, diarrhea, anemia, peripheral neuropathy, kidney degeneration, hepatic cirrhosis, and dermal keratosis.[108]

Kreiss et al., studying a population exposed to arsenic-contaminated well water, reported that total urine arsenic concentrations increased in proportion to the amount of arsenic ingested. Some of those exposed had abnormal nerve conduction, but this abnormality was not clearly related to the amount of arsenic ingested or excreted.[112] In another study, electromyographic changes were seen in persons drinking well water with greater than 0.05 ppm arsenic.[113]

The bone marrow, skin, and peripheral nervous system may become involved after acute or chronic exposure. Other effects include toxicity to liver parenchyma, resulting clinically in jaundice in early stages and in cirrhosis and ascites later.[108]

Elevated arsenic levels have been reported in the fetus born to a mother with actue ingestion of arsenic. Inorganic arsenic is teratogenic after administration of large dosages in rodents.[108]

2. *Carcinogenicity*

Cancer of the lung may occur among workers exposed to inhaled arsenic at manufacturing facilities. Unlike other human carcinogens, arsenic has not been shown to be carcinogenic to laboratory animals to date. Furthermore, mutagenesis tests have been essentially negative.[108] Basal cell cancer and squamous cell cancer have been reported after prolonged occupational exposure. The Occupational Safety & Health Administration has linked arsenic to cancer of the skin, lungs, lymph glands, and bone marrow. It has also been associated with bladder, kidney, skin, prostate, lung, and liver cancer.[114]

For drinking water regulations, the U.S. EPA designates arsenic as a category II substance, with equivical evidence for carcinogenicity from ingestion.

L. CHROMIUM

Chromium is generally present in air, water, and food at low levels. Annual world production of chromite is approximately seven times 10^6 tons, roughly three orders of magnitude greater than natural rates of environmental mobilization.[115] Large amounts of chromium are introduced into the environment through the use of chromium-containing phosphate fertilizers, discharge of industrial and sewage wastes, landfill dumping of chromium-containing sewage sludge and consumer products, and atmospheric emissions of chromium dust and aerosols.[115] Industrial wastewaters can contain total amounts of chromium ranging from 0.005 to 525 mg/L, with levels of hexavalent chromium in the range of 0.004 to 335 mg/L.[116]

The National Sewage Sludge Survey conducted by the EPA detected chromium 231 times in the wastewater plants sampled, with the mean concentration 258.5 mg/kg and range from 2.0 to 3750 mg/kg.[50]

1. *Toxicity*

Although the major source of chromium exposure and consequent health effects is occupational (inhalation and dermal), ingestion of chromium is a more important route of exposure for the general population. When ingested, hexavalent chromium is rapidly reduced to the poorly absorbed trivalent form by components of saliva and gastric juice.[116] Uptake of the trivalent form of chromium from the intestines is slow; however, some hexavalent chromium is absorbed there. Absorbed hexavalent ions are transported through the portal vein to the liver or are rapidly taken up by liver cells.[116] The liver, spleen, kidney, and testes accumulate the majority of chromium after exposure. Although trivalent chromium ions are rapidly cleared from the blood, hexavalent ions are retained longer through internalization by blood cells.[116] Following clearance from the blood, chromium is excreted principally in the urine.

Chromium toxicity varies with the particular chromium compound. The majority of the biologic effects following chromium exposure are due to the hexavalent chromium species. Metallic chromium, divalent, and trivalent chromium are relatively nontoxic. Hexavalent chromium compounds are hazardous following acute large exposures and may also be hazardous following chronic exposure to lower concentrations. Hexavalent chromium is corrosive and causes ulceration of the nasal passages and skin. It also induces hypersensitivity reactions of the skin. It can induce renal tubular necrosis from acute exposure.[117] Chronic low-level exposure to hexavalent chromium can result in gingivitis, periodontitis, ocular damage and eye lesions, conjunctivitis, and keratitis.[116] Chromium compounds can also be immunosuppressive.[116]

2. Carcinogenicity

Chromium is a human carcinogen, inducing lung cancers among workers exposed to it. The IARC classifies hexavalent chromium compounds as Group 1: agents that are carcinogenic in humans.[60]

Carcinogenicity is generally attributed to the hexavalent Cr^{6+}, which is corrosive and water insoluble, but both forms are active clastogens in intact cells.[116] It has been suggested that Cr^{6+}, which is more readily taken up by cells, converts to Cr^{3+} intracellularly. The trivalent chromium ion, although highly reactive with enzymes and macromolecules (e.g., nucleic acids), is relatively inert toward intact cells, and by itself is relatively unreactive with macromolecules and enzymes *in vitro*. Chang explains this by the fact that hexavalent chromium, *in vivo*, forms an oxyanion similar to the structure of that of sulfate oxyanions.[42] The chromium oxyanion can be transported into cells via the same channels and carriers as sulfates. Once inside the cell, the hexavalent chromium may then be released from the oxyanion complex and be reduced to the trivalent form, which is highly reactive with macromolecules (e.g., DNA) for the induction of toxicity and possible initiation of the carcinogenic process.[42,118] This suggests that the ultimately active agent is Cr^{3+}, which is dependent on entry as Cr^{6+} and then reduced to the lower valence state.[118] Most of the hexavalent chromium compounds are known to be mutagenic in cell systems.[116]

VI. TOXIC ORGANIC CHEMICALS: THE PUBLIC HEALTH CONTEXT

Chlorinated hydrocarbon compounds have left their imprint in the environmental matrix, being the basic compound source for numerous chemical substances produced and used after World War II. Chlorinated hydrocarbon insecticides, primarily DDT, not only contributed a fundamental resource for developing production agriculture after World War II, but also contributed to the initial effectiveness of a global public health initiative against infectious diseases, particularly malaria, that was probably the single most important factor in world population growth after the war.[119] Because many of the chlorinated hydrocarbon insecticides and other chlorinated hydrocarbons were not acutely toxic, a naive and indiscriminate use of these chemicals developed in industrial societies. While the environmental cost of this use was immediately noticed by some researchers, it was not until 1962 with the publication of Rachel Carson's *Silent Spring*[120] that public concern began to grow.

Another class of toxic organic chemicals that has played a fundamental part of post-World War II economic development is the halogenated polyaromatic hydrocarbons. Three classes of these chlorinated compounds are widely regarded as harmful. They have been the subjects of innumerable articles and monitoring programs. Polychlorinated biphenyls (PCBs), chlorinated dibenzo-p-dioxins (CDDs), and chlorinated debenzofurans (CDFs), altogether including some 419 possible compounds (209 of which are PCB congeners), are now widely dispersed in the environment.[121] CDDs and CDFs were never intentionally manufactured but were incidently generated as unwanted by-products, or waste. Concern about these compounds has increased during the past 15 years.[121]

Hall concludes that ambient levels for some of these toxic chemicals, which persist for decades in the environment, have reached levels high enough to affect the health of children.[122] The chlorinated hydrocarbons (e.g., PCBs, DDT, and the dioxin family) accumulate in human adipose tissue. Pregnant women pass the contamination to their fetuses, where it is believed that the developing nervous system is the most vulnerable. Neurobehavioral deficits, including short-term memory loss, have been detected in children born to mothers at the high end of the distribution curve of chlorinated hydrocarbon contamination.[122] Interpretation of these studies is complicated by the fact that the possibility of heavy metal exposures, which can produce similar toxic effects, was not investigated.

Wolf et al. recently found a statistically significant association between levels of the DDT metabolite DDE stored in the blood and risk of breast cancer from environmental exposures in New York City women.[123] Garabrant et al. observed a statistically significant (though with a wide confidence interval) association between DDT and pancreatic cancer in an occupational cohort with heavy and prolonged exposure.[124] Brown has studied workers employed at chlorinated hydrocarbon pesticide manufacturing plants. There was a statistically significant increase in liver and biliary cancer in one plant. These findings are difficult to interpret because the study was limited by the small number of deaths and lack of adequate exposure data.[125]

A. PREVALENCE OF TOXIC ORGANICS IN SLUDGE

The potentially hazardous toxic organics that may be present in wastewater and sludges number in the thousands, since they will reflect the wastewater stream that comprises their industrial, commercial, and residential sources. During the 1980s there was increasing interest in the behavior and fate of organic contaminants during the sewage treatment process. Compared to the metals, less is known about the range and concentration of synthetic organic chemicals that enter sewage treatment works, the efficiency with which they are degraded or removed during treatment, and their typical concentrations in municipal sludges.[126] The chlorinated aromatics, particularly polychlorinated biphenyls, are the most commonly analyzed and reported toxic organics. DEHP, a phthlate ester, is generally the single most frequently detected toxic organic in municipal sludges and phthlate esters, the most frequently detected class.[126]

An important difference between trace metal and trace organic additions to soil is the time that each may persist there. The half-life of the most persistent organics, e.g., PCBs, has been estimated at 10 or more years, while the persistence of most metals is estimated to be a few thousand years. Studies of trace organic behavior in soils must also consider assimilation mechanisms such as degradation (biotic and abiotic) and volatilization, in addition to factors such as solubility, adsorption/desorption, leaching, and plant uptake.[127,128] While these additional mechanisms make trace organic studies more difficult, they also lend themselves to management alternatives not available for trace metals. Long-term application programs with organics (e.g., petroleum wastes) have demonstrated the soil's ability to assimilate wastes over time.[127]

Based on the prevalence of organics in sludges and potential loadings to soils, agronomic or environmental risk due to the application of domestic sewage sludge to agricultural soils appears to be minimal.[127] In addition, many organics will be bound by soil organic matter and biologically degraded by soil microorganisms. However, persistent compounds like PCBs and the chlorinated pesticides could accumulate in soils from repeated sludge applications and could be a concern for food crop production. Many of the chlorinated hydrocarbon pesticides have been banned in the U.S. since the 1970s. Their persistence in the environment and their ability to induce tumors (particularly liver) in rodent experiments led the U.S. EPA to place severe restrictions on their use in agriculture.

The toxic organics of most concern for potential public health risk are those designated "priority pollutants" by the U.S. EPA. Data from available surveys appear to indicate that concentrations of these toxic organics in municipal sludge are usually relatively low.[126] Toxic organic concentrations are similar to or lower than routine pesticide soil concentrations. Even at high sludge application rates (100 to 1000 metric tons/ha), most of the toxic organics will be present in the soil-sludge mixture at concentrations of 1.0 to 10.0 mg/kg or less.[126]

The behavior of these compounds in the environment is based on their chemical properties. The lipophilic, nonpolar, nonvolatile compounds remain absorbed in the upper soil horizons within the root zone of most plants. Small, polar molecules pass through root tissues and are translocated throughout the plant.[129] These compounds are also the most likely to be lost from the soil through leaching and volatilization. Nonpolar molecules are thought to be prevented from movement into the plant by adsorption to root surfaces.[129] Leaves potentially can be contaminated by these nonpolar compounds when they volatilize from the soil and subsequently adsorb onto plant tops.

Two classes of nonpolar compounds common in sludge have been most often tested for their presence in plants, the polycyclic aromatic hydrocarbons and the polychlorinated biphenyls.[129] In general, root crops have higher concentrations of organic compounds than leafy vegetables. Vegetables will generally have higher concentrations of organic compounds than fruits.[129]

The principal pathways for potential human contamination from sludge-amended soil include: (1) uptake of toxic organics by plant roots, transfer to edible portions of plants, and consumption by humans; (2) direct application of toxic organics to edible plant parts as sludge, or as dust or mud, and consumption by humans; (3) direct ingestion of soil and sludge by children (pica); (4) uptake of toxic organics by plants used as animal feed, transfer to animal food products, and consumption by humans; and (5) direct ingestion of soil and sludge by grazing animals and transfer to animal food products, and consumption by humans.[130]

Pathways 1 and 4 are generally considered the most likely to lead to any contamination of the human food chain. Although direct ingestion of toxic organics (pathway 3), especially by children, can occur when sludge is used in home gardens, Dean and Suess concluded that this is likely to be a minor route of exposure, as is the inhalation of dusts or vapors.[130]

B. PLANT UPTAKE

Plant uptake/contamination, degradation, volatilization, and leaching are the pathways that could potentially lead to human exposure by toxic organics in sludge-amended soil. Most toxic organics are strongly absorbed to soils and its constituents, especially soil organic matter, leading to limited leaching and plant uptake. PCBs and some chlorinated hydrocarbon insecticides, such as lindane and dieldrin, can volatilize readily when surface applied, although soils and sludge itself can reduce these volatilization losses.[127] Most toxic organics undergo microbial degradation. Persistent toxic organics and some of the readily degraded components that can break down to toxic environmental metabolites are of most concern in contamination of the food chain.

A brief review of important sources, toxicity, and carcinogenic properties of illustrative examples of the three main classes of toxic organics is presented.

C. PHTHALATE ESTERS

Phthalate esters are a large group of chemicals, widely used and distributed throughout the world. They are used as plasticizers in polyvinyl chloride products and in the production of household articles, packages, cosmetics, and plant pesticides. They are ubiquitious in the environment (particularly as water pollutants) and are among the priority pollutants listed by the U.S. EPA. World production of phthalate esters is estimated to be several million tons a year.[131] Since phthalate esters are indispensable to this era, there is no reason to expect that the upward trend in their production and use will not continue. Menzer considers phthalate ester plasticizers to be general contaminants of virtually all soil and water ecosystems, although the concentrations are generally quite low.[64] They have been found complexed with the fulvic acid components of humic substances in soil and in both marine and estuarine waters.[64]

Ganning et al. have asked the two essential questions concerning potential environmental exposure to phthalate esters: What is the precise period of exposure required for toxic effects in humans with environmental levels of these esters? Will this time period be completed during a life span.[132] For example, some health researchers believe that bioaccumulation on the part of phthalate esters may take 30 to 40 years to reach a toxic level. Since widespread use of these compounds was introduced in the 1960s, they postulate that human toxicity may not occur until the turn of the century. Hence, at present the research evidence is not sufficient to evaluate the risk from lifetime environmental exposure to phthalate esters.[132]

DEHP (di(2-ethylhexyl) phthalate), a member of the phthalate ester group, is usually the most frequently detected and highly concentrated toxic organic in municipal sludges, with median contents in the 100 mg/kg dry weight range.[126] DEHP is moderately persistent and nonvolatile in soils. It is also possible that DEHP can accumulate in plants.[126]

O'Connor et al. could identify no field studies of crops in sludge-amended soils that reported DEHP contamination.[126] Greenhouses studies suggest that DEHP is taken up by plants but then metabolized to a variety of polar metabolites.[133] O'Connor et al. speculate that plants may, thus, serve as effective detoxifying barriers to toxic accumulation in the food chain. They conclude that the most prominent class of toxic organics in municipal sludges would appear to present little risk to the food chain.[126]

The National Sewage Sludge Survey conducted by the EPA detected bis(2-ethylhexyl) phthalate 189 times in the wastewater plants sampled, with the mean concentration 107.2 mg/kg and range from 0.5 to 89.1 mg/kg.[50]

1. Toxicity

The environmental transport and fate of phthalate esters are of concern because recent research indicates liver toxicity (in both animals and humans), renal toxicity in rodents, and some mutagenic, carcinogenic, and reproductive effects.[131,134,135] The phthalate esters produce toxic effects changing the structure of the liver by inducing peroxisomes, mitochondria, and enyzmes which participate in fatty acid transport and beta-oxidation.[132] Phthalate esters are well absorbed from the gastrointestinal tract and are widely distributed in the body, with the liver being the initial repository organ. Clearance from the body is rapid.

Chronic experiments with animals admininistered dosages comparable to human exposures indicate that phthalate esters have an accumulative effect on the liver. DEHP induces peroxisomal proliferation and concomitant oxidative stress.[136] It also decreases liver glutathione peroxidase, a selenoprotein. The bioavailability of selenium was decreased by DEHP incorporation in hepatocellular proteins.[136] Liver biopsies taken from dialysis patients and analyzed under the electron microscope show peroxisome proliferation indicating the possibility that human health may be influenced by these compounds.[137] At present, we do not know whether phthalate esters are metabolized and the products are excreted completely or if accumulation in certain tissues occurs. It will be difficult to definitively establish adverse effects of low-dose exposure to phthalate esters on human health due to the time element involved.

The biochemical evidence available suggests that the pattern of changes in membranes and enzymes that result from DEHP dosages involves several intracellular compartments. The principle effects observed are membrane destabilization and an increase in phospholipid fatty acid fluidity, properties that are of essential importance to membrane function.[132] Phthalates increase the level of dolichol in lysosomal membranes, leading to destabilization and a possible increase in permeability. If some of the lysosomal contents, such as hydrolytic enzymes are released, there is the possiblity of severe effects on the cell.[132]

As an acute toxicant, DEHP is relatively nontoxic with a single dose LD_{50} in the 30,600 to 33,900 mg/kg range in experimental animals.[131] To receive an equivalent dose, an adult human would have to consume about 4.5 lb of DEHP. Single oral doses of up to 10 g DEHP have not been lethal to humans.[131] Because phthalate esters do not demonstrate acute toxicity, most animal experiments employed in present toxicology are less effective in evaluating the risks associated with chronic exposure to these compounds.

The most apparent results of long-term, low-dose exposure to phthlate esters are effects that are initially limited, but increase in a continuous, almost linear manner with prolonged exposure.[132] Consequently, after sufficiently long exposure, even a very low dose of a plasticizer could give the same effects as a high dose for a short period of time. According to this argumentation, no threshold values for phthalates exist; that is, any level of intake continued over a sufficiently long period of time can have adverse effects.

DEHP and DBP have been shown to cause renal cysts in rodents and they also produce renal peroxisome proliferation.[138] There are no data to causally connect the two phenomena.[134]

2. Carcinogenicity

Phthalate esters are known to cause hepatic peroxisome proliferation in rodents and, after prolonged administration, hepatocarcinogenesis.[134] Peroxisome proliferators as a group, including a number of chlorinated hydrocarbons, tend to be hepatocarcinogenic, particularly in rodents.

It is unlikely that so many adverse effects are caused by a single agent; thus, an important task in future research on phthalate ester toxicity will be to analyze the effects of individual metabolites.[139] Dirven et al. studied the potential mutagenic effects of 4 metabolites of the rat liver carcinogen DEHP using the Ames assay and found concentration of these compounds up to 1000 micrograms per plate were negative with all tester strains.[140]

Recent studies have demonstrated the carcinogenicity of DEHP in Fischer 344 rats and B6C3F1 mice.[141] At both the maximally tolerated dose and one half of the maximally tolerated dose, liver tumors were produced in both sexes. A conference on phthlate esters evaluated these results and concluded that the weight of evidence on the carcinogenicity of DEHP was very strong. It has also been pointed out that much more testing is desirable on this class of compounds.[64]

Electron micrographs have revealed that, as in the rat, peroxisome proliferation in human liver does occur.[132] Phthalate esters have been tested in various *in vitro* systems, but insufficient information about the toxicity of these plasticizers has been obtained.

D. POLYCHLORINATED BIPHENYLS (PCBS)

Polychlorinated biphenyls, which belong to the class of polyhalogenated aromatic hydrocarbons, have been manufactured and used industrially since the early 1930s. The industrial grade Aroclors, sold by Monsanto, consists of mixtures of PCB congeners. Their thermal stability, chemical stability (e.g., general inertness to oxidation and hydrolysis), fire resistance, and dielectric properties have made them very useful in a variety of industrial applications. Although the U.S. EPA banned the production and commercial use of PCB in the late 1970s, human exposure continues. Resistance to degradation and their wide use have resulted in worldwide PCB contamination involving rivers, lakes, the atmosphere, fish, wildlife, and land-based hazardous waste sites. Like DDT residues, PCBs bioaccumulate and are concentrated in the food chain. Concern about the environmental persistence and human accumulation of PCBs is based on their ability to cause liver neoplasms in rodents (at high dosages), reproductive/developmental problems in humans and other species, evidence of subtle neurotoxic/behavioral effects, and alterations in immune function in certain species.

About 1.2 times 10^9 kg of PCBs have been manufactured in the last 50 years.[142] In the mid-1960s PCBs were determined to have widespread environmental persistence, with approximately one third of the total world production present in mobile reservoirs, including atmospheric, fresh, and marine waters and sediments, and sewage sludges. Polychlorinated biphenyls have been found in environmental and biological samples (e.g., human adipose tissue, milk, and serum) in many locations and tend to accumulate in sediments, soils, and biota. Because of the ubiquity, persistence, and possible toxicity of PCBs, their mobility in the soil-water environment is of concern.

Both environmentally and metabolically induced changes in congener patterns from the initial Aroclor contaminations may occur. Environmental degradation involves a number of processes, including differential solubility of lightly chlorinated congeners and anaerobic degradation.[142] Anaerobic degradation of PCBs as an environmentally acceptable means of remediating PCB-contaminated sediments and landfills may result in different potential health consequences. Anaerobic dechlorination results in a two- to sixfold increase in the concentrations of di- and trichlorinated congeners such and 2,2′-dichlorobiphenyl and 2,6,2′-trichlorobiphenyl. These congeners have a greater ability to alter neurochemical functions.[142]

The primary routes of exposure to PCBs are oral and pulmonary. PCBs are highly lipid-soluble compounds. PCBs are transported by the blood stream to the liver and muscle, where they are redistributed to adipose tissue.[143] Those PCBs that accumulate in adipose tissue are more highly chlorinated and less readily metabolized. PCBs can act as substrates for P-450 enzyme-catalyzed oxidation and are metabolized in the liver to form polar hydroxylated phenolic compounds, most likely through highly reactive arene oxide intermediates that are capable of forming covalent bonds with cellular constituents.[144] The more readily metabolized compounds are transformed to more polar compounds and excreted, primarily through urine. PCBs can also be eliminated from the body, unmetabolized, in association with agents having a high lipid content (i.e., milk, eggs). Because PCBs readily cross the placenta, they can be present in the fetus.

Environmental exposure occurs primarily through contaminated foods. The average PCB serum level for nonoccupationally exposed persons is approximately 7 ppb with a range up to 30 ppb. Adipose tissue levels are generally less than 3 ppm. Levels in the general population have been decreasing.[145]

Of the PCB congeners measured by the National Sewage Sludge Survey, PCB-1248 was detected 23 times with a mean concentration of 0.7 mg/kg and a range of 0.04 to 5.2 mg/kg. PCB-1254 was detected 13 times with a mean concentration of 1.8 mg/kg and a range of 0.3 to 9.3 mg/kg. PCB-1260 was detected 20 times with a mean concentration of 0.7 mg/kg and a range of 0.03 to 4.0 mg/kg. The other PCB compounds measured were below detection levels.[50]

1. *Toxicity*

PCBs are not generally acute toxicants. The LD_{50} ranges from 500 to 11,300 mg/kg depending on PCB congener and animal species.[143] Dermatologic effects are the most common clinical findings. Subacute and chronic exposures in animals lead to a variety of effects such as enzyme induction and inhibition, progressive weight loss, chloracne, alopecia, skin edema, swelling around the eyes, decreased immunocompetence, bone marrow depression, enlargement of the liver, and reproductive dysfunction.[143] The mechanism of toxic action is thought to be initiated by the interaction of specific PCB isomers with the aryl hydrocarbon (Ah) receptor, a cytosolic receptor. This PCB-receptor can translocate into the nucleus where it can induce an increased activity of a number of enzymes in the cell, resulting ultimately in the manifestations of toxicity.[146] This may also lead to altered gene expression. A current research initiative is determining which PCB congeners induce biochemical and toxic responses that do not act through the Ah receptor, since many of the congeners found in commercial mixtures do not.[144]

Current research concerns about human exposure to PCBs involve endocrinological, immunological, neurological, and reproductive/developmental effects. Of particular importance are investigations into possible neurobehavioral effects in children exposed *in utero*. Rodent studies indicate 0.25 mg/kg/day as the lowest observed effect level on the basis of developmental delays in growth and behavioral functions.[147]

In the late 1970s PCBs were discovered in sewage sludge used for fertilizer in Bloomington, IN. Mean serum PCB levels were 17.4 in the 89 sludge users (gardeners), 75.1 ppb in 18 workers with occupational exposure, and 24.4 ppb in 22 community residents without unusual exposure to PCB. PCB serum levels were not significantly associated with the amount of sludge used or the duration of exposure. Chloracne or systemic symptoms of PCB toxicity were not induced, nor were significant correlations found between PCB levels and tests of hematologic, hepatic, or renal function. Plasma triglyceride levels increased significantly. The authors concluded that the data indicated that PCBs may alter lipid metabolism at levels of exposure and bioaccumulation insufficient to produce overt symptoms.[148] No causal connection could be determined.

A more recent report by Emmett et al. did not find a number of the biochemical effects attributed to PCBs found in other studies, but did find evidence of possible subtle metabolic effects by PCB exposure.[149] Increased risk of cardiovascular disease, a major health effect that would be predicted from abnormal lipid levels, has not been shown to be increased in mortality studies of PCB-exposed workers.[150,151] The small number of deaths in both these studies makes the results difficult to interpret.

Neurological effects have been reported in laboratory mice and monkeys following prenatal, neonatal, or early postnatal exposures to PCBs. Exposure in adults generally does not produce measurable toxicity except for a few effects which can be reversible.[143]

Research on the subtle health effects of PCBs is still in its initial stage. A more detailed characterization of the long-term sequelae of perinatal and adult human exposure to PCBs on central nervous system function, a more complete analysis of the relationship between PCB structure and neurological activity, the determination of the biochemical mechanisms by which PCBs alter nervous system function, and the increased understanding of the types of interactions between PCBs and other neurotoxicants remain important goals.

2. Carcinogenicity

PCBs can induce neoplasia and adenofibrosis in the rodent liver. Animal studies show that PCBs can also act as tumor promoters or inhibitors. PCBs have been shown to promote hepatocellular tumors and preneoplastic lesions following ingestion of N-nitrosoamines.[152] They also inhibit these same tumors if the animals are treated with PCBs prior to carcinogen exposure.[153] See Silberhorn, et al.[152] and Safe[153] for recent reviews of the carcinogenicity and mutagenicity of PCBs. Brown in an updated mortality report found a statistically significant excess of deaths from cancer of the liver (primary and unspecified), gall bladder, and biliary tract among workers exposed through dermal and inhalation exposure, but again the small number of deaths makes the findings difficult to interpret.[154] Since PCBs have induced liver tumors in animal experiments, these findings are of particular interest.[154]

The IARC has classified polychlorinated biphenyls as Group 2A, agents that are probably carcinogenic to humans. The epidemiological evidence for carcinogenicity to humans is limited. The experimental evidence for carcinogenicity to animals is sufficient.[60]

E. POLYCYCLIC AROMATIC HYDROCARBONS (PAHS)

Polycyclic aromatic hydrocarbons (PAHs) are ubiquitous in the environment, being formed during the process of incomplete combustion or pyrolysis of organic matter. There are over 100 different PAHs known. Concentrations in the environment are dominated by inputs derived from burning fuels. Sewage sludges frequently contain a few compounds in this category. Values for individual PAHs typically are in the 1 to 10 mg/kg range. Total PAH contents of sewage sludges usually range from 10 to 50 mg/kg, although some sludges are found to contain up to 200 to 350 mg kg.[126]

Because of their widespread environmental presence, known carcinogenicity to animals, and mutagenicity, some researchers have identified PAHs as especially important with respect to public health risks from sludges. Some PAHs, particularly the higher molecular weight species, can persist in soils, making them a concern for plant uptake. The lower molecular weight species are more water soluble, suggesting easier assimilation into plant tissue than higher molecular weight species.[126]

PAHs are present in foods as a result of biosynthesis, adsorption of particulates on leafy surfaces from atmospheric fallout, or, more significantly, as a result of the processing or cooking of foods prior to consumption. Concentrations are higher on plant surfaces compared with internal tissue, and above ground plants have much higher concentrations than root crops. Broad leaf plants contain more PAHs than thin-leafed ones. Most of this PAH contamination is from atmospheric deposition. Concentrations of benzo[a]pyrene (BaP) in vegetation range from 0.1 to 150 μg/kg. Soil concentrations worldwide range from 100 to 1000 μg/kg, with total PAH typically about 10 times the BaP concentration.[155] Some terrestrial plants can take up and translocate PAHs through roots and leaves. The rate of uptake is dependent on physical factors of PAH deposition, soil type and condition, and plant species.[156]

BaP, usually used as an index compound, was detected seven times in the U.S. EPA's National Sewage Sludge Survey of 209 randomly selected wastewater treatment plants. The mean concentration was 10.8 mg/kg with a range from 0.7 to 24.7.[50]

1. Toxicity

There is little information available on distribution in humans. They are not generally acute toxicants, although little information is available. Ingestion of PAHs appears to be low in humans. Ingestion by animals indicates a relatively low acute toxicity, in part because PAHs are often poorly absorbed from the gastrointestinal tract.[157] Orally absorbed PAHs are rapidly and widely distributed in rats. Qualitative similarities in distribution among species suggest that distribution in humans would also be wide. Metabolism and excretion are probably relatively similar in humans and animals, but variability in specific activities of enzymes will alter the metabolic profiles among the species. PAH metabolites can be more toxic than the original compounds.[157] PAHs are metabolized by hydrocarbon hydroxylase, and the resulting phenolic compounds are conjugated and excreted in the urine or bile.[158] Placental transfer of PAHs in rodents appears to be limited. It is thought that human fetuses may be exposed to PAHs, but that levels would not be as high as maternal levels.[159]

Target organs of PAHs have been identified, primarily the skin and those organs with rapidly proliferating tissue such as the hematopoietic, lymphoid, and reproductive systems. PAHs are an example of indirect-acting reproductive toxicants that require metabolic activation to form a more reactive metabolite that exerts the toxic action.[160] BaP has been the most extensively studied PAH and thus many of the adverse effects of other PAHs must be inferred from these studies. This may underestimate or overestimate the health risks associated with the various PAHs.

2. *Carcinogenicity*

Several PAHs are considered to be carcinogenic.[159] PAHs can be activated by the hepatic microsomal enzyme system to forms that bind covalently to DNA.[158] The carcinogenic PAHs are an example of precarcinogens that require conversion through bioactivation to become ultimate carcinogens, either directly or through an intermediary stage, the proximate carcinogens. It is currently believed that the toxic and carcinogenic effects of PAHs are mediated by reactive diol-epoxide intermediates that interact directly with DNA and RNA, producing adducts. The formation of these adducts leads to neoplastic transformation as well as interfering with the normal functioning of rapidly proliferating tissues. These reactive intermediates are formed when PAHs are biotransformed by the P-450 enzymes.[159]

In a recent series of articles, Beech has proposed an alternative epigenetic mechanism for PAH carcinogenic promotion involving PAH and cationic Ah receptors inducing permanent cell membrane lesions that are inherited by daughter cells at mitosis.[161-65]

The IARC has classified benzo[a]pyrene as Group 2A, agents that are probably carcinogenic to humans. The epidemiological evidence for carcinogenicity to humans is limited. The experimental evidence for carcinogenicity to animals is sufficient.[60]

VII. NITRITE AND NITRATE TOXICITY

Nitrate is a normal constituent of foodstuffs as a nitrogen source. Plants absorb nitrogen in the form of nitrate or ammonium ions from the soil. Animal manure and/or inorganic fertilizers are used to replace the nitrogen absorbed by plants. The application of ever greater amounts of nitrogen fertilizer for improved crop yields has become a common agricultural practice. Data indicates that only 40 to 60% of nitrogen fertilizer applied to arable land is removed by crops, while about 25 to 30% is lost to the groundwater system. Evidence suggests that increased use of nitrogen fertilizer has resulted in higher nitrate levels in surface and groundwater above the standard recommended level of 10 mg/L NO_3-N (also expressed as 45 mg/L NO_3^- when measured as nitrate). Diffuse nitrate pollution of groundwater has been recognized as one of the most serious impacts of farming activities on the groundwater system.[166,167]

Vegetables are the main dietary source of nitrate for humans, contributing more than 70% of the total nitrate intake. Water, the second most important source, provides approximately 21% of total dietary nitrate. Meat and meat products contribute about 6.3% due to the use of sodium nitrate as a preservative and color enhancing agent in cured meats.[168] The World Health Organization (WHO) has recommended a daily nitrate and nitrite intake of 1.1 and 0.09 mg/kg, respectively.[169]

Water is the main source of nitrate for infants since most of their diet contains a higher proportion of water. The first case of fatal acquired methemoglobinemia in an infant due to ingestion of nitrate contaminated well water was reported in Iowa in 1945.[170] The Freshwater Foundations estimates that since that time about 2000 similar cases of acquired methemoglobinemia in young infants have been reported worldwide, with an estimated 10% case fatality rate.[171] The 2000 cases reported in the literature do not provide the actual estimate of the incidence rate or public health significance of this preventable disease, since it is generally not a notifiable/reportable disease. Recently, within the past 2 years, some states have adopted mandatory reporting of cases of methemoglobinemia to public health officials. Clearly, nitrate poisoning does contribute to national infant mortality statistics.[172] The most recent case of infant mortality due to methemoglobinemia in the U.S. was reported by Johnson and his associates in South Dakota in 1986.[173]

A. NITRATE SOURCES

1. Nitrate in Water

The Environmental Protection Agency's (EPA) national survey report on pesticides in drinking water wells estimates that about 1.2% of the community water system wells and about 2.4% of the rural domestic wells nationwide contain nitrate exceeding the recommended health advisory and Maximum Contaminant Level (MCL) of 10 mg/L NO_3 measured as nitrogen.[174] The survey also estimated that nearly 1.5 million people, including about 22,500 infants, are served by rural domestic drinking water wells that deliver water with levels of nitrate that exceed recommended health levels. Among those receiving water from community water supply wells, the survey estimated that approximately 3.0 million people, including 43,500 infants, use well water that exceed health advisory limits. In a similar survey of 686 private rural wells conducted in the agricultural state of Iowa, about 18.3% of the rural domestic wells contain nitrate above the health standard.[175] Public water supplies using surface water have reported nitrate levels exceeding health advisory levels, especially during spring melt conditions in agricultural areas. It is also estimated that 1 to 2% of the U.S. population using public water systems are consuming nitrate in excess of the EPA limit of 10 mg/L NO_3-N.[176]

Past reports have summarized various nitrate data collected from the same well, water system, or surface water sites over time, with some analyses dating to the early 1900s.[177-179] These data have clearly shown that nitrate concentrations in groundwater, and perennial surface waters that are fed by groundwater discharge, have increased dramatically, primarily since the early 1960s, and that these increases are regional in character. The regional increases occurred concurrently with the increased acreage of row-cropping and increased rates of use of nitrogen fertilizers in Iowa. Similar data have been reported across the corn-belt and in intensive agricultural regions worldwide.[180-183]

To reiterate, the concern with nitrate contamination in Iowa, and many agricultural regions, is not that the problem was unknown until recent years, nor that high concentrations were not present locally. The problem is now compounded by the regional increases in nitrate concentrations and its extension to greater depths in groundwater systems. Nitrate has become a nearly ubiquitous contaminant in the shallow groundwater system.

Considerable monitoring has also been undertaken in other states.[179] Statistical surveys limited to farm wells have been conducted in Nebraska and Kansas. Both states are less intensively farmed than Iowa, with lower pesticide and fertilizer use.

In 1985, the Nebraska Department of Health conducted a statistical sampling of 451 rural farm wells in 21 counties in central Nebraska.[184] Approximately 8% of the wells had NO_3-N concentrations >10 mg/L, while 4% showed pesticide detections (3.2% with atrazine). A similar statewide design survey was carried out in Kansas of 104 wells sampled in 50 counties.[185] In this study, 28% of farms had NO_3-N levels >10 mg/L. Pesticides were detected in 9% of the sample (4.5% for atrazine). In a second phase study, which targeted older and more shallow wells, 32% of farm wells showed NO_3-N concentration of >10 mg/L, and 11% had pesticide detections.[186]

The Wisconsin Department of Agriculture, Trade and Consumer Protection conducted a statistical survey of dairy well-water supplies. Water samples were collected from a random sample of 534 wells from dairy farms in Wisconsin from August 1988 through February 1989. This population-based survey found that 10% of the wells exhibited NO_3-N levels >10 mg/L and 13% of dairy farms had wells with pesticide detections (12% with atrazine).[187]

The nitrate concentration of public drinking water in Europe varies from country to country. In the U.K. public water supplies have to comply with a rolling three monthly mean not exceeding 17.8 mg/L nitrate, and no individual sample exceeding 22.2 mg/L. Approximately 85% of the population in Sweden receives surface water with very low concentrations of nitrate (about 0.5 mg/L), while groundwater rarely exceeds 2 mg/L nitrate.[188,189]

Human activities such as intensive livestock production with the attendant waste disposal problems, installation of substandard septic systems, and increased use of nitrogen fertilizers, among others, all contribute to the increased nitrate levels in groundwater. Shallow domestic wells (less than 100 ft deep) with deteriorating or improperly designed well casings are increasingly vulnerable to nitrate contamination. Well water contamination is more common and more serious today than it was a decade ago. Shallow wells that were constructed several years ago may have tested safe initially, but may now contain greater concentrations or even unsafe levels of nitrate or other environmental contaminants. Water from contaminated and poorly constructed systems are the most common cause of acquired infantile methemoglobinemia in rural areas. All public water supplies are required to regularly monitor their systems for nitrate concentrations and must notify the public via the media should nitrate concentrations exceed the health standard. However, private domestic wells used by rural populations are not tested routinely (often only every few years) for chemical, especially nitrate, and microbial contamination.

2. *Other Sources of Nitrate*

Although seldom a source of acute nitrate toxicity resulting in methemoglobinemia, more than 70% of the nitrate in a typical human diet is derived from vegetables such as spinach, celery, lettuce, cabbage, and root vegetables. Deliberate abuse of volatile nitrites (amyl, butyl, and isobutyl nitrites) as psychedelics or aphrodisiacs also occurs.[190-193]

Nitrate or nitrite exposure may occur from medications. Topical application of benzocaine as an anesthetic has resulted in reported cases of methemoglobinemia.[194,195] The use of silver nitrate in burn therapy (especially in infants and children), quinone derivatives (antimalarial), nitroglycerin, bismuth subnitrite (antidiarrheal), ammonium nitrate (diuretics), amyl and sodium nitrite which are used in cyanide and hydrogen sulfide poisoning antidotes and isosorbide di-, and tetra-nitrates constituents of vasodilators used in coronary artery disease therapy have been implicated in cases of nitrate or nitrite toxicity. Other sources of exposure include sodium nitrite used as an anticorrosive agent in cleaning fluids, accidental ingestion of ammonium nitrate found in cold packs, smoke inhalation, potassium nitrate/saltpeter, and nitrous gases in arc welding.[196-199] Serious poisoning and death has occurred when sodium nitrite, mistaken for table salt, was ingested with food.[200]

B. BIOLOGICAL FATE OF INGESTED NITRATE

In the human body, nitrate is rapidly absorbed from the proximal small intestine and distributed throughout the body.[201,202] Nitrite is absorbed by diffusion across the gastric mucosa and also through the wall of the intestinal tract.[203] Approximately 60 to 70% of an oral nitrate dose is excreted in urine in the first 24 h.[202] Twenty-five percent of ingested nitrate is excreted in saliva through an active blood nitrate transport system.[204,205] Nitrate enters the large bowel from the bloodstream and is rapidly converted to highly reactive nitrite partly by fecal microorganisms. In excess amounts, nitrite reacts with hemoglobin to form methemoglobin, thus significantly reducing the capacity of hemoglobin to transport oxygen to tissues.[206]

Another important biological fate is the conversion of nitrate to nitrite and then to N-nitroso compounds which are suspected to be direct-acting mutagens. Nitrate may also interact with pesticides such as atrazine, aldicarb, and carbofuran which contain secondary amine structures that may react with nitrite at low pH to form N-nitroso compounds.[207] In the laboratory Weisenburger and his associates synthesized N-nitrosoatrazine (NNAT) and found it to be readily formed from atrazine and nitrite at acid pH. NNAT is a weak mutagen in the Ames test; however, it is a strong mutagen in the Chinese hamster V-79 assay, producing revertants 3.4 times the dimethylnitrosamine control.[208] This potential chemical conversion has lead to an

Table 1 Causes of Methemoglobinemia

Acquired[a]	Hereditary
Direct oxidants, e.g., well water (Nitrate), foods high in Nitrate. Indirect oxidants, e.g., aniline dye derivatives, Benzocaine, Dapsone.	Hemoglobin M NADH methemoglobin Reductase deficiency

[a] Infants, 1 to 4 months old, are particularly susceptible to inducers of acquired methemoglobinemia because of reduced NADH reductase activity as a result of incomplete pediatric development of the enzyme.

interest in the role of these processes in the etiology of gastric, bladder, esophagal, and intestinal cancers in animals.[209]

C. HEALTH EFFECTS OF NITRATE TOXICITY

Unless conditions exist for reducing nitrate to nitrite during digestion, nitrate is metabolized and excreted without apparent adverse effects. However, the health effects of nitrate toxicity become manifest upon conversion to nitrite. Two factors influence the differences in individual susceptibility and variations in severity of nitrate toxicity: (1) The rate of development of the primary enzyme, NADH methemoglobin reductase system with age, and (2) The rate of bacterial conversion of nitrate to nitrite, and coliform infections reported to cause an increase in nitrite production without increased nitrate intake.

The health effects of nitrite are the same whether it enters the body through ingestion, inhalation, absorption through the skin, or is produced by bacterial conversion of nitrate during digestion. These effects include interference with normal oxygen carrying capacity of hemoglobin, depression of blood pressure, and interference with other physiologic processes.

1. Methemoglobinemia

Methemoglobinemia is a syndrome of elevated methemoglobin level, high blood nitrate, and frequently associated with acute diarrhea. The condition generally presents with few clinical signs other than cyanosis, but can result in coma and ultimately death depending on the extent of the hypoxia, which is the inability of red blood cells to carry oxygen to tissues.[210]

There are two main types of methemoglobin, the first is associated with the *normal* hemoglobin molecule. High methemoglobin may be present either because there is an increased production (acquired methemoglobinemia) or due to a hereditary deficiency of one of the methemoglobin reductase systems (Table 1).[211,212] The second type is associated with hereditary *abnormal* hemoglobin (hemoglobin M) which resists enzymatic reduction.[192,213-215]

Table 2 Reported Inducers of Methemoglobinemia

Agent	Source
Nitrate	
Sodium, potassium, calcium nitrate	Contaminated well water, fertilizer, food preservatives, vegetables — spinach, carrot juice
Bismuth subnitrate	Antidiarrheal
Ammonium nitrate	Fertilizer, diuretic
Silver nitrate	Topical burn therapy
Isosorbide di-, tetra-nitrate	Vasodilators
Nitrite	
Amyl nitrite	Lily cyanide antidote, vasodilator, abused inhalant
Sodium nitrite	Cyanide antidote, food preservative, anticorrosive
Butyl/isobutyl nitrite	Room deodorizer propellants
Ethyl nitrite	Folk medicine
Nitroglycerin	Oral, sublingual, or transdermal pharmaceuticals for treatment of angina, explosives
Aniline/aminophenols	Laundry ink, dyes, varnish, paints
Nitrobenzene	Industrial solvents, polishes
Benzocaine	Topical anesthetic
Lidocaine	Local and IV anesthetic, antiarrhythmic
Prilocaine	Local, caudal, epidural anesthesia
Sulfonamides	Antibacterial
Phenazopyridine	Pyridium
Chloroquine, Primaquine	Antimalarial
Dapsone	Dermatologic, antimalarial
Naphthalene	Mothballs, deodorizers
Copper sulfate	Fungicide
Methylene blue	Medical dye, methemoglobin therapy
Chlorates	Explosives, matches, toothpaste
Nitrogen oxide	Fires

Adapted from References 209 and 216.

The high sensitivity of infants less than 4 months of age, which is the segment of population at the most risk to nitrate toxicity, is due to a combination of factors:

1. Infants normally have a higher gastric pH than older children and adults. The higher gastric pH enhances bacterial invasion of the stomach and subsequent conversion of ingested nitrate to more potent nitrite
2. A large proportion of hemoglobin in infants is in the fetal hemoglobin form (hemoglobin F), which is more readily oxidized by nitrite to methemoglobin than adult hemoglobin (hemoglobin A)
3. In infants, NADH-dependent methemoglobin reductase, the enzyme responsible for reduction of induced methemoglobin back to normal hemoglobin, has only about half the activity which occurs in adults.[192]

Acquired methemoglobinemia can occur in infants due to subtle exposures such as use of diapers marked with aniline-containing inks or rinsed in oxidizing agents, wearing of shoes dyed with aniline compounds, and the use of rectal suppositories containing benzocaine or similar local anesthetics (Table 2).[209,216] There is presently no conclusive evidence to support earlier hypothesis that breast-fed infants may develop methemoglobinemia when their mothers consumed nitrate contaminated

Table 3 Signs and Symptoms Associated with Methemoglobinemia

Methemoglobin concentration (%)	Symptoms and clinical findings
0 3	Normal level
3–15	Slate gray cutaneous coloration
15–20	Central cyanosis of limbs/trunk, usually asymptomatic
20–50	Headache, dizziness, fatigue, lethargy, syncope, dyspnea
50–70	Dysrhythmia, seizures, CNS depression, coma
>70	High risk of mortality

Adapted from References 192 and 216.

well water. So far, available data from experimental animals indicate that neither nitrate nor nitrite accumulate or concentrate in the mammary gland or milk.[217,218]

Frequently pregnant and lactating women are advised not to consume water with nitrate concentrations above the MCL of 10 mg/L NO_3-N. To determine if the nitrate content of breast milk is influenced by the nitrate content of drinking water consumed, 20 women breast-feeding at least twice per day infants less than 6 months of age were enrolled in a 3-day crossover study. The women were asked to drink a minimum of 1500 cubic centimeters (cc) of water which contained 0 mg/L NO_3-N on day 1, 10 mg/L NO_3-N on day 2, and 22.2 mg/L NO_3-N on day 3. A daily diet record was kept of all foods and beverages consumed to determine daily dietary nitrate intake. Infants breast-fed on days 1 and 2, but received an alternate milk source on day 3. Mothers collected and froze samples of breast milk and urine at the end of each day. Mean results of the 20 subjects showed the following:

Time Period	Daily Nitrate Intake, mg		Concentration of NO_3-N	
of Study	Diet	Water	Breast Milk	Urine
Day 1	10.4	0.0	1.0	8.0
Day 2	15.2	22.2	1.1	14.7
Day 3	12.9	47.6	1.2	18.6

Women consuming water with nitrate concentration greater than or equal to the MCL do not produce breast milk with elevated nitrate levels (p-value >0.05).[219]

Nitrite is able to oxidize ferrous iron in hemoglobin to ferric iron and convert hemoglobin (the blood pigment that carries oxygen from the lungs to the tissues) to methemoglobin. Since methemoglobin is incapable of binding molecular oxygen, the physiologic effect is oxygen deprivation or suffocation with cyanosis. This may result in progressive central nervous system effects ranging from mild dizziness and lethargy to coma and convulsions, cardiac dysrhythmia, and circulatory failure.

In the early stages of symptomatic nitrate intoxication or mild toxicity, the methemoglobin levels may range from 10 to 20% of total hemoglobin. The lips and mucous membranes of persons with nitrate/nitrite toxicity usually have more a brownish than a bluish cast, normally referred to as "chocolate cyanosis". At low levels of methemoglobin the skin may remain pale, becoming bluish to purplish or blue-brown with increasing methemoglobin levels. The extent of these effects depends on the percentage of total hemoglobin that is in the oxidized form (Table 3).[192,216]

2. Reproductive and Developmental Effects

In an Australian epidemiological study, statistically significant risk increases occurred specifically for birth defects of the central nervous system and musculoskeletal system in a population of women consuming water containing about 3 mg/L NO_3-N.[220] However, experimental animal studies using the same drinking water source failed to demonstrate the presence of a teratogenic agent in the water.[221] A Canadian case control study comparing baseline nitrate levels of less than 0.1 mg/L NO_3-N to exposures of nitrate of about 6.0 mg/L NO_3-N from private well water sources estimated a moderate increased risk, though not statistically significant (risk odds ratio = 2.30), for central nervous system birth defects.[222] Animal and human data have so far failed to provide any conclusive evidence of teratogenic effects attributable to nitrate or nitrite ingestion.

The methemoglobin level of asymptomatic pregnant women with normal gestation may increase from the normal of 0.5 to 2.5% of total hemoglobin to a maximum of about 10.5% at the 30th week of pregnancy. The methemoglobin level subsequently declines and returns to normal after delivery. Pregnant women may therefore be more sensitive to the induction of clinical methemoglobinemia by nitrite at approximately the 30th week of pregnancy.[208]

3. Carcinogenic Effects

High nitrate levels in groundwater have been associated with elevated rates of non-Hodgkins lymphoma (NHL, cancer of lymphoid tissues) in a Nebraska study. Weisenburger observed that the incidence of NHL is increased in many counties in eastern Nebraska. Histologic analysis has revealed a twofold increase in the clinically aggressive, diffuse large cell subtype of NHL. A population-based case-control study was conducted in eastern Nebraska in 1985 to investigate the possible association between NHL and agricultural exposures. Telephone interviews were conducted with 201 men having histologically confirmed NHL and 705 controls. Among men, the use of the herbicide 2,4-D was associated with a 50% increased risk of NHL (OR 1.5). Personal exposure to 2,4-D for more than 20 days per year increased the risk threefold (OR 3.3). Several classes of insecticides were also associated with increased risk: organophosphates (OR 1.9), carbamates (OR 1.8),and chlorinated hydrocarbons (OR 1.4). As a result of intense agrichemical use, extensive contamination of shallow groundwater by nitrate and atrazine has occurred in eastern Nebraska. A twofold increased incidence of NHL is present in counties with greater than 20% of the wells contaminated by nitrate (>10 ppm) and in counties with intense fertilizer use. The findings by Weisenburger suggest that NHL in eastern Nebraska may be related to the use of pesticides and nitrogen fertilizers.[223]

Although several reviews have tended to associate nitrate intake with gastric cancer,[224] the National Academy of Sciences concluded that "evidence implicating nitrate, nitrite, and nitroso-compounds in the development of cancer in humans is circumstantial."[225] The WHO, after summarizing available evidence in 1985, also concluded that "no convincing evidence of a relationship between gastric cancer and consumption of drinking-water containing nitrate levels up to 10 mg/L NO_3-N has emerged. Furthermore, no firm epidemiological evidence has been found linking gastric cancer and drinking water containing higher levels of nitrate, but a link cannot be ruled out due to the inadequacy of the data available. Very few studies have considered human cancers other than gastric cancer in relation to nitrates, and none of them provides convincing evidence that nitrate ingestion influences cancer incidence at other sites." Currently, there is no conclusive evidence that either nitrate or nitrite at normal exposure levels cause cancer in experimental animals or humans.[226-230]

4. Mutagenicity

Both sodium nitrate and sodium nitrite have not demonstrated mutagenicity in test animals.[231,232] The ingestion of meals containing different levels of nitrate, nitrite, and nitrosamines resulted in increased production of unscheduled DNAs (UDS), a response to DNA damage. However, since metabolic activation of nitrosamines to genotoxic derivatives occurs mainly in the liver, such short-lived reactive metabolites may not have much opportunity to damage leucocyte DNA.[189]

5. Other Potential Health Effects

There is preliminary evidence to associate autosomal recessive hereditary methemoglobinemia with mental retardation. Though rare, the syndrome is reported to be caused by a deficiency of enzyme NADH cytochrome b5 reductase.[233-235]

Widespread contamination of waters by runoffs of nitrate applied to farm lands and disposal of sewage in large bodies of water may lead to denitrification with production of nitrous oxides, which are reported to cause nitrous-oxide-induced ozone depletion.[236]

D. PUBLIC HEALTH IMPLICATIONS OF NITRATE TOXICITY

In cases of mild acquired methemoglobinemia, no treatment may be required other than avoiding consumption of the nitrate-contaminated water or food. The provision of treatment alone in acute nitrate poisoning is not adequate. It is important to identify and permanently eliminate the nitrate source. Normal prenatal care for rural patients should include a recommendation to have private wells tested at regular intervals for nitrate. If excess nitrate concentration is determined, the well water should not be used in preparing infant formula or otherwise consumed by infants, particularly those less than 6 months of age.

Clues to potential nitrate/nitrite exposure may be obtained by considering family location and occupation, drinking water source, nutritional status (especially infants), complete medical history, including even topical medications, and any recent cases of gastrointestinal disorders.[236] Infants with acute diarrhea occasionally exhibit high

Table 4 Regulatory and Recommended Nitrate Levels

		Level (mg/L)			
Agency	**Focus**	**NO_3-N**	**NO_3^-**	**NO_2-N**	**Comments**
U.S. EPA[a]	Water	10.0[b]	45	1	Regulatory and Health Advisory Level(HAL)
Germany[241]	Water	4.4	20		Regulatory/HAL
S.Africa[241]	Water	4.4	20		Regulatory/HAL
Denmark[242]	Water	5.6	25		Guideline

[a] EPA = Environmental Protection Agency; [b] mg/L in water is equivalent to ppm.

methemoglobin levels even when fed diets which contain less than the EPA recommended nitrate standard, since diarrhea usually results in endogenous *de novo* synthesis of nitrate and nitrite.[237-240] It is therefore especially important to assess both the nutritional and health status of the infant.

In situations where well water is the sole potable water, information on well depth, location, construction (e.g., type of casing), and frequency of microbiological and nitrate testing is useful in assessing possible nitrate exposure. The conventional practice of boiling water to destroy microorganisms before mixing infant formula, though a good measure against microbial infection, is not a safe practice with nitrate-contaminated water. Boiling increases the nitrate concentration in the water. Alternative sources of safe water should therefore be found. These may include water from a well that has been tested safe, bottled water, water from a new and deeper well, or water from a monitored public water supply. Water treatment technologies such as ion exchange resins or reverse osmosis to remove nitrate from water are available, but are not commonly used.

The following determinants are important for effective evaluation of nitrate toxicity:

- **Health standards in water** — Water tests for nitrate can be obtained from any public health laboratory which utilizes approved EPA or WHO methods, such as the cadmium reduction method. Care should be taken in interpreting the results correctly for the units of nitrate concentration provided by the laboratory. The nitrate limit in drinking water was established as a safeguard against infantile methemoglobinemia. All the agencies in the U.S., both regulatory and monitoring, have recommended and/or regulated similar levels of nitrate in potable water. Regulatory or monitoring maximum contaminant levels of nitrate in water differ in other countries (Table 4).
- **Nitrate and nitrite in body fluids** — The urinary and salivary systems together excrete approximately 85 to 95% of nitrate from the body. It is therefore significant to determine nitrate and nitrite in these systems. Automated systems for the analysis of nitrate and nitrite in urine, saliva, deproteinized plasma, gastric juice, and milk by reduction with a high-pressure cadmium column, rapid ion-exchange liquid chromatography and conductivity, and microanalytical gas chromatograph method may be employed for the detection of nitrate and/or nitrite in various samples including body fluids.[243-246]

- **Blood nitrate level** — High nitrate concentrations are normally observed in patients with infantile methemoglobinemia, partly as a result of the reaction of nitrites with oxyhemoglobin.[247] A correlation between blood nitrate and methemoglobin concentrations can be established. However, in about 20% of cases of infants with acute diarrhea, moderately higher blood nitrate concentrations could be measured although no elevated methemoglobin levels are observed. The relationship between blood nitrate and methemoglobin is not normally linear at lower nitrate concentration. A certain minimum amount of nitrite needs to enter the blood stream before a measurable increase in methemoglobin concentration can be detected.[238]
- **Blood methemoglobin level** — The methemoglobin and hemoglobin levels in blood are the most useful tests for the evaluation of nitrate toxicity.[248] A high methemoglobin level is a positive indicator of severe nitrate toxicity. The filter paper technique is a simple and adequate method to initially evaluate methemoglobinemia, or "chocolate-brown cyanosis". A drop of blood suspected of higher methemoglobin is placed on a piece of filter paper alongside a drop of blood of a normal individual. When dry, the blood containing methemoglobin (>15%) will turn a deep chocolate-brown or slate-gray color. In a tube, methemoglobin-containing blood will not turn red when shaken in air or when oxygen is bubbled through it, while blood that is dark because of a high content of normal deoxyhemoglobin will turn red on oxygenation.[190]
- **Arterial blood gases** — Methemoglobinemia may result in reduced measured blood oxygen saturation, but calculated oxygen saturation from partial pressure of oxygen (PO_2) levels may remain normal in the presence of methemoglobinemia.[190,249]

E. PRESENT AND FUTURE OF NITRATE TOXICITY RESEARCH

Well water of untested origin used in preparing formulas for infants has been a primary suspect in most cases of infantile methemoglobinemia. However, high nitrate levels in water supplies may suggest the presence of other agricultural chemicals and bacterial contaminants which may have more serious health consequences, especially for infants and pregnant women (potential risk to the fetus).

There is still much work to be done to assist in our understanding of all the health hazards of nitrate toxicity and associated microorganisms. To date the main focus has been on treatment of acute methemoglobinemia and strategies for the prevention of nitrate toxicity in infants. The long range impact of persistent methemoglobinemia on human health remains relatively unstudied. It is therefore significant to reevaluate epidemiological investigations and improve rapid methods for determining the effects of nitrate toxicity. Other areas of further investigation may include:

- **Nitrate toxicity and pregnancy** — The effects of concern are related to pregnant women consuming water containing high nitrate levels. Since elevated methemoglobin levels normally occur during pregnancy the possible increased sensitivity of pregnant women and their fetus to anoxic (oxygen deficiency) conditions is a concern.
- **Nitrate toxicity and mental retardation** — Evidence in the literature indicates a hereditary manifestation of mental retardation due to hereditary deficiency of

cytochrome b5 reductase. The combined effect of exposing individuals with this deficiency to elevated nitrate in their food and water on mental retardation requires further investigation.

- **Nitrate/nitrite and other cancers** — Even though the body of data summarized by both the WHO and NAS is presently inconclusive for a direct link between nitrate consumption and cancer, there continues to exist epidemiologic evidence associating increased nitrate use and possible exposure with certain cancers. It is therefore very essential to extensively investigate this association. The production of nitrous oxide (N_2O) during denitrification of nitrate by soil and aquatic bacteria is currently an area of interest because of the role of N_2O in the destruction of stratospheric ozone, and the consequent increased risk of skin cancer.[236]
- **Nitrate induced "allergic" arthritis** — Epstein reported hypersensitivity to sodium nitrate in the form of recurrent arthritic attacks.[250] The observation that exposure to nitrate can result in rheumatoid-like arthritis is a potential concern to the health of farmers who are occupationally exposed to nitrate in fertilizers for long periods.[251]
- **Nitrate and neurotoxicity** — Since nitrate toxicity results in anoxia, the potential exists for permanent neurotoxic effects following acute methemoglobinemia conditions. Similarly, repeated low levels (subclinical and asymptomatic) of anoxia induced by chronic methemoglobinemia may result in neurotoxic effects, particularly in infants and children. Animal studies to investigate the role that nitrate may play in neuro-behavioral and developmental disorders are being considered by the National Institute of Environmental Health Sciences.[252]
- **Nitrate/pesticide mixtures and birth defects** — Research focus on nitrate and pesticide mixtures in water is continuing. Animal reproductive toxicity studies using the "Iowa Cocktail" mixture of nitrate and common herbicides, such as atrazine, alachor, metolachor, metribuzin, among others, were conducted by NIEHS' National Toxicology Program. Preliminary results indicate no statistically significant reproductive toxicity associated with the mixtures. On-going studies of second generation animals will be concluded in late 1992.[252]
- **Nitrate/atrazine and leukemia** — Because of the known production of N-nitrosoamines in the human gastrointestinal tract and the synthesis of nitrosoatrazine in the laboratory, a study is being designed to use a leukemia cell transplant model[253] and a fish bioassay to evaluate a mixture of nitrate and atrazine. The study is to be performed by NIEHS in Research Triangle Park, NC.[254]
- **Nitrate and developmental disorders** — The authors have heard several anecdotal reports from physicians in Iowa about children who survived acute methemoglobinemia as infants and experienced learning disabilities in later life during their school years. Case study reports should be developed for these children to determine if a pattern exists.

ACKNOWLEDGMENT

Financial support is gratefully acknowledged from Pioneer Hi-Bred International, Inc., Des Moines, IA. Much of the information regarding nitrate toxicity used to generate this chapter was assembled at their request to assist in formulating their corporate environmental policy statement regarding the environmental significance of nitrate. Additional original research presented in this chapter was supported by the University of Iowa Center for Health Effects of Environmental Contamination, and the Environmental Health Sciences Research Center, funded by the National Institute of Environmental Health Sciences.

REFERENCES

1. Truhaut, R., A survey of the dangers of the chemical era: impacts on human and environmental health, in *Environment and Health: A Holistic Approach*, Krieps, R., Ed., Gower Publ., Brookfield, VT, 1989, 33.
2. Tarr, J., The search for the ultimate sink: urban air, land, and water pollution in historical perspective, *Rec. Columbia Hist. Soc.,* 51, 1, 1984.
3. National Research Council, *Environmental Epidemiology: Public Health and Hazardous Wastes*, National Academy Press, Washington, D.C., 1991.
4. Burby, R.J. and Okun, D.A., Land use planning and health, *Annu. Rev. Public Health,* 4, 47-67, 1983.
5. Landrigan, P.J., Prevention of toxic environmental illness in the twenty-first century, *Environ. Health Perspect.,* 86, 197, 1990.
6. Dixon, R.L. and Nadolney, C.H., Problems in demonstrating disease causation following multiple exposure to toxic or hazardous substances, in *Environmental Impacts on Human Health: The Agenda for Long-Term Research and Development*, Draggon, S.D., Cohrssen, J.J., and Morrison, R.E., Eds., Praeger, New York, 1987, 117.
7. National Toxicology Program, *Sixth Annual Report on Carcinogens*, U.S. Department of Health and Human Services, Research Triangle Park, NC, 1991.
8. National Research Council, *Toxicity Testing: Strategies to Determine Needs and Priorities*, National Academy Press, Washington, D.C., 1984.
9. Pimentel, D., Dazhong, W., and Giampietro, M., Technological changes in energy use in U.S. agricultural production, *Agroecology: Researching the Ecological Basis for Sustainable Agriculture*, Springer-Verlag, New York, 1990.
10. Miller, G.T., Jr., *Living in the Environment*, Wadsworth, Belmont, CA, 1988.
11. Hallberg, G.R., From hoes to herbicides: agriculture and groundwater quality, *J. Soil Water Conserv.,* Nov/Dec, 357, 1986.
12. National Research Council, *Alternative Agriculture*, National Academy Press, Washington, D.C., 1989.
13. Bogardi, I. and Kuzelka, R.D., Eds., *Nitrate Contamination: Exposure, Consequence, and Control*, Springer-Verlag, New York, 1991.
14. United States Environmental Protection Agency, *Development of Risk Assessement Methodology for Land Application and Distribution, and Marketing of Municipal Sludge*, U.S. Environmental Protection Agency, Washington, D.C., 600/6-89/001, 1989.
15. Donnelly, K.C., Brown, K.W., and Chisum, C.P., Mutagenic potential of municipal sewage sludge and sludge amended soil, in *Chemical and Biological Characterization of Municipal Sludges, Sediments, Dredge Spoils, and Drilling Muds*, Lichtenberg, J.J., Winter, J.A., Weber, C.I., and Fradkin, L., Eds., American Society For Testing and Materials, Philadelphia, PA, 1988, 288.

16. Briuns, R.J.F., Fradkin, L., Stara, J.F., Peirano, W.B., Molak, V., and Lomnitz, E., Analytical methods necessary to implement risk-based criteria for chemicals in municipal sludge, in *Chemical and Biological Characterization of Municipal Sludges, Sediments, Dredge Spoils, and Drilling Muds*, Lichtenberg, J.J., Winter, J.A., Weber, C.I., and Fradkin, L., Eds., American Society for Testing and Materials, Philadelphia, PA, 1988, 465.
17. Straub, T.M., Pepper, I.L., Gerba, C.P., Hazards from pathogenic microorganisms in land-disposed sewage sludge, *Rev. Environ. Contamination Toxicol.,* 132, 55, 1993.
18. Pahren, H.R., Microorganisms in municipal solid waste and public health implications, *CRC Crit. Rev. Environ. Control,* 17, 187, 1987.
19. Scarpino, P.V., Fradkin, L., Clark, C.S., Kowal, N.E., Lomnitz, E., Baseheart, M., Peterson, J.M., Ward, R.L., and Hesford, M., Microbiological risk assessment for land application of municipal sludge, in *Chemical and Biological Characterization of Sludges, Sediments, Dredge Spoils, and Drilling Muds*, Lichtenberg, J.J., Winter, J.A., Weber, C.I., and Fradkin, L., Eds., American Society For Testing and Materials, Philadelphia, PA, 1988, 480.
20. Strauch, D., Improvement of the quality of sewage sludge: microbiological aspects, in *Sewage Sludge Treatment and Use: New Developments, Technological Aspects and Environmental Effects*, Dirkzwager, A.H. and L'Hermite, P., Eds., Elsevier, New York, 1989, 160.
21. Williams, C., Assessing risks: putting toxicity in perspective, *Environ. Protection*, 12, 1993.
22. Gochfeld, M., Toxicology: principles of toxicology, in *Maxcy-Rosenau-Last Public Health and Preventive Medicine*, 13th ed., Last, J.M. and Wallace, R.B., Eds., Appleton & Lange, Norwalk, CT, 1992, 315.
23. Office of Technology Assessment, *Neurotoxicity: Identifying and Controlling Poisons in the Nervous System*, Van Nostrand, New York, 1992.
24. Hattis, D., The value of molecular epidemiology in quantitative health risk assessment, in *Environmental Impacts on Human Health: The Agenda for Long-Term Research and Development*, Draggon, S.D., Cohrssen, J.J., and Morrison, R.E., Praeger, New York, 1987, 89.
25. Hansen, L.G., and Chaney, R.L., Environmental and food chain effects of the agricultural use of sewage sludges, *Rev. Environ. Toxicol*., 1, 103, 1984.
26. Chaney, R.L., Bruins, R.J., Baker, D.E., Korcak, R.F., Smith, J.E., and Cole, D., Transfer of sludge-applied trace elements to the food chain, in *Land Application of Sludge: Food Chain Implications*, Page, A.L., Logan, T.J., and Ryan, J.A., Eds., Lewis Publishers, Chelsea, MI, 1987, 67.
27. Younos, T.M., Ed., Role of trace elements, in *Land Application of Wastewater Sludge: A Report of the Task Committee on Land Application of Sludge*, American Society of Civil Engineers, New York, 1987, 30.
28. Elinder, C. G. and Kessler, E., Toxicity of metals, in *Utilization of Sewage Sludge on Land: Rates of Application and Long-Term Effects of Metals*, Berglund, S., Davis, R.D., and L'Hermite, P.D., Eds., Reidel, Boston, 1984, 116.
29. Goyer, R.A., Toxic effects of metals, in *Casarett and Doull's Toxicology: The Basic Science of Poisons*, Amdur, M.O., Doull, J., and Klaassen, C.D., Eds., Pergamon Press, New York, 1991, 623.
30. Chang, A.C., Warneke, J.E., Page, A.L., and Lund, L.J., Accumulation of heavy metals in sewage-sludge-treated soils, *J. Environ. Qual*., 13, 87, 1984.

31. Holmgren, G.G.S., Meyer, M.W., Chaney, R.L., and Daniels, R.B., Cadmium, lead, zinc, copper, and nickel in agricultural soils of the United States of America, *J. Environ. Q.*, 22, 335, 1993.
32. Dean, J.G., Bosqui, F.L., and Lanouette, K.H., Removing heavy metals from waste water, *Environ. Sci. Technol.*, 6, 518, 1972.
33. Naylor, L.M. and Loehr, R.C., Priority pollutants in municipal sewage sludge, *BioCycle*, 23, 18, 1982.
34. Corey, R.B., King, L.D., et al., Effects of sludge properties on accumulation of trace elements by crops, in *Land Application of Sludge: Food Chain Implications*, Page, A.L., Logan, T.J., and Ryan, J.A., Eds., Lewis Publishers, Chelsea, MI, 1987, 25.
35. Chaney, R.L., Twenty years of land application research: part I, *BioCycle*, 31, 54, 1990.
36. Singhal, R.L. and Merali, Z., Biochemical toxicity of cadmium, in *Cadmium Toxicity*, Mennear, J.H., Ed., Marcel Dekker, New York, 1979, 61.
37. Chaney, R.L., Public health and sludge utilization: part II, *BioCycle*, 31, 68, 1990.
38. Crites, R., Land use of wastewater and sludge, *Environ. Sci. Technol.*, 18, 140A, 1984.
39. Chang, A.C., Hinesly, T.D., et al., Effects of long-term sludge application on accumulation of trace elements by crops, in *Land Application of Sludge: Food Chain Implications*, Page, A.L., Logan, T.J., and Ryan, J.A., Eds., Lewis Publishers, Chelsea, MI, 1987, 53.
40. Coogan, T.P., Latta, D.M., Snow, E.T., and Costa, M., Toxicity and carcinogenicity of nickel compounds, *CRC Crit. Rev. Toxicol.*, 19, 341, 1989.
41. Friberg, L., Norberg, G.F., and Vouk, V.B., Eds., *Handbook on the Toxicology of Metals*, Elsevier, Amsterdam, 1986.
42. Chang, L.W., The concept of direct and indirect neurotoxicity and the concept of toxic metal/essential element interactions as a common biomechanism underlying metal toxicity, in *The Vulnerable Brain and Environmental Risks, Volume 2: Toxins in Food*, Isaacson, R.L. and Jensen, K.F., Plenum Press, New York, 1992, 61.
43. Minnema, D., Neurotoxic metals and neuronal signalling processes, in *The Vulnerable Brain and Environmental Risks, Volume 2: Toxins in Food*, Isaacson, R.L. and Jensen, K.F., Plenum Press, New York, 1992, 83.
44. Murray, M.J. and Thomas, P.T., Toxic consequences of chemical interactions with the immune system, in *Principles and Practice of Immunotoxicology*, Miller, K., Turk, J.L., and Nicklin, S., Eds., Blackwell Scientific, Boston, 1992, 65.
45. Dean, J.H. and Murray, M.J., Toxic responses of the immune system, in *Casarett and Doull's Toxicology: The Basic Science of Poisons*, Amdur, M.O., Doull, J., and Klaassen, C.D., Eds., Pergamon Press, New York, 1991, 282.
46. Burger, E.J., Tardiff, R.G., and Bellanti, J.A., Eds., *Environmental Chemical Exposures and Immune System Integrity*, Princeton Scientific Publishing, Princeton, NJ, 1987.
47. Lewis, R., Sr., *Reproductively Active Chemicals: A Reference Guide*, Van Nostrand Reinhold, New York, 1991.
48. Friberg, L., Elinder, C.G., Kjellstrom, T., and Norberg, G.F., *Cadmium and Health: A Toxicological and Epidemiological Appraisal*, CRC Press, Boca Raton, FL, 1986.
49. Davis, R.D., Cadmium — a complex environmental problem: part II, cadmium in sludges used as fertilizer, *Experientia*, 40, 117, 1984.
50. Kuchenrither, R.D. and McMillan, S.I., Preview analysis of National Sludge Survey, *BioCycle*, 31, 60, 1990.

51. Ryan, J.A., Pahren, H.R., and Lucas, J.B., Controlling cadmium in the human food chain: a review and rationale based on health effects, *Environ. Res.,* 28, 251, 1982.
52. Tanaka, K., Min, K., et al., The origin of metallothionein in red blood cells, *Toxicol. Appl. Pharmacol.,* 78, 63, 1985.
53. Fowler, B.A., Mechanisms of kidney cell injury from metals, *Environ. Health Perspect.,* 100, 57, 1992.
54. Nordberg, G.F., Cadmium carcinogesis and its relationship to other health effects in humans, *Scand. J. Work Environ. Health,* 19(Suppl. 1), 104, 1993.
55. Hazen-Martin, D., Todd, J.H., et al., Electrical and freeze-fracture analysis of the effects of ionic cadmium on cell membranes of human proximal tubule cells, *Environ. Health Perspect.,* 101, 510, 1993.
56. Waalkes, M.P., Effects of dietary zinc deficiency on the accumulation of cadmium and metallothionein in selected tissues of the rat, *J. Toxicol., Environ. Health,* 18, 301, 1986.
57. Iwata, K., Saito, H., Moriyama, M., and Nakano, A., Renal tubular function after reduction of environmental cadmium exposure: a ten-year follow-up, *Arch. Environ. Health,* 48, 157, 1993.
58. Agency for Toxic Substances and Disease Registry, *Toxicological Profile for Cadmium (Update),* U.S. Department of Health & Human Services, Washington, D.C., TP-92/06, 1993.
59. Thomas, P.T., Tatajczak, H.V., et al., Evaluation of host resistance and immune function in cadmium-exposed mice, *Toxicol. Appl. Pharmacol.,* 80, 446, 1986.
60. International Agency for Research on Cancer, *IARC Monographs on the Evaluation of Carcinogenic Risks to Humans: An Updating of IARC Monographs Volumes 1 to 42, Supplement 7,* International Agency for Research on Cancer, Lyon, France, 1987.
61. United States Environmental Protection Agency, Cadmium, *Rev. Environ. Contamination Toxicol.,* 107, 25, 1989.
62. Waalkes, M.P., Rehm, S., et al., Cadmium carcinogenesis in the male Wistar rats: dose-response analysis of tumor induction in the prostate and testes and at the injection site, *Cancer Res.,* 48, 4656, 1988.
63. Waalkes, M.P., Rehm, S., et al., Cadmium carcinogenesis in the male Wistar rats: dose-response analysis of effects of zinc on tumor induction in the prostate and in the testes and at the injection site, *Cancer Res.,* 49, 4282, 1989.
64. Menzer, R.E., Water and soil pollutants, in *Casarett and Doull's Toxicology: The Basic Science of Poisons,* Amdur, M.O., Doull, J., and Klaassen, C.D., Eds., Pergamon Press, New York, 1991, 872.
65. Hammond, P.B. and Dietrich, K.N., Lead exposure in early life: health consequences, *Rev. Environ. Contamination Toxicol.,* 115, 91, 1990.
66. Grandjean, P., Lyngbye, T., and Hansen, O.N., Lessons from a Danish study on neuropsychological impairment related to lead exposure, *Environ. Health Perspect.,* 94, 111, 1991.
67. Fischbein, A., Occupational and environmental lead exposure, in *Environmental and Occupational Medicine,* Rom, W.N., Ed., Little, Brown, Boston, 1992, 735.
68. Mahaffey-Six, K. and Goyer, R.A., Experimental enhancement of lead toxicity by low dietary calcium, *J. Lab. Clin. Med.,* 76, 933, 1970.
69. Riess, J.A. and Needleman, H.L., Cognitive, neural, and behavioral effects of low-level lead exposure, in *The Vulnerable Brain and Environmental Risks,* Vol. 2, Isaacson, R.L. and Jensen, K.F., Eds., Plenum Press, New York, 1992, 111.

70. Goldstein, G.W., Developmental neurobiology of lead toxicity, in *Human Lead Exposure*, Needleman, H.L., Boca Raton, FL, 1992, 125.
71. Goyer, R.A., Transplacental transport of lead, *Environ. Health Perspect.,* 89, 101, 1990.
72. Pirkle, J.L., Schwartz, J., Landis, J.R., and Harlan, W.R., The relationship between blood lead levels and blood pressure and its cardiovascular risks, *Am. J. Epidemiol.,* 121, 246, 1985.
73. United States Environmental Protection Agency, Nickel, *Rev. Environ. Contamination Toxicol.,* 107, 103, 1989.
74. Agency for Toxic Substances and Disease Registry, *Toxicological Profile for Nickel (Update)*, U.S. Department of Health & Human Services, Washington, D.C., TP-92/14, 1993.
75. Smialowicz, R.J., Riddle, M.M., et al., Immunologic effects of nickel: II. Suppression of natural killer cell activity, *Environ. Res.,* 36, 56, 1985.
76. Sunderman, F.W., Jr., Search for molecular mechanisms in the genotoxicity of nickel, *Scand. J. Work Environ. Health,* 19(Suppl. 1), 75, 1993.
77. Costa, M., Molecular targets of nickel and chromium in human and experimental systems, *Scand. J. Work Environ. Health,* 19(Suppl. 1), 71, 1993.
78. Prasad, A.S., Essentiality and toxicity of zinc, *Scand. J. Work Environ. Health,* 19(Suppl. 1), 134, 1993.
79. McClain, C.J., McClain, M.L., Boosalis, M.G., and Hennig, B., Zinc and the stress response, *Scand. J. Work Environ. Health,* 19(Suppl. 1), 132, 1993.
80. Agency for Toxic Substances and Disease Registry, *Toxicological Profile for Zinc*, U.S. Department of Health & Human Services, Washington, D.C., TP-89/25, 1989.
81. Sandstead, H.H., Zinc requirements, the recommended dietary allowance and the reference dose, *Scand. J. Work Environ. Health,* 19(Suppl. 1), 128, 1993.
82. Chandra, R.K., Excessive intake of zinc impairs immune responses, *JAMA,* 252, 1443, 1984.
83. Earl, F.L. and Vish, T.J., Teratogenicity of heavy metals, in *Toxicity of Heavy Metals in the Environment: Part 2*, Oehme, F.W., Ed., Marcel Dekker, New York, 1979, 617.
84. Beale, A.M., Fasulo, D.A., and Craigmill, A.L., Effects of oral and parenteral selenium supplements on residues in meat, milk and eggs, *Rev. Environ. Contamination Toxicol.,* 115, 125, 1990.
85. Combs, G.F., Jr., Essentiality and toxicity of selenium with respect to recommended dietary allowances and reference doses, *Scand. J. Work Environ. Health,* 19(Suppl. 1), 119, 1993.
86. Lockitch, G., Jacobson, B., et al., Selenium deficiency in low birth weight neonates: an unrecognized problem, *J. Pediatr.,* 114, 865, 1989.
87. Hill, K.E., Burk, R.F., and Lane, J.M., Effect of selenium depletion and repletion on plasma glutathione and glutathione-dependent enzymes in the rat, *J. Nutr.,* 117, 99, 1987.
88. Medinsky, M.A., Cuddihy, R.G., et al., Projected uptake and toxicity of selenium compounds from the environment, *Environ. Res.,* 36, 181, 1985.
89. Alexander, J., Risk assessment of selenium, *Scand. J. Work Environ. Health,* 19(Suppl. 1), 122, 1993.
90. Longnecker, M.P., Taylor, P.R., et al., Selenium in diet, blood, and toenails in relation to human health in a seleniferous area, *Am. J. Clin. Nutr.,* 53, 1288, 1991.

91. Shoham, J., The effect of nutrition on the immune system, in *Principles and Practice of Immunotoxicology*, Miller, K., Turk, J.L., and Nicklin, S., Eds., Blackwell Scientific, Boston, 1992, 161.
92. Birt, D.F., Julius, A.D., et al., Enhancement of BOP-induced pancreatic carcinogenesis in selenium-fed Syrian golden hamster under specific dietary conditions, *Nutr. Cancer,* 11, 21, 1988.
93. Milner, J.A., Effect of selenium on virally induced and transplantable tumor models, *Fed. Proc.,* 44, 2568, 1985.
94. Salonen, J.T., Alfthan, G., Huttunen, J.K., and Puska, P., Association between serum selenium and the risk of cancer, *Am. J. Epidemiol.,* 120, 342, 1984.
95. Willett, W.C., Polk, B.F., and Morris, J.S., Prediagnostic serum selenium and risk of cancer, *Lancet*, 2, 130, 1983.
96. Schrauzer, G.N., White, D.A., and Schneider, C.J., Cancer mortality correlation studies. III. Statistical associations with dietary selenium intakes, *Bioinorg. Chem.,* 7, 23, 1977.
97. Salonen, J.T., Salonen, R., et al., Risk of cancer in relation to serum concentrations of selenium and vitamin A and E: matched case-control analysis of prospective data, *Br. Med. J.,* 290, 417, 1985.
98. Virtamo, J., Valkeila, E., et al., Serum selenium and risk of cancer: a prospective follow-up of nine years, *Cancer*, 60, 145, 1987.
99. Clark, L.C., The epidemiology of selenium and cancer, *Fed. Proc.,* 44, 2584, 1985.
100. Wilkenfeld, M., Metal Compounds and rare earths, in *Environmental and Occupational Medicine*, Rom, W.N., Ed., Little, Brown, Boston, 1992, 815.
101. Owen, C.A., Jr., *Physiological Aspects of Copper: Copper in Organs and Systems*, Noyse, Park Ridge, NJ, 1982.
102. Owen, C.A., Jr., *Copper Deficiency and Toxicity: Acquired and Inherited, in Plants, Animals, and Man*, Noyse, Park Ridge, NJ, 1981.
103. Carson, B.L., Ellis, H.V., and McCann, J.L., *Toxicology and Biological Monitoring of Metals in Humans: Including Feasibility and Need*, Lewis Publishers, Chelsea, MI, 1986.
104. Abdulla, M., Copper levels in Indian childhood cirrhosis, *Lancet*, 2, 246, 1979.
105. Tanner, M.S., Bhave, S.A., et al., Early introduction of copper-contaminated animal milk feeds as a possible cause of Indian childhood cirrhosis, *Lancet*, 2, 992, 1983.
106. Adelson, J.W., Indian childhood cirrhosis is a result of copper hepatotoxicity — in all likelihood, *J. Pediatr. Gastroenterol. Nutr.,* 6, 491, 1987.
107. Rowe, D.M., Solomayer, J.A., and Zenz, C., Other metals, in *Occupational Medicine, Principles and Practical Application*, Zenz, C., Ed., Year Book Medical Publishers, Chicago, IL, 1988, 639.
108. Agency for Toxic Substances and Disease Registry, *Toxicological Profile for Arsenic (Update)*, U.S. Department of Health & Human Services, Washington, D.C., TP-92/02, 1993.
109. Tseng, W.P., Effects and dose-response relationships of skin cancer and Blackfoot disease with arsenic, *Environ. Health Perspect.,* 19, 109, 1977.
110. National Research Council, *Drinking Water and Health*, National Academy of Sciences, Washington, D.C., 1977.
111. Crecelius, E.A., Changes in the chemical speciation of arsenic following ingestion by man, *Environ. Health Perspect.,* 19, 147, 1977.

112. Kreiss, K., Zach, M.M., et al., Neurologic evaluation of a population exposed to arsenic in Alaskan well water, *Arch. Environ. Health,* 38, 116, 1983.
113. Hindmarsh, J.T., McLetchie, O.R., et al., Electromyographic abnormalities in chronic environmental arsenicalism, *J. Anal. Toxicol.,* 1, 270,1977.
114. Chen, C.L., Kuo, T.L., and Wu, M.M., Arsenic and cancers, *Lancet,* 1, 414, 1988.
115. Outridge, P.M. and Scheuhammer, A.M., Bioaccumulation and toxicology of chromium: implications for wildlife, *Rev. Environ. Contamination Toxicol.,* 130, 31, 1993.
116. Cohen, M.D. and Costa, M., Chromium compounds, in *Environmental and Occupational Medicine,* Rom, W.N., Ed., Little, Brown, Boston, 1992, 799.
117. Agency for Toxic Substances and Disease Registry, *Toxicological Profile for Chromium (Update),* U.S. Department of Health & Human Services, Washington, D.C., TP-92/08, 1993.
118. Jennette, K.W., The role of metals in carcinogenesis: biochemistry and metabolism, *Environ. Health Perspect.,* 40, 233, 1981.
119. Ecobichon, D.J. and Joy, R.M., *Pesticides and Neurological Diseases,* CRC Press, Boca Raton, FL, 1982.
120. Carson, R., *Silent Spring,* Houghton Mifflin, Boston, 1962.
121. Alford-Stevens, A.L. and Budde, W.L., Determination of polychlorinated compounds (dioxins, furans, and biphenyls) by level of chlorination with automated interpretation of mass spectrometric data, in *Chemical and Biological Characterization of Municipal Sludges, Sediments, Dredge Spoils, and Drilling Muds,* Lichtenberg, J.J., Winter, J.A., Weber, C.I., and Fradkin, L., Eds., American Society For Testing and Materials, Philadelphia, 1988, 204.
122. Hall, R.H., A new threat to public health: organochlorines and food, *Nutr. Health,* 8, 33, 1992.
123. Wolf, M.S., Toniolo, P.G., et al., Blood levels of organochlorine residues and risk of breast cancer, *J. Natl. Cancer Inst.,* 85. 648, 1993.
124. Garabrant, D.H., Held, J., et al., DDT and related compounds and risk of pancreatic cancer, *J. Natl. Cancer Inst.,* 84, 764, 1992.
125. Brown, D.P., Mortality of workers employed at organochlorine pesticide manufacturing plants — an update, *Scand. J. Work Environ. Health,* 18, 155, 1992.
126. O' Connor, G.A., Chaney, R.L., and Ryan, J.A., Bioavailability to plants of sludge-borne toxic organics, *Rev. Environ. Contamination Toxicol.,* 121, 129, 1991.
127. Jacobs, L.W., O'Connor, G.A., Overcash, M.A., and Rygiewicz, M.J., Effects of trace organics in sewage sludges on soil-plant systems and assessing their risk to humans, in *Land Application of Sludge: Food Chain Implications,* Page, A.L., Logan, T.J., and Ryan, J.A., Lewis Publishers, Chelsea, MI, 1987, 101.
128. Jury, W.A., Winer, A.M., Spencer, W.F., and Focht, D.D., Transport and transformations of organic chemicals in the soil-air-water ecosystem, *Rev. Environ. Contamination Toxicol.,* 99, 119, 1987.
129. Connor, M.S., Monitoring sludge-amended agricultural soils, *BioCycle,* 1, 47, 1984.
130. Dean, R.B. and Suess, M.J., The risk to health of chemicals in sewage sludge applied to land, *Waste Manage. Res.,* 3, 251, 1985.
131. Agency for Toxic Substances and Disease Registry, *Toxicological Profile for Di [2-Ethylhexyl] Phthalate (Update),* U.S. Department of Health & Human Services, Washington, D.C., TP-92/05, 1993.
132. Ganning, A.E., Brunk, U., et al., Effects of prolonged administration of phthlate ester on the liver, *Environ. Health Perspect.,* 73, 251, 1987.

133. Aranda, J., O'Connor, G.A., and Eiceman, G., Effects of sewage sludge on DEHP uptake by plants, *J. Environ. Q.*, 18, 45, 1989.
134. Woodward, K.N., Phthalate esters, cystic kidney disease in animals and possible effects on human health: a review, *Hum. Exp. Toxicol.*, 9, 397, 1990.
135. See *Environ. Health Perspect.*, 65, 1986 for an issue devoted to the spectrum of phthalate ester toxicity.
136. Garberg, P. and Hogberg, J., Selenium metabolism in isolated hepatocytes: inhibition of incorporation in proteins by mono(2-ethylhexyl)phthalate, a metabolite of the peroxisome proliferator di(2-ethylhexyl)phthalate, *Carcinogenesis*, 12, 7, 1991.
137. Ganning, A.E., Brunk, U., and Dallner, G., Phthalate esters and their effect on the liver, *Hepatology*, 4, 541, 1984.
138. Crocker, J.F., Safe, S.H., and Acott, P., Effects of chronic phthalate exposure on the kidney, *J. Toxicol. Environ. Health*, 23, 433, 1988.
139. Albro, P.W., Absorption, metabolism, and excretion of di(2-ethylhexyl) phthalate by rats and mice, *Environ. Health Perspect.*, 65, 293, 1986.
140. Dirven, H.A., Theuws, J.L., Jongeneelen, F.J., and Bos, R.P., Nonmutagenicity of 4 metabolites of di(2-ethylhexyl)phthalate (DEHP) and 3 structurally related derivatives of di(2-ethylexyl)adipate (DEHA) in the Salmonella mutagenicity assay, *Mutation Res.*, 260, 121, 1991.
141. Kluwe, W.M., Carcinogenic potential of phthalic acid esters and related compounds: structure — activity relationships, *Environ. Health Perspect.*, 65, 271, 1986.
142. Seegal, R.F. and Shain, W., Neurotoxicity of polychlorinated biphenyls: the role of ortho-substituted congeners in altering neurochemical function, in *The Vulnerable Brain and Environmental Risks, Volume 2: Toxins in Food*, Isaacson, R.L. and Jensen, K.F., Eds., Plenum Press, New York, 1992, 169.
143. Agency for Toxic Substances and Disease Registry, *Toxicological Profile for Selected PCBs (Update)*, U.S. Department of Health & Human Services, Washington, D.C., TP-92/16, 1993.
144. Safe, S., Toxicology, structure-function relationship, and human and environmental health impacts of polychlorinated biphenyls: progress and problems, *Environ. Health Perspect.*, 100, 259, 1992.
145. Mes, J., PCBs in human populations, in *PCBs and the Environment*, CRC Press, Boca Raton, FL, 1987, 39.
146. Safe, S., Polychlorinated biphenyls (PCBs) and polybrominated biphenyls (PBBs): biochemistry, toxicology and mechanism of action, *CRC Crit. Rev. Toxicol.*, 13, 319, 1984.
147. Golub, M.S., Donald, J.M., and Reyes, J.A., Reproductive toxicity of commercial PCB mixtures: LOAELs and NOAELs from animal studies, *Environ. Health Perspect.*, 94, 245, 1991.
148. Baker, E.L., Jr., Landrigan, P.J., et al., Metabolic consequences of exposure to polychlorinated biphenyls (PCB) in sewage sludge, *Am. J. Epidemiol.*, 112, 553, 1980.
149. Emmett, E.A., Maroni, M., et al., Studies of transformer repair workers exposed to PCBs: II. Results of clinical laboratory investigations, *Am. J. Ind. Med.*, 14, 47, 1988.
150. Gustavsson, P., Hogstedt, C., and Rappe, C., Short-term mortality and cancer incidence in capacitor manufacturing workers exposed to polychlorinated biphenyls (PCBs), *Am. J. Ind. Med.*, 10, 341, 1986.
151. Bertazzi, P.A., Riboldi, L., et al., Cancer mortality of capacitor manufacturing workers, *Am. J. Ind. Med.*, 11, 165, 1987.

152. Silberhorn, E.M., Glauert, H.P., and Robertson, L.W., Carcinogenicity of polyhalogenated biphenyls: PCBs and PBBs, *Crit. Rev. Toxicol.,* 20, 440, 1990.
153. Safe, S., Polychlorinated biphenyls (PCBs): mutagenicity and carcinogenicity, *Mutation Res.,* 220, 31, 1989.
154. Brown, D.P., Mortality of workers exposed to polychlorinated biphenyls — an update, *Arch. Environ. Health,* 42, 333, 1987.
155. Shabad, L.M., Cohan, Y.L., et al., The carcinogenic hydrocarbon benzo(a)pyrene in the soil, *J. Natl. Cancer Inst.,* 47, 1179, 1971.
156. Kowal, N.E., *Health Effects of Land Application of Municipal Sludge*, U.S. Environmental Protection Agency, Washington, D.C., EPA 600/1-85 015, 1985.
157. Pucknat, A.W., Ed., *Health Impacts of Polynuclear Aromatic Hydrocarbons*, Noyes Data Corporation, Park Ridge, NJ, 1981.
158. Schwarz-Miller, J., Rom, W.N., and Brandt-Rauf, P.W., Polycyclic aromatic hydrocarbons, *Environmental and Occupational Medicine*, Rom, W.N., Ed., Little, Brown, Boston, 1992, 873.
159. Agency for Toxic Substances and Disease Registry, *Toxicological Profile for Polycyclic Aromatic Hydrocarbons (Update)*, U.S. Department of Health & Human Services, Washington, D.C., Draft for Public Comment, 1993.
160. Mattison, D.R., The mechanisms of action of reproductive toxins, *Am. J. Ind. Med.,* 4, 65, 1983.
161. Beech, J.A., Polycyclic aromatic hydrocarbons initiate carcinogenesis by forming permanent membrane lesions of increased permeability. Other non-viral carcinogens may also form similar lesions, *Med. Hypotheses,* 36, 1, 1991.
162. Beech, J.A., Cancer promotion in cells initiated by polycyclic aromatic hydrocarbons and other non-viral carcinogens, *Med. Hypotheses,* 36, 4, 1991.
163. Beech, J.A., Cell proliferation and carcinogenesis may share a common basis of permeable plasma membrane clusters, *Med. Hypotheses,* 38, 208, 1992.
164. Beech, J.A., The membrane potential theory of carcinogenesis, *Med. Hypotheses,* 29, 101, 1989.
165. Beech, J.A., Mitosis stimulation near wounds by temporary membrane lesions of increased permeability, *Med. Hypotheses,* 35, 295, 1991.
166. Benes, V., Pekny, V., Skorepa, J., and Vrba, J., Impact of diffuse nitrate pollution sources on groundwater quality — Some examples from Czechoslovakia, *Environ. Health Perspect.,* 83, 5, 1989.
167. Hallberg, G.R., Pesticide pollution of groundwater in the humid United States, *Agric. Ecosyst. Environ.,* 26, 299, 1989.
168. Vogtmann, H. and Biedermann, R., The nitrate story — No end in sight, *Nutr. Health,* 3(4), 217, 1985.
169. WHO, Specifications for the identity and purity of food additives and their toxicological evaluation: food colours and some antimicrobials and antioxidants. Eighth Report of the Joint FAO/WHO Expert Committee on Food Additives, *WHO Tech. Rep. Ser.,* 309, 5, 1965.
170. Comly, H.H., Cyanosis in infants caused by nitrates in well water, *JAMA,* 257, 2877, 1987.
171. Freshwater Foundation, Nitrates and Groundwater: A public health concern. Navarre: Freshwater Foundation, 1988.
172. Johnson, C.J. and Kross, B.C., Continuing importance of nitrate contamination of groundwater and wells in rural areas, *Am. J. Ind. Med.,* 18(4), 449, 1990.

173. Johnson, C.J., Bonrud, P.A., Dosch, T.L., et al., Fatal outcome of methemoglobinemia in an infant, *JAMA,* 257(20), 2796, 1987.
174. U.S. EPA, National pesticide survey: summary results of pesticides in drinking water wells, Office of Pesticides and Toxic Substances, U.S. Environmental Protection Agency, Washington, D.C., 1990.
175. Kross, B.C., Hallberg, G.R., Bruner, D.R., et al., The Iowa State-Wide Rural Well-Water Survey — Water-Quality Data: Initial Analysis, Tech. Info. Series 19, Iowa Dept. Natural Resources, 1990.
176. Craun, G.F., Greathouse, D.G., and Gunderson, D.H., Methemoglobin levels in young children consuming high nitrate well water in the United States, *Int. J. Epidermiol.,* 10(4), 309, 1981.
177. McDonald, D.B. and Splinter, R.C., Long-term trends in nitrate concentration in Iowa water supplies, *J. Am. Water Works Assoc.,* 74, 437, 1982.
178. Johnson, G., Kurz, A., Cerny, J., et al., Nitrate levels in water from rural Iowa wells, a preliminary report, *J. Iowa State Med. Soc.,* 2, 1946.
179. Hallberg, G.R., Nitrate in groundwater in the United States, in *Nitrogen Management and Groundwater Protection,* Follett, R.F., Ed., Elsevier, Amsterdam, 1989, 35.
180. Keeney, D.R., Sources of nitrate to groundwater, *CRC Crit. Rev. Environ. Control,,* 16(3), 257, 1986.
181. Powers, J.F. and Schepers, J.S., Nitrate contamination of groundwater in North America, in *Effects of Agriculture on Groundwater,* Bouwer, H. and Bowman, R.S., Eds., *Agric. Ecosys. Environ.,* 26, 165, 1989.
182. Strebel, O., Duynisveld, W.H.M., and Bottcher, J., Nitrate pollution of groundwater in Western Europe, in *Effects of Agriculture on Groundwater,* Bouwer, H. and Bowman, R.S., Eds., *Agric. Ecosys. Environ.,* 26, 189, 1989.
183. Juergens-Gschwing, S., Ground water nitrate in other developed countries (Europe) — relationships to land use patterns, in *Nitrogen Management and Groundwater Protection,* Follett, R.F., Ed., Elsevier, Amsterdam, 1989, 75.
184. Nebraska Dept. of Health, Domestic well water sampling in central Nebraska: Laboratory findings and their implications, Dept. of Health, Lincoln, NE, 1985.
185. Steichen, J., Koelliker, J., and Grosh, D., Kansas farmstead well survey for contamination by pesticides and volatile organics, *Proc. Agric. Impacts Ground Water Conf.,* Natl. Water Assoc., Dublin, OH, 1986, 530.
186. Snethen, D. and Robbins, V., Farmstead well contamination study, Kansas State Dept. Health Environ., Bureau Water Protection, Topeka, KS, 1987.
187. LeMasters, G. and Doyle, D.J., Grade A dairy farm well water quality survey, Wisconsin Dept. Agric. Trade Consumer Protection and Wisconsin Agric. Stats. Serv., Madison, WI, 1989.
188. Walker, R., Nitrates, nitrites and N-nitrosocompounds: a review of the occurrence in food and diet and the toxicological implications, *Food Addit. Contamination,* 7(6), 717, 1990.
189. ECETOC, *Nitrate and Drinking Water,* Ecetoc Technical Report No. 27, European Chemical Industry Ecology and Toxicology Centre, Brussels, 1988.
190. Ellenhorn, M.J. and Barceloux, D.G., Airborne toxins, in *Medical Toxicology — Diagnosis and Treatment of Human Poisoning,* Elsevier, New York, 1988, 845.
191. Forsyth, R.J. and Moulden, A., Methemoglobinemia after ingestion of amyl nitrite, *Arch. Dis. Child,,* 66(1), 152, 1991.

192. Goldfrank, L.R., Price, D., and Kirstein, R.H., Nitroglycerin (Methemoglobinemia), in *Goldfrank's Toxicologic Emergencies,* 4th ed., Goldfrank, L.R., Flomenbaum, N.E., Lewin, N.A., et al., Eds., Appleton and Lange, Norwalk, 1990, 391.
193. Linden, C.H., Volatile substances of abuse, *Emerg. Med. Clin. North Am.,* 8(3), 559, 1990.
194. Linares, L.A., Peretz, T.Y., and Chin, J., Methemoglobinemia induced by topical anesthetic (benzocaine), *Radiother. Oncol.,* 18(3), 267, 1990.
195. Grum, D.F. and Rice, T.W., Methemoglobinemia from topical benzocaine, *Cleveland Clin. J. Med.,* 57(4), 357, 1990.
196. Gowans, W.J., Fatal methemoglobinemia in a dental nurse. A case of sodium nitrite poisoning, *Br. J. Gen. Pract.,* 40(340), 470, 1990.
197. Challoner, K.R. and McCarron, M.M., Ammonium nitrate cold pack ingestion, *J. Emerg. Med.,* 6, 289, 1988.
198. Hoffman, R.S. and Sauter, D., Methemoglobinemia resulting from smoke inhalation, *Vet. Hum. Toxicol.,* 31(2), 168, 1989.
199. Sporer, K.A. and Mayer, A.P., Saltpeter ingestion, *Am. J. Emerg. Med.,* 9(2), 164, 1991.
200. Kaplan, A., Smioth, C., Promnitz, D.A., et al., Methaemoglobinaemia due to accidental sodium nitrite poisoning. Report of 10 cases, *S. Afr. Med. J.,* 77(6), 300, 1990.
201. Hartman, P.E., Nitrates and nitrites: ingestion, pharmacodynamics and toxicology, in *Chemical Mutagens—Principles and Methods for Their Detection,* de Serres, F.J. and Hollaender, A., Eds., Plenum Press, New York, 7, 211, 1982.
202. Bartholemew, B. and Hill, M.J., The pharmacology of dietary nitrate and the origin of urinary nitrate, *Food Chem. Toxicol.,* 22, 789, 1984.
203. U.S. EPA, *Health Effects Criteria Document for Nitrate/Nitrite,* Criteria and Standards Division. Office of Drinking Water, U.S. Environmental Protection Agency, Washington, D.C., 1985.
204. Spiegelhalder, B., Eisenbrand, G., and Preussman, R., Influence of dietary nitrate on nitrite content of human saliva: possible relevance to *in vivo* formation of N-nitroso compounds, *Food Cosmet. Toxicol.,* 14, 545, 1976.
205. Tannenbaum, S.R., Weisman, K., and Fett, D., The effect of nitrate intake on nitrite formation in human saliva, *Food Cosmet. Toxicol.,* 14, 549, 1976.
206. Parks, N.J., Krohn, K.A., Mathis, C.A., et al., Nitrogen-13-labeled nitrite and nitrate: distribution and metabolism after intratracheal administration, *Science,* 212, 58, 1981.
207. Anonymous, Nitrate: rerun of an old horror, *Health Environ. Dig.,* 1(12), 1, 1988.
208. Weisenburger, D.D., Joshi, S.S., Hickman, T.I., Walker, B.A., and Lawson, T.A., Mutagenesis tests of atrizine and nitrosoatrizine: compounds of special interest to the Midwest, *Proc. AACR,* 28, 103, 1988.
209. Donovan, J.W., Nitrates, nitrites, and other sources of methemoglobinemia, in *Clinical Management of Poisoning and Drug Overdose,* 2nd ed., Haddad, L.M. and Winchester, J.F., Eds., W.B. Saunders, Philadelphia, 1990, 1419.
210. Caudill, L., Walbridge, J., and Kuhn, G., Methemoglobinemia as a cause of coma, *Ann. Emerg. Med.,* 19(6), 677, 1990.
211. Hafsia, R., Meddeb, B., Mtimet, B., et al., Congenital cyanosis due to methemoglobin reductase deficiency: first reported Tunisian case, *Nouv. Rev. Fr. Hematol.,* 31(5), 371, 1989.
212. Prchal, J.T., Borgese, N., Moore, N.R., et al., Congenital methemoglobinemia due to methemoglobin reductase deficiency in two unrelated American black families, *Am. J. Med.,* 89(4), 516, 1990.

213. Wimberley, P.D., Nielsen, I.M., Blanke, S., and Olessen, H., Hematoglobinopathies in Danish families, *Scand. J. Clin. Lab. Invest. Suppl.,* 194, 45, 1989.
214. Office of Technology Assessment, *The Role of Genetic Testing in the Prevention of Occupational Disease,* OTA Congress of The United States (Library of Congress Catalog #83-600526), Washington, D.C., 1983, 90.
215. Harris, J.W. and Kellermeyer, R.W., Red cell metabolism and methemoglobinemia, in *The Red Cell — Production, Metabolism, Destruction: Normal and Abnormal,* Rev. ed., Harvard University Press, Cambridge, MA, 1970, 447.
216. Dabney, B.J., Zelarney, P.T., and Hall, A.H., Evaluation and treatment of patients exposed to systemic asphyxiants, *Emerg. Care Q.,* 6(3), 65, 1990.
217. Fan, A.M., Willhite, C.C., and Book, S.A., Evaluation of the nitrate drinking water standard with reference to infant methemoglobinemia and potential reproductive toxicity, *Regul. Toxicol. Pharmacol.,* 7(2), 135, 1987.
218. U.S. EPA, *Nitrate/Nitrite,* Health Advisory. Office of Drinking Water, U.S. Environmental Protection Agency, Washington, D.C., 1987.
219. Dusdeiker, L.B., Stumbo, P., Kross, B.C., Murph, W., and Dungy, C., Abstract: Are nitrates concentrated in human milk, *Am. J. Dis. Child,* 146, 501, 1992.
220. Dreosti, I.E., McMichael, A.J., and Bridle, T.M., Mount Gambier drinking water and birth defects, *Med. J. Austr.,* 141, 409, 1984.
221. Dorsch, M.M., Sragg, R.K.R., McMichael, A.J., et al., Congenital malformations and maternal drinking water supply in rural South Australia: a case-control study, *Am. J. Epidermiol.,* 119(4), 473, 1984.
222. Arbuckle, T.E., Gregory, G.J.S., Corey, P.N., et al., Water nitrates and CNS birth defects: a population-based case-study, *Arch. Environ. Health,* 43(2), 162, 1988.
223. Weisenburger, D.D., Environmental epidemiology of Non-Hodgkin's lymphoma in Eastern Nebraska, *Am. J. Ind. Med.,* 18(3), 303, 1990.
224. Fraser, P., Chilvers, C., Beral, V., and Hill, M.J., Nitrate and human cancer: A review of the evidence, *Intl. J. Epidemiol.,* 9(1), 3, 1980.
225. National Academy of Sciences, *The Health Effects of Nitrite, Nitrate and N-Nitroso Compounds,* National Academy Press, Washington, D.C., 1981.
226. Lijinsky, W., Induction of tumors in rats by feeding nitrosatable amines together with sodium nitrite, *Food Chem. Toxicol.,* 22, 715, 1984.
227. Shank, R.C. and Newberne, P.M., Dose-response study of the carcinogenicity of dietary sodium nitrite and morpholine in rats. *Food Cosmet. Toxicol.,* 14, 1, 1976.
228. Greenblatt, M. and Mirvish, S.S., Dose-response studies with concurrent administration of piperazine and sodium nitrite to strain A mice, *J. Natl. Cancer Inst.,* 50, 119, 1972.
229. Andrews, A.W., Fornwald, J.A., and Lijinsky, W., Nitrosation and mutagenicity of some amine drugs, *Toxicol. Appl. Pharmacol.,* 52, 237, 1980.
230. Knight, T.M., Forman, D., Pirastu, R., et al., Nitrate and nitrite exposure in Italian populations with different gastric cancer rates, *Int. J. Epidemiol.,* 19(3), 510, 1990.
231. Whong, W.Z., Speciner, N.D., and Edwards, G.S., Mutagenicity detection of *in vivo* nitrosation of dimethylamine by nitrite, *Environ. Mutagenesis,* 1, 277, 1979.
232. Couch, D.B. and Friedman, M.A., Interactive mutagenicity of sodium nitrite, dimethylamine, methylurea and ethylurea, *Mutat. Res.,* 31, 109, 1975.
233. Golterman, L.K. and Maaswinkel-Mooy, P.D., An infant with hereditary methemoglobinemia, *Tijdschr. Kindergeneeskd.,* 57(2), 60, 1989.
234. Fialkow, P.J., Browder, J.A., Sparkes, R.S., and Motulsky, A.G., Mental retardation in methemoglobinemia due to diaphorase deficiency, *N. Engl. J. Med.,* 273, 840, 1965.

235. Junien, C., Leroux, A., Lostanleu, D., et al., Prenatal diagnostic of congenital enzymopoenic methemoglobinemia with mental retardation due to generalized cytochrome b5 reductase deficiency: first report of two cases, *Prenat. Diagn.*, 1(1), 17, 1981.
236. National Academy of Sciences, *Nitrates: An Environmental Assessment,* National Academy of Sciences, Washington, D.C., 1978.
237. Smith, M.A., Shah, N.R., Lobel, J.S., and Hamilton, W., Methemoglobinemia and hemolytic anemia associated with *Campylobacter jejuni* enteritis, *Am. J. Pediatr. Hematol. Oncol.,* 10(1), 35, 1988.
238. Hegesh, E. and Shiloah, J., Blood nitrates and infantile methemoglobinemia, *Clin. Chim. Acta,* 125, 107, 1982.
239. Dagan, R., Zaltzstein, E., and Gorodischer, R., Methaemoglobinaemia in young infants with diarrhea, *Eur. J. Pediatr.,* 147, 87, 1988.
240. Wettig, K., Schulz, K.R., Scheibe, J., et al., Endogenous nitrate synthesis in selected infectious diseases and ulcerative colitis, *Neoplasm,* 38(3), 337, 1991.
241. Hesseling, P.B., Toens, P.D., and Visser H., An epidemiological survey to assess the effect of well-water nitrates on infant health at Rietfontein in the northern Cape Province, South Africa, *S. Afr. J. Sci.,* 87, 300, 1991.
242. WHO, *Health Hazards from Nitrates in Drinking-Water.* Report on a WHO Meeting, March 5–9, 1984, World Health Organization, Copenhagen, 1985.
243. Green, L.C., Wagner, D.A., Glogowski, J., et al., Analysis of nitrate, nitrite, and [^{15}N]Nitrate in biological fluids, *Anal. Biochem.,* 126, 131, 1982.
244. Cortas, N.K. and Wakid, N.W., Determination of inorganic nitrate in serum and urine by kinetic Cadmium-Reduction method, *Clin. Chem.,* 36(8), 1440, 1990.
245. Lippsmeyer, B.C., Tracy, M.L., and Moller, G., Ion-exchange liquid chromatographic determination of nitrate and nitrite in biological fluids, *J. Assoc. Off. Anal. Chem.,* 73(3), 457, 1990.
246. Dunphy, M.J., Goble, D.D., and Smith, D.J., Nitrate analysis by capillary gas chromatography, *Anal. Biochem.,* 184, 381, 1990.
247. Schneider, N.R. and Yeary, R.A., Measurement of nitrite and nitrate, *Am. J. Vet. Res.,* 34, 133, 1973.
248. Hegesh, E., Gruener, N., Cohen, S., et al., A sensitive method for the determination of methemoglobin in blood, *Clin. Chim. Acta,* 30, 679, 1970.
249. Hoffman, R.S. and Garay, S.M., Pulmonary principles, in *Goldfrank's Toxicologic Emergencies,* 4th ed., Goldfrank, L.R., Flomenbaum, N.E., Lewin, N.A., et al., Eds., Appleton and Lange, Norwalk, 1990, 141.
250. Epstein, S., Hypersensitivity to sodium nitrate: a major causative factor in case of palindromic rheumation, *Ann. Allergy,* 27, 343, 1969.
251. Panush, R.S., Food induced ("allergic") arthritis: clinical and serologic studies, *J. Rheumatol.,* 17, 291, 1990.
252. Heindell, J., personal communication, 1992.
253. Dieter, M.P., Jameson, C.W., French, J.E., et al., Development and validation of a cellular transplant model for leukemia in Fischer rats: a short-term assay for potential anti-leukemic chemicals, *Leukemia Res.,* 13(9), 841, 1989.
254. Dieter, M.P., personal communication, 1992.

CHAPTER 7

Ameliorating Effects of Alternative Agriculture

W. D. Pitman

TABLE OF CONTENTS

0-87371-860-7/95/$0.00+$.50
© 1995 by CRC Press, Inc.

I. INTRODUCTION

Two distinct assumptions are evident from the title of this chapter. The first is that improvements can be made from what is, or has been, considered the normal or standard approach in agricultural production. The second is that there are viable alternatives. There are limitations yet to be overcome in the use of the available land, biological resources, soil amendments, and agricultural chemicals despite the tremendous success of agriculture in the U.S. in meeting the food and fiber needs of this nation and beyond. Production choices or alternatives are continually encountered in agricultural enterprises.

The word "alternative" often brings to mind some very specific concepts in relation to agriculture. While some of the common concepts and practices may characterize the available options in some situations, the alternatives now available and rapidly being developed for enhancement of agricultural production extend substantially beyond the aspects of earlier conservation efforts or present concepts of organic farming. Even a cursory survey of the current literature reveals continuing improvements in use of the common conservation farming methods for specific situations along with a great diversity of novel concepts representing potential alternatives for specific agricultural practices.

II. EVOLVING AGRICULTURAL PHILOSOPHY

"Alternative agriculture" and the commonly interchanged term "sustainable agriculture" are often used in reference to a philosophy of agricultural production rather than a set of practices or particular method of farming. The prevailing philosophy or paradigm of agricultural production has evolved throughout the history of the U.S. and in most other nations as well. Upon initial settlement of the vast area of the U.S., the prevailing paradigm of agriculture was characterized by the idea of conquering the wilderness and taming the frontier. Nutrients released from storage in the large standing biomass of forests initially and, subsequently, in the dense grasslands of the prairie were exploited for short-term crop production with a seemingly endless supply of new land for future exploitation. The fact that the supply of new land was not unlimited was only gradually recognized and acknowledged. The depletion of nutrients by and susceptibility to erosion of the prevailing farming practices at the time were not immediately appreciated nor remedied.

By the mid-1800s, loss of agricultural productivity was recognized as an economic constraint in the eastern states. While there were still extensive frontier areas for development farther west, the search for improved farming methods and especially restoration of soil fertility became a major agricultural concern in the eastern states.[1] Even as the adverse consequences of overly exploitative methods of farming were becoming widely recognized, a number of trends developed shortly after World War I which delayed acceptance and general adoption of conservation practices. The availability and adoption of farm machinery, increases in farm size, reduction in farm population and consequent increase in productivity required per farm worker, and the emergence of federal credit for agriculture[2] increased the scale of farm operations at the expense of attention to detail.

The drastic environmental consequences of the Dust Bowl experience in the Great Plains in the 1930s became the compelling force which propelled a shift in agricultural paradigms. The frontier had been conquered and the new paradigm was based on the need for conservation and the wise use of natural resources in agricultural production. Such practices as contour farming, crop rotations involving high residue crops and soil-building green manure crops, strip cropping, planting of perennial grasses on highly erodible lands, etc. became common approaches of the leading farmers.

As apparently unlimited opportunities for agricultural development initially delayed recognition of limitations involved with early methods of farming and new opportunities for increased scale of operations delayed general acceptance of conservation measures just before the dry 1930s, circumstances soon overwhelmed emphasis on the conservation philosophy. World War II brought a short period of demand for maximum production. This was followed by years of continuing emphasis on production efficiency as agricultural production exceeded demand for food.[3] Scientific advances provided genetically improved plants and animals, chemicals for soil amendments and pest control, larger and more efficient machinery, and numerous other contributions. Reduced vulnerability to damage by plant pests was provided by plant genetic advances and pesticides, increased production potential was provided by additional plant genetic advances and soil amendments, and mechanical advances allowed improvements in planting, cultivating, and harvesting. Dramatic increases in crop productivity and product quality, along with reduced labor requirements, led to rapid acceptance of these developments by agricultural producers. Increased quantities, variety, and quality of food items at generally decreasing cost relative to total income led to endorsement of agricultural developments by the consuming public.

While the idea was prevalent that agricultural productivity had to be continually increased to keep pace with population increases, surpluses of many agricultural products characterized the period following World War II.[3,4] The prevailing philosophy of agriculture was one of economic survival. The conservation of resources was relegated to the status of a concession of necessity rather than an esteemed philosophy. Economic survival remains the paradigm of modern agriculture in the U.S. Scientific advances which allow increased production efficiency and enhanced profit potential with minimal risk are quickly adopted. Most agricultural enterprises respond to a market economy with a rather narrow margin of profit. Often government agricultural programs have been keys to profit potential and production decisions. These programs along with federal crop insurance programs and emergency disaster relief payments have served to reduce the need for diversified agricultural operations, thus leading to highly specialized, sometimes single-crop, enterprises. Economics dictate to a substantial degree the alternatives chosen in agriculture with government programs often accentuating the resulting trends. The paradigm of economic survival prevails in agriculture, as in other modern industries, simply because those who base decisions on other criteria typically do not remain in the business for long. Agriculture in the U.S. has evolved to a large extent from essentially "a way of life" characterized by independence and self-reliance to an economic industry where profit is essential for existence and public policy increasingly influences profit potential.

III. THE NEED FOR ALTERNATIVES

A. OVERVIEW

Alternatives are desirable when present approaches fall short of the goal. The following goal of agriculture in a national context was stated in the 1948 Yearbook of Agriculture:[5] "Our goal is permanency in agriculture — an agriculture that is stable and secure for farm and farmer, consistent in prices and earnings; an agriculture that can satisfy indefinitely all our needs of food, fiber, and shelter in keeping with the living standards we set. Everybody has a stake in a permanent agriculture." While recent concepts expand the goal to include effects beyond the farm, "sustainable agriculture" and "a permanent agriculture" represent a continuing concept. Certainly, agriculture must be responsible for its effects beyond the farm to provide sustainability of the contingent environment as well.

Although the specific topic of this book is soil amendments, alternative agriculture involves the full range of activities involved in the production of food and fiber from the land. Therefore, this chapter will include aspects of the broader area of the need for and availability of alternatives in agricultural production.

There appears to be a growing public perception that the simple life on the family farm with its characteristic concern for the land and high value system has been replaced by an impersonal, profit-oriented, corporate agricultural industry which only accepts responsibility for its social and environmental effects to the extent required by law and its active enforcement. However, agriculture in the U.S. developed and has evolved with the production of essential goods, resulting in a rather high degree of environmental modification. Depleted, eroded farmland is more generally a characteristic of past agriculture than current approaches. Sustained, and in many cases continually increasing, yields indicate that the conventional agriculture of the past 30 years is not a completely unsustainable agriculture. Sustained yields do, however, generally depend on regular inputs of mineral fertilizers, often repeated applications of various pesticides, continuing genetic improvement of crop cultivars, and sometimes even the routine mechanical modification of land scarred by recurring soil erosion.

Several aspects of this characteristically high-input, conventional agriculture demand scrutiny from a society becoming increasingly aware of the environmental consequences of various human activities. Nonrenewable resources are extensively used to provide fuel and fertilizer. Short-term economics, rather than long-term availability of these resources or even projected immediate need for the agricultural products, determine use of these resources. Despite increasing awareness of limitations in water supply, the area under irrigation in the U.S. has more than doubled since the 1940s. Profitability, especially with such high-value crops as fruits and vegetables, is the major reason for this increase.[6] Effects of irrigation on both amount and quality of water are locally important issues with concerns ranging from depletion of supplies to increased salinity of water and soil and salt-water intrusion of coastal aquifers. The movement of soil and associated soil amendments and agricultural chemicals from agricultural lands due to wind and water remains widespread despite continuing efforts from individual land owners and operators and government soil and water conservation programs for more than 50 years.

B. SOIL EROSION

Bentley[7] indicated that almost a billion dollars a year have been spent on soil conservation efforts in this country since 1940. While scale of operation, short-term profitability, management requirements, and other immediate concerns are contributing factors to the failure to further control soil erosion, the predominant factors may well be that the effects on productivity are often not immediately obvious and the degree of off-site damage by soil particles from such a diffuse source is difficult to associate with an individual field or farm. Quantification of the adverse effects of soil erosion is essentially impossible. Although government agencies often attempt to estimate levels and extent of soil erosion and associated sedimentation, such estimates are generalized and subjective. Nationwide surveys of soil erosion in the U.S. conducted in 1977 and 1982 by the Soil Conservation Service provide an overview of soil erosion in this country.[8,9] Soil erosion is a continuing problem primarily on cropland, with the highest erosion rates reported for the Appalachian states and the Corn Belt.[8] Damaging erosion can be so subtle as to consist of only the redistribution of 2 or 3 cm of topsoil over short distances within a field immediately after crops have been planted. Resulting uneven plant emergence and poor stands of precision-planted crops can make the difference in ultimate profit or loss. Even more obvious erosion resulting in removal of topsoil from sloping areas and deposition on bottom lands can affect only individual farms or even fields. Such erosion would certainly not be detected by monitoring the suspended sediments of major water systems. It is also difficult to distinguish between natural levels of erosion and accelerated erosion due to agricultural activity in some situations. Tolerance levels, based on sustaining crop yields, are projected for purposes of monitoring and evaluating soil erosion.[10]

In 1939, it was estimated that 40,500,000 of the 167,000,000 ha used for crop production in the U.S. had been so severely damaged by erosion that they were no longer suitable for crop production.[11] Loss of topsoil by sheet erosion and wind damage and extensive gully development on some of the most productive farm lands seemed to be recognized only after the fact. Despite an apparent public commitment to remedy the situation and substantial government programs on soil conservation over the following 30 years, it was suggested in 1970 that the Mississippi River carried 390 million Mg per year of soil to the Gulf of Mexico.[12] This was described as equivalent to all topsoil on 192,000 ha of land. Damage from sediment throughout the nation in 1970 was estimated at $500,000,000.[12] Serious erosion rates, based on sediment removal exceeding the threshold, were reported to occur on 400,000 km^2 or 15% of the area of the 11 western states in 1975.[13] Associated mineral contamination of surface waters added 80 to 90 million Mg of salt to substantially degrade water quality. Nationwide in 1975, almost 2.7 billion Mg of soil were lost from cropland,[14] an average of 16 Mg per hectare. This was suggested as almost double the acceptable rate of soil loss, with most of the cropland in the eastern two thirds of the country exceeding the tolerable limits. From monitorings of the nation's rivers from 1974 to 1981, suspended sediment concentrations increased in river basins predominated by cropland.[15]

Description and characterization of actual or potential erosion under specific conditions can also be of value in assessing the problem. In 1981, Olson[16] suggested, "Soil erosion is probably the most destructive process that acts to reduce production from the land". A recent assessment[6] of water quality determined that, "Water pollution may be the most damaging and widespread adverse environmental effect of agricultural production". A 1989 report from the National Research Council[17] identified agriculture as the nation's largest source of nonpoint water pollution. Pollution resulting from erosion of agricultural lands involves sediments, nitrate, phosphate, and some agricultural pesticides.[6] The previously mentioned monitoring of the nation's rivers[15] indicated that total phosphorus concentrations were closely associated with suspended sediments. Losses of fertilizer phosphorus apparently increased as cropland soil erosion increased. Nitrate concentrations in the nation's rivers increased over the 1974 to 1981 sampling period, except in the western portion of the country. These nitrate increases appeared to be associated with fertilization and livestock concentration. Recent information[18] indicates that, at least in some situations, farming methods and systems in addition to fertilization practices may be closely associated with levels of nitrate loss. Although predominantly carried in runoff water, pollutants from soil erosion can also be transported by wind. Wind erosion in the Red River Valley of North Dakota has recently been suggested as a potential source of nitrate and salinity for surface and groundwater.[19]

Occurrence of soil erosion is highly event-driven with heavy spring rains immediately following seedbed preparation being particularly damaging. While chemical components of soil erosion and excessive runoff such as nitrates, phosphorus, and pesticides will generally be immediately evident only upon appropriate chemical analyses, movement of soil particles and resulting sedimentation are particularly widespread and noticeable. Rural roadsides, farm ponds, streams, lakes, rivers, and flood control structures are often recipients of large quantities of sediment from eroding agricultural lands. The effects of sediments are highly complex and extend from structural damage of water bodies to drastic alteration of aquatic communities.[20] While accumulations of water-transported soil particles are readily evident, all movement of soil on agricultural lands by wind and water is not due to inappropriate management. Soil erosion or the movement of soil particles by wind and water is a natural process and is even a soil-building process. However, detrimental acceleration of this process as a consequence of agricultural activity must be acknowledged.

In contrast to the consequences of pollution from nutrients and pesticides, effects of soil erosion and sedimentation on the most highly developed human societies of history have been studied and documented.[11,16] Ancient societies in both the Middle East and the Americas developed major cities whose populations were supported by well-developed agriculture and even intricate irrigation systems. These highly developed societies typically exploited soil resources for short-term benefits, with a decline of the society corresponding with decreased agricultural productivity of eroded soils and sedimentation of irrigation systems, city water reservoirs, and other structures. Certainly their experiences, if not those of our own nation, graphically illustrate the consequences to society of failure to recognize and appropriately modify activities which adversely affect the basic resources supporting the society.

C. ADVERSE EFFECTS OF AGRICULTURAL CHEMICALS

In addition to the adverse effects of soil erosion, which can be anticipated and to some extent predicted from prevailing agricultural practices, chemical modification of the environment is also a recognized hazard of present agriculture.

Effects of agricultural chemicals are much less predictable. Information regarding the extent and degree of chemical contamination, which is generally of low or even negligible degree but rather wide extent,[6,20,21] is from recent investigations. Consequences of the reported levels of chemicals in the environment are, perhaps, minimal but not definitely known. Nitrates and phosphorus in runoff water can contribute to accelerated eutrophication of surface waters. Nitrates and pesticides in water supplies can present hazards to human health as well as that of other species. While some such aspects of nutrient enrichments and chemical contaminants are understood, uncertainties regarding many aspects of chemical contamination prevail. Although the direct effects of pesticides are typically rather thoroughly evaluated prior to approval for use under current procedures, not all potential circumstances, effects, and interactions can be anticipated. The various agricultural enterprises with associated soils, chemical treatments, varying weather conditions, and management factors provide a tremendous diversity of potential hazards. Characterization of these hazards in terms of actual risk is currently progressing.[6,22,23] Some aspects of the use of chemicals in agriculture are simply not known, as opposed to those which can be predicted with some degree of probability based on an existing information base. Strategies for dealing with this true uncertainty are also being proposed.[24]

Unintended effects of agricultural chemicals and soil amendments can be as dramatic and far-reaching as those of DDT[25] or so subtle as to be indistinguishable from natural processes. Improved analytical methods and increased instances of sampling or monitoring have revealed that agricultural chemicals occur in waters, wetlands, and surface-water sediments.[6,20] While isolated instances of contamination are often due to accidents or inappropriate use of a product, extensive detection indicates the need for evaluation of utilization procedures. The most alarming current detections are of the persistent pesticides which are no longer used but remain in the environment from past use. Some currently used insecticides, although not possessing the combination of persistence and accumulation progressively up the food chain of DDT, are toxic to some aquatic organisms such as species of Crustacea.[20] Herbicides moved off-site to surface waters may reduce primary productivity through toxicity to phytoplankton.[20] Although not extensively used now, arsenate-containing pesticides can add to the low natural concentrations of this element in localized environments. Pesticide properties which present hazards include mobility, persistence, and volatility (especially when subject to codistillation), with the latter property allowing redepositing of a pesticide in fog or rain to nontarget areas.[26] Nitrate has been labeled as the most widely occurring chemical contaminant of the world's aquifers.[27] In the U.S., incidence of groundwater exceeding the maximum contaminant level for nitrate-N in drinking water of 10 mg/L has primarily been associated with intensively managed, irrigated cropland on well-drained soils in the western portion of the country and localized areas where high rates of manure or commercial fertilizer have been applied to well-drained soils in the eastern states.[27]

D. UNINTENDED EFFECTS OF AGRICULTURAL PRACTICES ON ECOSYSTEM FUNCTIONS

Disturbance of existing systems through either major disruptions or only minor alterations affects the interactions of organisms and life processes throughout the natural community. Agricultural practices often are applied for a specific effect on a target species without an understanding of other effects on the environment. In the process of accelerated eutrophication mentioned previously, nutrients targeted for enhancement of crop growth can be transported by runoff to lakes and reservoirs. In these bodies of water, complex aquatic communities exist in a dynamic, interactive balance. Nutrient limitations normally restrict populations of algae in fresh water. Additions of large quantities of the limiting nutrient, which is often phosphorus but can also be nitrogen in some situations, may lead to bursts of growth or blooms by populations of individual algal species. Such blooms may restrict light penetration to adversely affect plants at lower strata in the water. Oxygen depletion and death of fish are frequent effects of excessive nutrient enrichment of surface waters. Such polluted waters are greatly diminished in value for recreational use, municipal water supplies, and most other uses. More subtle effects of nutrient enrichment can alter aquatic communities at lower levels of contamination.[6] Actual extent of occurrence of such damage to fresh water and especially determination of specific causes among many potentially contributing factors are not well documented.

Numerous other ecological aspects of agricultural practices are potentially detrimental to various species, natural biological systems, and even the global environment. Release of carbon into the atmosphere from stored forms such as forests which have been cleared, soil organic matter which is often depleted, and petroleum which is processed into fuel and fertilizer contribute to some extent to the concern over global warming. Extensive production of a limited number of crops over large areas with clean cultivation, large fields, and chemical control of weeds, insects, and plant pathogens characterize some of the most productive agricultural enterprises. A very restricted environment is produced by these practices. Elimination of essentially all noncrop plants is often desired to prevent competition with crops. Lack of plant diversity, even along field edges or borders, and pesticide effects on nontarget insect, microbial, and other species result in very limited food supplies and habitats for the various animal populations which previously occupied these areas. The various species of these complex ecosystems include many that are beneficial to agriculture.

Nitrogen-fixing soil bacteria and natural predators of agricultural pests are, perhaps, the most obvious. Tillage and cropping systems in such enterprises have often contributed to depletion of soil organic matter. At the extreme, complex interactive biological systems have been reduced to allow maximum control of living organisms by mechanical and chemical treatments. Other factors, such as interactions among remaining noncrop species and natural chemical and physical processes including those affecting resource loss (nitrification, infiltration, leaching, erosion, etc.), may not bc adequately considered. Such systems sacrifice the potential benefits of many available biological processes, which are apparently assumed to be inconsequential in relation to the merits of optimizing control of the production environment.

Loss of buffering capacity in agricultural systems can occur within individual crops as well as for the system as a whole. Vulnerability of crops increases with extensive planting of monocultures of the same crop and can be even greater with extensive use of a superior cultivar or germplasm source used to produce hybrid cultivars. Vigilance in this respect is a continuing requirement with intensive crop production of a small number of genotypes. Pathogen populations are typically so dynamic that highly virulent forms of some species can develop and reach epidemic proportions in less time than resistant cultivars can be developed. Such an experience with the Southern Corn Leaf Blight in 1970[28] had nationwide repercussions and devastating effects on some individual farmers and entire communities. A less extensive, but locally critical epidemic of Northern Corn Leaf Blight occurred across Texas in 1992 due to extensive use of susceptible hybrids and favorable weather for the disease.[29]

Along with the sacrificed potential benefits directly to crop production from various components of a more diverse environment, other sacrifices are made with consequences beyond agricultural production. Populations of individual wildlife species may be greatly diminished. Low soil organic matter levels can limit infiltration rates and water storage capacity, increasing potential for soil erosion and downstream flooding. These soil conditions can also reduce moisture and nutrient storage and increase vulnerability to drought, while lack of ground cover after crop harvest predisposes extensive areas to accelerated rates of soil erosion. Limitations to natural predators of crop pests produce an enhanced dependence on pesticides and a vulnerability if such applications are not appropriately made. Thus, the increased control over many agricultural systems sacrifices the potential benefits of diverse natural interactions among species and enhances risks of disaster due to reduced buffering of the system.

The complexity of interactions involved in manipulation of biological systems is well illustrated by responses to pesticides where invertebrate pests and their natural enemies are involved. Resurgence is a phenomenon where pest populations decrease immediately after pesticide application but subsequently build to greater levels than those initially present.[30] For pests which are relatively sessile and protected by waxy coverings such as scale insects or those protected by leaf cuticle such as leaf miners, Waage[30] suggested that the natural enemies may suffer substantially greater mortality from pesticide applications than do these pests. Population resurgence may lead to an economic loss from such pests even when they were initially below the economic threshold and not the target pests of the original pesticide treatment. Pesticide control can be sufficiently effective to locally eliminate a specific natural enemy due to an inadequate supply of the specific susceptible stage of the pest for a critical period. Possible interactions among pests, pesticides, and natural enemies include resurgence of the pest, similar effects of the pesticide on pest and enemy for no net response, additive effects of natural enemy and pesticide on the pest, and even synergistic effects of the two limiting factors on the pest population. Selective pesticides and appropriate timing and rate of application of broad-spectrum pesticides can be critical to manipulation of interacting populations of invertebrate species. While information for such specific application of pesticides is not available for many situations, recognition and acknowledgment of possible effects beyond those intended are essential steps for improvement. Development of pesticide-resistant pest populations has become an increasingly common response to repeated applications of a pesticide which must be considered in development of pest control programs.

The control of crop growth sought through stringent cultural and chemical treatments in many agricultural systems comes at the expense of potentially beneficial interactions among naturally occurring species and processes. Loss of buffering capacity of these natural components of the ecosystem can increase susceptibility to disaster. Regular application of mineral fertilizers can sustain crop production on many soils, at least for some period, even with continuing soil erosion. Available tillage equipment can readily mask the effects of substantial soil erosion. Continuing occurrence of soil erosion, sedimentation, nutrient enrichment of surface waters, leaching of nitrates to groundwater, movement of pesticide residues to surface and groundwater, and an awareness of the complex biological interactions involved dictate that undesirable effects of agricultural production must be acknowledged as valid concerns of society. Regardless of the degree of immediate danger or urgency of the situation, alternatives which contribute to a sustainable environment and allow acceptable levels of production and profit are more appropriate and responsible than are those prone to degradation of the environment.

IV. ALTERNATIVES NOW AVAILABLE

A. TRADITIONAL SOIL CONSERVATION MEASURES

Despite the continuing, pervasive problem of soil erosion in the U.S., the technology to greatly reduce levels of soil loss are now available. In fact, extensive areas which were reported to be essentially destroyed by soil erosion in the 1930s[11] are now productive agricultural lands. Adequate protection for most erosion-prone soils can be readily provided by appropriately managed permanent grass cover. In the southeastern portion of the country, Bermuda grass (*Cynodon dactylon* L.) has been widely used to convert unstable, eroded farmland into productive pasture and hayland. Thus, the initial alternative or choice required for development of a sustainable agriculture is that of land use. An extensive information base including soil maps, land capability classification, identification of factors limiting particular land uses, and technical expertise to interpret and assist in application of appropriate soil conservation measures are available to agricultural producers, without charge, throughout the U.S. A number of options are available for most soils and agricultural enterprises to permit development of productive farming systems which are not predisposed to excessive rates of soil erosion.

An extension of the idea of proper land use to identification of appropriate crops for particular soils or land areas can also contribute to erosion control. Just as some erosion-prone lands are best suited for use under permanent grass cover, other land areas may be suitable for cropland use only when close-grown crops such as small grains are the primary crop. Other areas may be suitable for row crops as long as sufficient amounts of crop residue are produced and effectively used to protect the soil from erosion. This may require crop rotations including some high residue-producing crops, the regular use of cover crops following cash crops, or extended periods for growth of sod-forming grass crops. Thus, land use, cropping systems, and crop residue management are keys to erosion control. Contour farming, especially in combination with other practices, can contribute to reduced rates of runoff and less vulnerability to erosion in some situations. Strip cropping has been an effective means of using close-grown crops interspersed with row crops in contour farming patterns to reduce the adverse effects of clean-tilled land over long slopes.

A recent national initiative to increase awareness of the soil conservation benefits of crop residues emphasizes the merits of this option.[31] Much of the recent research involving management of crop residue involves conservation tillage systems,[32–34] which are discussed later. Aspects of crop residue management from much of this research are not necessarily limited to specific tillage systems. Protection of plant pathogens is one of the detrimental effects of surface residue. Increased incidence of disease with surface residue has not necessarily resulted in reduced yields in all cases, as illustrated with spring barley (*Hordeum vulgare* L.) in Oregon.[34]

Interactions of residue with plant nutrients illustrate the complexity of effective residue management. Decomposition of legume residue was reported to be greater and phosphorus loss was more rapid when residues were buried than when left on the soil surface.[32] Nitrogen loss from crop residue was only minimally less on the surface than when buried.[35] Timing nitrogen fertilizer application with crop needs and residue mineralization can enhance effectiveness of legume residues.[36] Placement of fertilizer nitrogen in relation to surface residue can greatly affect plant response.[37] Effects of residue and potassium rate and placement on corn (*Zea mays* L.) in Ohio included lack of a response to banding with surface residue, although a response to banding was obtained with no surface residue.[38] This differential response was attributed to greater root development for more efficient nutrient uptake under the residue. Effects of residue management on plant pathogens and nutrient cycling in various crops and tillage systems will be critical for the continued use of some recent strategies.

On sloping land, fertile soils are often capable of production of crops which cannot be adequately managed by manipulation of cropping systems and cultural practices. A number of landscape modifications have been successfully used to deliver runoff water from the land with greatly reduced impact. The construction of graded terraces has been the most widely used landscape modification for erosion control in the humid and subhumid croplands of the U.S.

Terraces have also been a source of frustration to many farmers and concentration of water for enhanced development of gully erosion where use and maintenance were not adequate. Terrace layout during the 1940s and 1950s was based on contour of the land to allow a gradual slope of terrace channels within narrow tolerances. These design criteria produced odd-shaped land areas with point rows and other limitations to farming operations. As equipment size increased in the 1960s and 1970s, many of the expensively developed terrace systems were plowed down to accommodate large equipment. Parallel terrace systems, with somewhat greater tolerance limits for slope of terrace channels, replaced the old cumbersome and rejected systems during this time. Terrace systems typically represent a substantial public commitment to soil conservation since most such systems have been constructed with cost-sharing and technical expertise provided by government programs.

Most terrace systems require construction of suitable outlets with capacity to carry the volume of water safely to natural drainage areas. Grassed waterways or other outlet structures are, thus, typically components of most graded-terrace systems. Additional landscape modifications such as diversion terraces to safely dispose of up-slope water accumulations and drop-structures to stabilize active gullys are options for controlling or reducing rates of soil erosion due to runoff water. In semi-arid areas such as the western Great Plains, level terraces have been extensively constructed with the dual purpose of reducing erosion from runoff and for water conservation by retaining most of the typically limited rainfall on the land.

As with control of soil erosion by water, management of crop residue is critical for control of soil erosion by wind in susceptible areas. Cropping systems to provide ground cover by either crop residues or growing crops during critical seasons can often be effective. Plantings of wind breaks can be especially effective at reducing soil loss by wind. However, the amount of moisture required to support trees for effective wind breaks in semi-arid areas can reduce soil moisture for a distance up to three times the height of the trees.[39] Alternate strips or barriers of grain crops planted perpendicular to the direction of the prevailing wind can provide protection from wind erosion for leeward strips of fallow ground or new establishing crops. In some situations, management of surface roughness of bare ground can be a primary means of controlling wind erosion. Bedding land and plowing when soil is moist to produce clods are means of increasing surface roughness. A smooth, bare surface over an extended distance is typically susceptible to soil loss by wind. Even minor disruption of a surface crust can greatly reduce soil movement by wind. Such factors as surface roughness become especially important in the drier, western portion of the Great Plains where either the crop grown or moisture limitations may not allow production of sufficient crop residue to effectively control wind erosion.

B. CONSERVATION TILLAGE SYSTEMS

In addition to the preceding traditional soil-conserving practices, the more recently promoted conservation tillage systems have recently been extensively adopted in some regions of the country.[40,41] These systems are generally based on reduced tillage and maintenance of crop residue at or near the soil surface. Conservation tillage systems encompass a range of practices including minimum tillage, no tillage, and other tillage approaches which are based on management of crop residue to reduce soil erosion. While their effective use in many areas has been dependent upon recent development of specialized planting machinery and herbicides for effective weed control, these systems are based on the practice of stubble mulching used in the Great Plains in the 1930s.[40] Although widely recommended and extensively used in some regions, considerable research is still in progress to improve and better understand various aspects of conservation tillage systems. Even basic aspects regarding crops to be included in these systems and frequency required for high-residue crop production are currently being evaluated in some regions.[42] Of particular concern are pest management, soil compaction, aeration, root growth, microbial populations, and, ultimately, crop yield.[41]

Recent efforts to further develop effective weed control in conservation tillage systems have included use of plant competition involving cover crops[43–46] along with herbicide effectiveness and weed ecology.[47] The need to understand ecological aspects of weed population dynamics in conservation tillage systems and then base weed control strategies on underlying causes was recently addressed by Swanton et al.[48] Effects of residue on nitrogen nutrition require consideration.[49] While surface-soil layers often, but not always, increase in bulk density and penetration resistance with initial use of conservation tillage systems,[50–52] Pikul et al.[53] suggested that such soil changes were inconsequential in their case because yield reductions did not result from such soil changes over a 19-year period. In contrast, soil changes with no-till corresponded with lower yields in a 15-year evaluation with continuous corn in a silt loam.[54] Where distinct traffic zones and row zones were maintained under conservation tillage for 20 years, high bulk density and penetration resistance were distinct characteristics primarily of the traffic zone.[55] These recent research results with aspects of weed control, plant nutrition, and compaction illustrate the complexity of the processes involved in conservation tillage systems. At least in some situations useful conservation tillage systems now available will likely undergo continuing improvements and refinements over the next several years.

C. NUTRIENT MANAGEMENT

Along with available soil conservation options, alternatives exist for many fertilization needs. Legumes present an alternative source of nitrogen, and organic wastes can provide nitrogen and other nutrients. These alternative sources of fertilizer nutrients can be used to substantially greater extents than current levels.

Legumes were important components of cropping systems in American agriculture until the availability of low-cost nitrogen fertilizer following World War II.[56] Increased production and profit potential, simplification of cropping systems, and increased reliability were advantages of nitrogen fertilizer over legumes. Legumes are available and continue to be viable alternatives for erosion control, nitrogen fixation, and high-quality forage. While legume production for nitrogen fixation alone will often not be an economically superior choice in many crop production systems, the combined benefits of erosion control and nitrogen fixation make legumes a viable option for some systems. In the southeastern U.S. where soil fertility is low and erosion hazard is generally high, there is considerable potential for use of legume cover crops in both conventional[57] and conservation tillage[58] systems. Legumes can be components of the most profitable cropping systems in some situations,[33,59] although as long as 10 years may sometimes be required for the benefits of legumes to be evident.[59] Despite a history of legume use prior to World War II and availability of adapted legumes for most regions, uncertainties remain regarding renewed legume use. Most adapted legumes now available have been developed for production of forage or grain crops. While some success can be attained with the available legumes, substantial advances can be expected from genetic improvement for nitrogen fixation and erosion control. Competitive ability with weeds, disease resistance, insect tolerance, and nitrogen fixation potential may be substantially improved through selection and traditional crop breeding programs.[58] Recent evaluations have identified suitable legume species for particular environments.[60] In drier regions where moisture limitations for crop establishment are critical, legume cover crops may deplete surface moisture supplies excessively in some cropping systems.[58] Additional considerations with legumes include their competitiveness with crops[61] and effects of legume residues on germination and seedling growth of subsequent crops.[62] Such interactions can be especially complicating with many tropical legumes,[63] which typically produce various secondary metabolic products.

Although erosion control is not generally a factor in use of legumes in grass pastures, the combination of nitrogen fixation potential and high-quality forage make legumes a desirable pasture component. A major limitation of legume-based pastures is the lack of persistence of many forage legumes.[64] Thus, potential nitrogen fixation rates[65] often exceeding 100 kg ha^{-1} yr^{-1} and sometimes even over 300 kg ha^{-1} yr^{-1} are generally not sustained in commercial pastures. Current economics often favor nitrogen fertilizer over legumes as nitrogen sources for grass pastures in humid and subhumid regions of the U.S. due to poor persistence of available legumes. Dependability, production potential, and ease of management are typically advantages of nitrogen fertilizer over legume-based pasture systems.[56,66] Use of legume-based pastures where the enhanced forage quality can be best utilized, such as for young growing livestock, provides opportunities for optimum benefits from the enhanced management required. Despite the potential of legumes for nitrogen fixation in pastures and cropping systems, dependability and production potential of legume-based systems are often inferior to those based on nitrogen fertilizer where such fertilizer is readily available.

Agricultural and municipal wastes are potential sources of nutrients which are often locally underutilized. While some use of these materials as sources of plant nutrients is made in agriculture, the variability of these waste materials in nutrient composition, handling difficulties, potential for disease transmission, introduction of weeds, salt accumulation, and heavy-metal content of some industrial components have been limitations to further use.[67] Disposal needs have added increasing initiative in recent years for generators of such wastes to develop sites and methods for safe disposal and recycling. Excessive nutrient loading of soils from disposal of wastes is also a current localized concern.[68–70] Aspects of handling and using various wastes have been evaluated over an extended period of time.[67,71] Efficiency of nitrogen utilization from livestock wastes has typically been only about half that obtained from nitrogen fertilizer due to nitrate leaching, denitrification, volatilization, and unavailable fractions of nitrogen in livestock wastes.[72] Although loss of phosphorus from wastes can be greater than that from fertilizer, phosphorus and potassium from livestock wastes are typically similar in availability to those in commonly used fertilizers.[72] Livestock wastes can also enhance soil structure, aeration, water intake, and stability of soil aggregates.[67]

Recent research with poultry *(Gallus gallus domesticus)* litter indicates that nitrogen can be effectively cycled through systems utilizing productive warm-season grasses with the primary environmental hazard from nitrogen being potential nitrate leaching in some soils.[68] However, phosphorus can accumulate in soils where litter application is based on nitrogen dissipation, resulting in potential for loss of phosphorus in surface runoff. Cool-season grasses with lower potential productivity may allow excessive accumulations of both nitrogen and phosphorus as well as some other elements with poultry litter application rates typical of those in some locations.[69] Potential losses of nutrients in runoff water and methods for minimizing these losses are areas of continuing research.[73]

Efforts to incorporate anticipated differences in nutrient availability from cattle *(Bos taurus)* manure due to changes in ambient temperature indicated that interacting factors were too complex for such an approach to be effectively used to predict responses to additional spring fertilizer.[74] Attempts to determine crop yield responses to nutrients from manure have shown some relationships, especially to soil nitrate or ammonia plus nitrate during the growing season with a soil nitrogen test after manure application recently suggested as a management tool.[75] Modeling approaches to determining the value of manure for use on depleted cropland indicated that transportation costs limit the economic range of feedlot manure distribution to approximately 55 km from the source for wheat (*Triticum aestivum* L.) production.[76] The more depleted the site, the greater the distance that could be economically justified. While crop yield responses to nutrients from applied manure are typical at the lower range of normal application rates, excessive quantities can reduce grain yields, especially when moisture is a limiting factor for crop growth.[77] The additional concern regarding heavy metal content of sewage sludge has led to evidence indicating that cadmium and lead uptake by plants may increase over time after application of contaminated sludge as pH decreases.[78] Excessive accumulations of copper and zinc in sewage sludge can substantially reduce soil microbial populations.[79] While composting has potential for enhancing biological characteristics of organic wastes, especially reducing pest hazards, nutrient availability and excesses of some elements will continue to be important management considerations with use of organic wastes as sources of plant nutrients.

Developments with chemical fertilizers and their use also provide several alternatives which can reduce the potential for loss of plant nutrients, especially nitrogen, from the area of application. Developments with fertilizers include the slow-release nitrogen fertilizers and the use of nitrification inhibitors with nitrogen fertilizers.[80] These advances are particularly appropriate where high rainfall or irrigation result in movement of substantial proportions of applied nitrogen from the rooting zone before it can be taken up by crops. Nitrification inhibitors have been reported to contribute to increased protein content of grain, increased grain yield, decreased incidence of disease in some grain crops, greater nitrogen use efficiency, decreased leaching, and decreased denitrification.[81] However, nitrification inhibitors did not consistently affect yields of potato (*Solanum tuberosum* L.) in Florida, even though a slow-release nitrogen source did effectively increase yields in most cases.[82] Urease-inhibitor amendments have increased nitrogen fertilizer use efficiency from urea applied in bands or broadcast.[83] Cost of slow-release nitrogen fertilizers relative to that of conventional soluble nitrogen sources is a limiting factor for many agricultural uses.[80] Appropriate crop sequences to effectively utilize residual nitrogen from slow-release sources may be critical to their profitable use as illustrated by response of a subsequent crop.[84] Recent results illustrate that simply using slow-release nitrogen fertilizer is not an adequate approach to management of high fertilizer rates on sites susceptible to nitrate leaching.[85] Effective plant uptake of nitrogen as it becomes available is essential to minimize nitrate leaching.

Additional alternatives which reduce potential for nutrient loss include the appropriate timing and placement of proper amounts of nutrients.[86,87] Band application rather than broadcasting fertilizer for row crops is often an appropriate consideration, with row configuration and placement of fertilizer bands of importance in some cases. Proximity of nutrients to seed especially under some environmental conditions can produce effects that persist through the growing season and affect yields.[88] Soil testing for various fertilizer elements has been widely used but often primarily to make sure levels were not below some target level. Although fertilizer amounts exceeding the optimum rates are not cost effective, the added expense of excess fertilizer is often incidental in comparison with crop value, especially for many fruit and vegetable crops. Considerable adjustment of fertilizer rates could be accomplished in some situations through use of soil and plant analyses and recently expanded information bases regarding optimal ranges. Excessively high fertilizer rates as insurance against insufficient nutrient availability can often have a higher environmental cost than the immediate economic cost. Thus, nutrient management should be based on providing appropriate ranges of nutrients for particular crop and soil combinations and not for simply preventing deficiencies. Nutrient cycling and nutrient pools in various forms can be effectively managed to reduce dependence on repeated applications of high fertilizer rates in many instances where soil organic matter, crop residues, and soil microbial populations are adequately considered in nutrient management. Recent interest in refining fertilizer recommendations to consider variations in soil and fertility within fields and fertilize accordingly for a particular yield goal may result in increased efficiency of fertilizer use and greater yields, but still not offset the increased cost.[89]

D. PEST MANAGEMENT

To some extent, because of the success of chemical pest control, other means of pest control have often been perceived as outdated. Even in situations where no chemical control has been developed, recognized alternatives have been overlooked or abandoned. This has been illustrated by Cook[90] in discussions of the pest-control benefits of crop rotation. Such cultural practices as crop rotation and residue management provide distinct challenges for integration into farming systems when short-term economics, government farm programs, crop residue management for erosion control, and pest management provide both benefits and limitations for each alternative available. Crop rotation for some agronomic crops may be a particularly beneficial approach to dealing with several problems including soil erosion, pest problems, and inefficient nutrient utilization. The concept of integrated pest management, which refers to combining the various means of pest control in an effective strategy, can be enhanced further by consideration of all aspects of the farming operation and surrounding environment. Such a comprehensive approach to integration of farming practices into an environmentally acceptable system involves use of alternatives currently being called "best management practices".

Increased emphasis on ecological approaches to pest control, even in integrated pest management systems, holds considerable potential for reducing costs and hazards to the environment. While many aspects of such approaches require further development prior to widespread application, steps in that direction can be taken in most instances. Use of scouting to monitor pest populations with pesticide application only at particular thresholds often minimizes the use of pesticides while still adequately controlling pest populations. The cost of monitoring pest populations may exceed the cost of the additional pesticide applications which comprise a program of fixed pesticide application frequency. In such cases, the benefits to the environment of less pesticide must be evaluated in light of the additional cost. Interactions of pests with biological and climatic factors typically produce fluctuating population dynamics. Appropriate timing of either chemical or cultural treatments based on the ecology of a particular pest can considerably enhance treatment effectiveness.

Many components of integrated pest management systems are effective primarily at levels greater than individual fields or farms. Highly mobile pests such as many insects, pathogens, and even weed species with effective seed dispersal mechanisms can often be more effectively managed with coordinated efforts across a community to limit habitat for the pest, conserve beneficial insects, and reduce opportunity for pest build-up through production of healthy, uniform crops of resistant varieties. Even coordination on a regional scale can be important in instances where crops are progressively earlier at lower latitudes. Pests often build up and follow the developing crops. Use of resistant varieties at the lowest latitudes of continuous production, the earliest crop, can often greatly limit some pests for a particular season. Management of insect resistance to insecticides through sequential use of different classes of insecticides and other appropriate pesticide application practices also requires coordinated efforts throughout a production area.[91]

The most widely publicized aspect of the ecological approach to pest management, biological control, certainly has tremendous potential for management of agricultural pests. Only a small portion of that potential has actually been developed to the extent to be available for routine use. In the greenhouse vegetable and ornamental industry, biological control was initiated in 1968 with continuing developments.[92] The 15 species of natural enemies of 18 pest species successfully used include insect parasites, arthropod predators, and pathogens.[92] Biological control of soil-borne plant pathogens has been obtained using commercially available microbial inoculates,[93] however, success with such products has not been universal. Natural occurrence of such biological control can also be manipulated for use as a component of integrated pest management systems.[93] The bacterium *Bacillus thuringiensis* Berliner has been used to control a range of insects,[94] primarily in forestry and vegetable production.[95] This bacterium accounted for over 90% of bio-insecticide sales internationally in 1991.[96] Bio-insecticides comprised the majority of biological control products for the year, which accounted for less than 0.5% of global pesticide sales. Available commercial products including pheromones, antifeedants, and growth regulators of plants and insects provide opportunities for increased use of natural product structures or bioregulators in pest control programs.[97] Unfortunately, this potential is currently available only for very specific situations, which do not represent a substantial proportion of agricultural pest management needs.

V. DEVELOPING TECHNOLOGY

Much of the publicity regarding recent and current agricultural research efforts focuses on biotechnology and approaches to reducing dependence on inorganic fertilizers and chemical pest control. These efforts will provide new products and directions for some aspects of agriculture. Based on the extent of dependence of most agricultural production on inorganic fertilizers and chemical pesticides and the extensive research and development programs the chemical industry has supported, substantial developments can be expected in these areas. Limitations of fertilizers and pesticides, which have contributed to environmental concerns, will be alleviated to some extent through improved formulations, development of more appropriate application procedures, rates, and conditions, and the introduction of new products developed with emphasis on compatibility with the environment. Emphasis on efficiency, rather than scale of operation, and sensitivity to unnecessary environmental modifications such as soil compaction and reductions in soil tilth, organic matter, and ground cover will contribute to continuing advances in farm equipment design.

A particularly distinct increase in research and development efforts with various biological control alternatives is apparent not only from an increase in reports of such research but also the recent initiation of several specialized journals in the field. Development of commercial biological control in the greenhouse industry may be the pattern to expect, but at a considerably slower pace and perhaps extent, in more extensive agriculture. Biological control has become a common component of integrated pest management in the European greenhouse industry as costs and registration regulations reduced chemical alternatives, pest resistance to chemicals increased, and the use of beneficial insects (insect pollinators) expanded.[92] The limitations of chemical pesticides provide opportunities or niches where demand exists for new alternatives. Powell and Jutsum[96] have suggested that these situations will represent the major uses of expanding biocontrol products in the near future. Although primarily expected to fit a niche market, biological control agents may be reasonably expected as alternative components of integrated pest management for major pests of major crops over extensive areas.[95]

Examples of potential commercial developments with biological control which are probable include additional uses of *Bacillus thuringiensis* as illustrated by Zuckerman et al.[98] Initial expectations from inoculant biocontrol agents for management of soil-borne plant pathogens has moderated, with use now expected to be primarily in integrated approaches along with practices to enhance natural biocontrol systems.[93] *Rhizobium* species appear to control some soil-borne fungal pathogens of several crop species.[99] The continued effectiveness of a host-specific virus of soybean looper *(Pseudoplusia includens)* in Louisiana 12 to 15 years after its introduction indicates that it could spread and persist from relatively small introductions.[100] Experience with this attempt at biological control in a row crop was suggested as a potential model for further similar attempts. The naturally occurring fungal entomopathogen *Beauveria bassiana* and its role in controlling European corn borer *(Ostrinia nubilalis)* have also been suggested as examples for development of biological control in row crops.[101] While research with sex pheromones indicates potential for direct biological control as illustrated by Shaver and Brown,[102] a more common use in the short term will likely be in insect monitoring and survey, where pheromone-baited traps enhance the effectiveness of scouting efforts.[103] Enhanced use of commercially available insect growth regulators will undoubtedly increase as new information enhances effectiveness of their use. Combinations of feeding stimulants and growth regulators provide such potential.[104] Characterization of available genetic resources, identification of ecotypic variation, matching individual ecotypes to specific target environments, knowledge of interactions of species, and even genetic manipulation to enhance adaptability and effectiveness of biological control organisms hold tremendous potential for enhancement of selected organisms and their effectiveness as described for mites (Phytoseiidea) in Oregon.[105] Effective use of sustained populations of biological control organisms is highly dependent upon interactions with pesticides in integrated pest management programs with responses differing for various chemicals available as illustrated by Jalali and Singh.[106]

Strategies for acceptable biological control of weeds have been limited by host specificity and lack of lethal effects of potential control agents.[107] Sands and Miller[107] have suggested a new strategy based on genetic manipulation of lethal, broad host-range organisms to limit host-range or long-term survivability. Such short-term survival could enhance commercialization through continuing marketability and increased environmental safety.[107] Potential appears to exist for use of nonpathogenic isolates of fungal pathogens to limit some plant diseases.[108] Foliar fungal pathogens are subject to natural suppression from microorganisms which provide potential for development of mycoparasites for biological control of some particular plant diseases.[109] Although environmental safety is a major benefit anticipated from use of biological control in place of extensive chemical pesticide use, warnings of potential effects of biocontrol agents on nontarget organisms[110] must serve as precautions for adequate preliminary evaluations.

Development of herbicide-resistant crops to allow more economical and safer chemical pesticide programs, while highly publicized and rather controversial, will likely be much more difficult and less extensively successful than originally anticipated.[111] Precautions to minimize opportunities for adverse effects are also recommended here.[112] Management of herbicide-resistant weeds, which can develop as a response of plant populations to repeated use of a herbicide over time, will likely be a continuing concern. Conservation tillage systems may be especially vulnerable, with both biological control efforts[96] and chemical pesticide advances[113] developing for satisfactory weed management. Several approaches for integration of ecological principles into weed management are being evaluated. Certainly the clean-tilled, weed-free picturesque field, common through many areas with extensive row-crop production, is not essential for economic crop production. Use of crop interference,[114] management of weed seed banks with microorganisms,[115] incorporation of weed succession into management strategies,[48] and use of modeling may become aspects of future weed management strategies. Identification of critical periods of weed control for individual crops may be a key aspect of ecological management.[116,117] Effects of timing of application[118] and novel formulations[119] continue to enhance safety of available herbicides. Soil solarization can be expected to be more widely used for management of weeds, nematodes,[120] and soil-borne diseases,[121] especially with high-value crops.

In contrast to expectations from biotechnology with herbicide-resistant crops, the transfer of genes for traits imparting insect resistance to crop plants appears to be progressing and on the verge of commercial use as illustrated by Micinski et al.[122] with cotton (*Gossypium hirsutum* L.). The rather extensive and successful use of integrated pest management approaches on insect pests in several cropping systems can be readily modified to incorporate new developments which will undoubtedly provide additional insect resistance from traditional plant breeding and biotechnological approaches, introduced biological control agents, management to enhance natural biological interactions, and additional aspects of a greater knowledge of insect-crop interactions. Aspects of plant resistance to disease, effects of chemical control of pathogens, and their interactions are being further explored for enhanced pest management as illustrated by the research of Pataky and Eastburn.[123] Basic knowledge of plant responses to disease, including the molecular basis of disease resistance and pathogen effects, has recently been expanded tremendously.[124] Information about pathogenesis-related proteins, their genetic control, and cloning techniques for resistance genes suggests that management for crop diseases, and perhaps insects and even some other stress responses, may well be approaching major milestones.

The complex ecology of soils is becoming an increasingly appreciated aspect of sustainable crop production. Along with pathogenic organisms, decomposers, organisms involved in nitrogen cycling, and numerous others, vesicular-arbuscular mycorrhizal fungi are suggested to be of special significance because of their roles in plant nutrition and development of soil structure.[125] While numerous evaluations of responses in plant growth to these fungi, especially in relation to rates of phosphorus, have been conducted, neither the basic ecology of these organisms nor means of effectively managing for their beneficial effects are adequately understood. Relationships of mycorrhizal fungi and crop management are currently being discerned for particular cropping systems, especially low-input systems.[126,127] Lack of practical propagation procedures for these fungi for widespread commercial use in responsive cropping systems appears to be a key limitation.

Even though legumes can be effectively used for biological nitrogen fixation, as pointed out by Henzell in 1988,[128] the substantial progress made in elucidating basic aspects of the nitrogen fixation process has not been reflected in new applications. Substantial efforts to more effectively utilize legumes for nitrogen fixation in low-input systems have not been overwhelmingly successful, although management recommendations for specific situations have been developed.[129] Expectations for major developments with associative nitrogen fixation of grasses have also failed to materialize, even though the rather low levels of nitrogen fixation reported[130] could allow selection of superior genotypes of some tropical grass crops for low-input systems.

Technology at levels beyond basic biological sciences and individual farming practices or even farming systems that will influence agriculture directly is also developing. Government policies and regional planning efforts to limit adverse effects of society on the environment will be enhanced by such developments as land information systems,[131] geographical information systems,[132] remote sensing,[133] and modeling of landscape or regional relationships of weather, hydrology, and various scales and details of land use.[134,135] Such information may allow development of economically acceptable approaches for use of pollution-interception techniques based on landscape planning even beyond the individual farm level. Vegetation buffer strips, which may consist of permanent vegetation managed for hayland, pasture, or forest, can be strategically located to protect water supplies from various sources of pollution.[136]

VI. IMPLEMENTING APPROPRIATE CHANGE

A. NATURE OF ANTICIPATED CHANGE

Just as the history of agriculture is one of change, agriculture remains dynamic. Several factors indicate that the rate of change may be greater at the present time than in most periods of the past. The economic disasters of some agricultural enterprises and even rural financial institutions and communities during the 1980s, uncertainty of export markets and product demand, imminent change in federal farm programs, and increasing influence of an environmentally conscious, urban society indicate lack of stability in agricultural policies. Change in the philosophy of agriculture is indicated at the national level by the programs and policies established and/or espoused by governmental bodies and key leaders. Numerous organizations, causes, and coalitions are proponents and aggressive advocates of change ranging from specific details of agricultural production and land use to complete alteration of ideas and attitudes. Some change is needed. Change is occurring. Agriculture is changing. The agricultural industry can provide leadership and influence this change in a favorable way, or agriculture can react to the changes instituted by a nonagricultural, somewhat activist-led society.

Increasing public awareness of the vulnerability of the global environment to adverse human activities has led to the pervasive perception that a new concept, a new approach, even a completely new paradigm for man's interaction with the world around him is required. This new paradigm will be based on the concept of sustainability. All aspects of our culture and commerce will eventually be affected. The nature of agriculture, the national economy, and the standard of living will be determined by the outcome of the currently evolving national philosophy. From an agricultural standpoint, the goal of sustainability is certainly not a point of contention. The definition, scope, and means to sustainability are highly controversial. Some strong proponents of sustainability in agriculture advocate a very narrow, localized scope of sustainability, emphasizing an almost self-contained concept of an individual farm. This philosophy attempts to lend essentially synonymous meanings to the terms sustainable agriculture, alternative agriculture, low-input farming, and organic farming. The principles of ecology and biology along with the concepts of regenerative processes are emphasized, often to the exclusion of essentially all inorganic, chemically synthesized, or even off-farm purchased inputs. This philosophy has recently progressed from an apparently minor, almost radical movement to approach the status of the only acceptable position for an individual or organization genuinely concerned with environmental sustainability. While agriculture must accept and embrace sustainability as a goal and guiding principle, does this imply that low-input, organic farming is the only alternative? Furthermore, even though organic farming can be viable on an individual farm basis, is there adequate information available to indicate that a nationwide, low-input, organic agriculture can dependably sustain the present and growing population without drastic alterations in the standard of living? What will be the costs for such a change to a low-input, organic agriculture on an extensive basis? These questions must be appropriately presented and addressed by those with the relevant technical expertise, because decisions will ultimately be made by a society with very limited expertise or even awareness of production agriculture. If the concepts of sustainability and low-input, organic farming are allowed to continue to converge in general use, an unknowing, but concerned, society can readily influence public policy and evolving regulatory mandates in an inappropriate direction for the public welfare.

In its 1989 assessment of the role of alternative farming methods in production agriculture, the National Research Council[17] chose to contrast what were called "alternative agriculture" and "conventional agriculture". While the report acknowledged lack of distinct differences in farming practices to characterize two separate systems, the philosophy of low-input alternative farming approaches or systems was characterized as more sustainable and innovative. This alternative agriculture was illustrated by individual case studies with a strong emphasis on organic farming. Certainly, many of the practices commonly used in organic farming should be considered appropriate potential alternatives in agricultural enterprises not currently using them. The implication that such an alternative agriculture based on organic farming practices is available to replace a rejected conventional, chemical-based agriculture has been questioned by a CAST assessment.[137] A key aspect of the CAST review was the point that many of the desirable aspects of a more ecologically based production agriculture, such as biological pest control and pollution-free sources of plant nutrients, are simply not available in most cases. Options for biological pest control are limited, and even organic sources of nutrients are potential sources of pollution.

In what appeared to be an obvious progression, Papendick et al.[138] presented an overview of the adverse environmental consequences of prevailing agricultural methods followed by presentation of potential improvement from use of specific practices suggested to characterize "alternative agriculture". Organic farming was used as an essentially interchangeable term with alternative agriculture. As has become somewhat characteristic, the implication was made that the philosophy of the alternative of organic farming, rather than simply the use of appropriate farming practices, was the means to enhanced sustainability. This implication, that the conventional approach is not sustainable and contrasts with a better alternative approach, has become so frequently used that it is difficult to perceive that such an immediate extensive change in practice or philosophy may not be appropriate for the general welfare. Certainly, practices that have been used to characterize low-input systems have potential to minimize groundwater contamination.[139] Some practices that are emphasized in biodynamic farming can enhance physical, biological, and chemical soil properties.[140] Aspects of organic farming present distinct benefits for conservation of resources including reduced erosion, reduced runoff, and enhanced nutrient recycling.[141] The benefits of specific practices, whether identified closely with a particular system or not, must be considered for their potential contribution regardless of the philosophy or system they appear to represent.

Since practices characterizing low-input farming, organic farming, and regenerative farming are acknowledged as viable alternatives for use in many agricultural enterprises across the country, why should the accompanying philosophy not be embraced by American agriculture? Such approaches are not appropriate for some specific commodities.[142] Current technologies, though progressing, are not adequate for an immediate, extensive substitution.[137,142,143] The scientific information for ecologically based agriculture is inadequate.[142]

These scientific and technical limitations suggest that attempts, though well-meaning, to immediately implement such a change in agriculture, could have drastic consequences on the quantity, quality, variety, and price of food available to the public. An additional aspect is the response to the subtle message that American agriculture, governing and regulatory entities, and the general public are gradually endorsing the philosophy of low-input, organic farming as the appropriate alternative for American agriculture. More environmentally acceptable chemical approaches can undoubtedly be developed by the extensive research and development programs of the agricultural-chemical industry. Decisions regarding such long-term, costly programs could be tempered by anticipated product marketability. Multinational corporations providing the fertilizers, pesticides, and other inputs that are integral components of American agriculture must evaluate the opportunities and constraints, both existing and anticipated, which affect profit potential. Existing and anticipated markets along with regulatory requirements and public opinion that indicate future controls are important considerations in investment decisions. As emphatically presented by Avery,[144] abandonment of the technical developments which support a highly productive agriculture could embody sacrifice of world leadership in addressing food production needs, national economic strength, and even the opportunity for environmental stability as reduced yields could lead to more extensive farming of marginal lands. Even the perception of such trends could have substantial effects on business decisions, which are continuing to move industry, jobs, products, and economic opportunity internationally to the more advantageous locations.

A philosophy or paradigm based on rejection of basic aspects of current production capabilities must at least evoke caution. Recognized limitations in the comprehensiveness of applicability of such a new philosophy should indicate a need to restrict its adoption. The extensive need for enhanced understanding of theoretical aspects of such a new philosophy should indicate its potential for the future rather than its appropriateness for the present. Recognition of the possibilities, or perhaps probabilities, that nationwide adoption of a low-input, organic farming philosophy could lead to undesired trends in food availability, quality, and price should cause grave concern. The disincentive for research and development of environmentally compatible chemicals that such a philosophy could produce must be considered. While the practices which characterize the various approaches to farming should be considered potential alternatives to present practices, the particular philosophy of a low-input, organic agriculture must be rejected as the basis for a new national paradigm of production agriculture for the present time.

Consideration of economic opportunity and economic stability may require a reassessment of the definition of sustainability. Sustainability may be addressed strictly in reference to natural resources and their conservation. Sustainable development may expand the concept from protection or conservation to effective and efficient use of available resources. Sustainability of human populations must be considered, as conditions of despair currently exist not only due to political and social inequities but also because of localized shortages of food, fuel, and clean water. A sustainable agriculture must provide for the present and maintain the resources for the future. It must be productive and conserve the resources of production, effectively and efficiently utilizing the resources available. The social, cultural, economic, and ecological dimensions of agricultural sustainability must be considered in a global context, which also addresses the concepts of economic entitlement and ecological obligation.[145] Economic entitlement involves the rights for an opportunity to earn a living and access to basic human necessities. Ecological obligation refers to the responsibility of those with opportunity and affluence to make responsible, rather than extravagant or selfish, use of resources. Predictions of continuing population growth, increased demand for food, and possibilities of major advances in food production capability through biotechnological developments[146] suggest that "ecological obligation" could extend to a responsibility, by those with the opportunity and expertise, to develop and use the necessary technology to sustain the resource base and the growing population. Dependence of high levels of agricultural production on technological developments of the past 50 years is illustrated by the suggestion that maintenance of per capita cereal production will require a doubling of chemical nitrogen use worldwide by the year 2030.[147]

B. LEADERSHIP NEEDED

Ecological obligation is a concept applicable to agriculture and its sustainability. This obligation must be more fully accepted by those making the decisions in agriculture. More widespread acceptance of this obligation will have an effect on both the sustainability of agriculture and the credibility of the agricultural industry. This credibility or public perception along with environmental performance in the immediate future will largely determine how change to a more sustainable agriculture proceeds. Two distinct forces will propel the unavoidable change in agriculture toward sustainability. Some decisions will be made on a voluntary basis by farmers and others directly involved in agriculture. Other decisions will be made through governmental and regulatory processes and introduced as public policy. Public policy can range from incentives such as tax breaks and direct subsidies to mandated regulations. Either extreme can lead to decisions not necessarily of preference. Agricultural interests have historically asserted their independence and accountability but have recently come to be perceived, to some extent, as dependent upon government programs and policies. A responsive and accountable agriculture will be able to influence public perception positively and maintain considerable independence. An unresponsive agriculture, or one so perceived, will be increasingly coerced by typically inefficient and often frustrating public policy on a rather incongruous path toward sustainability.

An example of restrictive public policy is legislation proposed in England to regulate the amount of fertilizer a farmer can use.[148] The numbers of statutes and regulations administered by the U.S. Environmental Protection Agency (EPA) are increasing to impact new targets including small companies.[149] This tremendous growth in government controls has been referred to as "pathological growth of regulations".[149] It is anticipated that public concern over potential water contamination could readily subject individual farmers to numerous regulations concerning use of agricultural chemicals and fertilizers, as well as livestock management constraints, due to potential nonpoint source pollution. The present philosophy expressed by the EPA provides opportunity for progress where collaborative efforts could be beneficial.

As stated by Browner,[150] "We are committed to emphasizing public-private partnerships for voluntary efforts *outside* regulatory frameworks". Apparently initiative from major agricultural entities could lead to development of voluntary efforts to monitor and improve problem situations before remedial practices and penalties are mandated by law.

Unfortunately, it appears that much of the discussion regarding the means to sustainability in agriculture revolves around public policy. The need to assess effects of existing public policy on decisions regarding agricultural sustainability has been effectively presented by the National Research Council.[17] Certainly the aggregate influence of federal policy has not pointed agriculture in the direction of sustainability. The Carnegie Commission[151] concluded that the government's environmental research and development effort suffers from lack of direction, focus, and coordination. Such inadequacies are reflected in environmental policies affecting agriculture. In an assessment of agriculture's role in water quality problems, CAST[6] emphasized the need for public policy to direct agriculture in a more sustainable approach. The Carnegie Commission has presented the need and suggested means for improved utilization of science and technology[152] and the regulatory process[153] in addressing the goals of society through more responsive and focused public policy. Public policy will undoubtedly set the overall framework within which a sustainable agriculture must develop. However, greater progress toward the goal of improved environmental performance by agriculture on its own initiative will surely reduce the need and public pressure for additional government involvement.

Such a direct approach has both advantages and limitations. Costs of some aspects of enhanced environmental sustainability are the rightful responsibility of a society demanding and enjoying such benefits as improved wildlife habitat. Any such public cost contributions involve considerable public policy and government involvement. In cases where costs are not prohibitive, agriculture can generally be more efficient and less cumbersome without the benefit of excessive public policy. Public policy is typically the result of the perception by society that there is an unmet need for action. While the failure of existing public policy to effect a sustainable agriculture is before the nation, agriculture has an opportunity to take the initiative and restore public confidence through independent, responsible stewardship. Or agriculture can irresponsibly optimize short-term benefits before impending public policy developments further restrict independence and freedom. Perhaps the point has not been reached where appropriate action will prevail. It has been suggested in this context that "only when systems break down do people do sensible things".[154] However, a trend toward more widespread use and acceptance of the farming practices that enhance sustainability has become apparent.[143] Agriculture, in the private sector, has the necessary organizational structure involving commodity associations and broader-based local organizations to institute effective educational and motivational efforts to enhance the image and adoption of sustainable alternatives.

Existing and developing successes in sustainable agriculture, and even acknowledgment of areas needing attention and being addressed, must be conveyed effectively to the public. A sensitivity to real environmental concerns must be projected by agriculture. Agriculture must overcome the image that "U.S. farmers are, in the aggregate, among the most anti-environmental of major social groups" as stated by Buttel et al.[142] A similar image of agriculture was depicted in the 1960s,[155] and deservedly so for some individuals, including personnel in land grant universities, the agricultural-chemical industry, and the U.S. Department of Agriculture. A recent National Research Council report has called for a combined effort from policymakers, researchers, and farmers to acknowledge shortcomings in environmental areas and to attempt to restore public confidence in agriculture.[148] To avoid counterproductive government efforts to stimulate progress, individual leaders in various agricultural commodity areas must assume responsibility for sensitizing the industry to environmental concerns.

Initiative by leadership of the agricultural industry is needed to assess potential and existing environmental shortcomings of various production systems for particular commodities. Industry leaders must also convey research and/or extension needs to appropriate agencies where needs are not being adequately met by existing programs. Failure to do so will eventually result in detrimental loss of flexibility and options as suggested in a recent NRC report.[148] Agricultural leadership across broad areas of influence must build consensus to accept and proclaim a commitment to sustainability and environmental responsibility. This consensus must be broad enough to allow the new evolving paradigm to succeed. It must include the means to sustain the population, conserve the natural resources, protect the environment, provide economic opportunity, and stimulate future progress. The new paradigm of sustainable agriculture must be broad enough to exert an obvious peer pressure on those hesitant to act responsibly. The alternative is an already developing complex maze of public policies intended to provide incentive, control, motivation, and limitation to numerous different aspects of agriculture as it affects or is perceived to affect the environment and human population.

REFERENCES

1. Kerr, N. A. *The Legacy, a Centennial History of the State Agricultural Experiment Stations*, Missouri Agricultural Experiment Station, University of Missouri, Columbia, MO, 1987.
2. Eisenhower, M. S., Ed., *Yearbook of Agriculture 1930*, United States Department of Agriculture, Washington, D.C., 1930.
3. USDA, *The Yearbook of Agriculture 1950–1951*, United States Department of Agriculture, Washington, D.C., 1951.
4. USDA, *The Yearbook of Agriculture 1966*, United States Department of Agriculture, Washington, D.C., 1966.
5. USDA, *Grass, the Yearbook of Agriculture 1948*, United States Department of Agriculture, Washington, D.C., 1948.
6. CAST, *Water Quality: Agriculture's Role*, Council for Agricultural Science and Technology, Ames, Iowa, 1992.
7. Bentley, O. G., Soil erosion and crop productivity: a call for action, in *Soil Erosion and Crop Productivity*, Follett, R. F. and Stewart, B. A., Eds., American Society of Agronomy, Madison, WI, 1985, 1.

8. McCracken, R. J., Lee, J. S., Arnold, R. W., and McCormack, D. E., An appraisal of soil resources in the U.S.A, in *Soil Erosion and Crop Productivity*, Follett, R. F. and Stewart, B.A., Eds., American Society of Agronomy, Madison, WI, 1985, 49.
9. Crosson, P., Temperate region soil erosion, in *Sustainable Agriculture and the Environment,* Ruttan, V. W., Ed., Westview Press, Boulder, CO, 1992, 11.
10. Larson, W. E., Pierce, F. J., and Dowdy, R. H., The threat of soil erosion to long-term crop production, *Science*, 219, 458, 1983.
11. Bennett, H. H., *Soil Conservation*, McGraw, New York, 1939.
12. Wadleigh, C. H. and Dyal, R. S., Soils and pollution, in *Agronomy and Health*, Blaser, R. E., Ed., American Society of Agronomy, Madison, WI, 1970, 9.
13. USDA, *Erosion, Sediment and Related Salt Problems and Treatment Opportunities*, United States Department of Agriculture, Soil Conservation Service, Special Projects Division, Golden, CO, 1975.
14. USDA, *Cropland Erosion*, United States Department of Agriculture, Soil Conservation Service, Washington, D.C., 1977.
15. Smith, R. A., Alexander, R. B., and Wolman, M. G., Water-quality in the nation's rivers, *Science*, 235, 1607, 1987.
16. Olson, G. W., *Soils and the Environment: A Guide to Soil Surveys and Their Applications*, Chapman & Hall, New York, 1981.
17. NRC, *Alternative Agriculture,* National Research Council, National Academy Press, Washington, D. C., 1989.
18. Keeney, D. R. and DeLuca, T. H., Des Moines River nitrate in relation to watershed agricultural practices: 1945 versus 1980s, *J. Environ. Qual.*, 22, 267, 1993.
19. Cihacek, L. J., Seeney, M. D., and Deibert, E. J., Characterization of wind erosion sediments in the Red River Valley of North Dakota, *J. Environ. Qual.*, 22, 305, 1993.
20. Cooper, C. M., Biological effects of agriculturally derived surface water pollutants on aquatic systems — A review, *J. Environ. Qual.,* 22, 402, 1993.
21. Liebman, M. and Dyck, E., Weed management, a need to develop ecological approaches, *Ecol. Appl.*, 3, 39, 1993.
22. Gallant, J. C. and Moore, I. D., Modelling the fate of agricultural pesticides in Australia, *Agric. Syst.*, 43, 185, 1993.
23. Varshney, P., Tim, U. S., and Anderson, C. E., Risk-based evaluation of ground-water contamination by agricultural pesticides, *Ground Water*, 31, 356, 1993.
24. Costanza, R. and Cornwell, L., The 4P approach to dealing with scientific uncertainty, *Environment*, 34 (No. 9), 12, 1992.
25. Dunlap, T. R., *DDT: Scientists, Citizens and Public Policy*, Princeton University Press, Princeton, NJ, 1981.
26. Gish, T. J. and Sadeghi, A., Agricultural water quality priorities: A symposium overview, *J. Environ. Qual.,* 22, 389, 1993.
27. Spalding, R. F. and Exner, M. E., Occurrence of nitrate in groundwater — A review, *J. Environ. Qual.,* 22, 392, 1993.
28. Tatum, L. A., The southern corn leaf blight epidemic, *Science*, 171, 1113, 1971.
29. Krausz, J. P., Fredericksen, R. A., and Rodrigues-Ballesteros, O. R., Epidemic of northern corn leaf blight in Texas in 1992, *Plant Dis.*, 77, 1063, 1993.
30. Waage, J., The population ecology of pest-pesticide-natural enemy interactions, in *Pesticides and Non-Target Invertebrates*, Jepson, P. C., Ed., Intercept Ltd., Dorset, England, 1989, 81.
31. Schertz, D. L. and Bushnell, J. L., USDA crop residue management action plan, *J. Soil Water Conserv.,* 48, 175, 1993.

32. Buchanan, M. and King, L. D., Carbon and phosphorus losses from decomposing crop residues in no-till and conventional till agroecosystems, *Agron. J.*, 85, 631, 1993.
33. Hanson, J. C., Lichtenberg, E., Decker, A. M., and Clark, A. J., Profitability of no-tillage corn following a hairy vetch cover crop, *J. Prod. Agric.*, 6, 432, 1993.
34. Smiley, R. W. and Wilkins, D. E., Annual spring barley growth, yield, and root rot in high- and low-residue tillage systems, *J. Prod. Agric.*, 6, 270, 1993.
35. Smith, S. J. and Sharpley, A. N., Nitrogen availability from surface-applied and soil-incorporated crop residues, *Agron. J.*, 85, 776, 1993.
36. Westermann, D. T. and Crothers, S. E., Nitrogen fertilization of wheat no-till planted in alfalfa stubble, *J. Prod. Agric.*, 6, 404, 1993.
37. Gordon, W. B., Whitney, D. A., and Raney, R. J., Nitrogen management in furrow irrigated, ridge-tilled corn, *J. Prod. Agric.*, 6, 213, 1993.
38. Yibirin, H., Johnson, J. W., and Eckert, D. J., No-till corn production as affected by mulch, potassium placement, and soil exchangeable potassium, *Agron. J.*, 85, 639, 1993.
39. Fryear, D. W. and Skidmore, E. L., Methods for controlling wind erosion, in *Soil Erosion and Crop Productivity*, Follett, R. F. and Stewart, B. A., Ed., American Society of Agronomy, Madison, WI, 1985, 443.
40. Allmaras, R. R., Unger, P. W., and Wilkins, D. W., Conservation tillage systems and soil productivity, in *Soil Erosion and Crop Productivity*, Follett, R. F. and Stewart, B. A., Ed., American Society of Agronomy, Madison, WI, 1985, 357.
41. Gebhardt, M. R., Daniel, T. C., Schweizer, E. E., and Allmaras, R. R., Conservation tillage, *Science*, 230, 625, 1985.
42. Lund, M. G., Carter, R. P., and Oplinger, E. S., Tillage and crop rotation affect corn, soybean, and winter wheat yields, *J. Prod. Agric.*, 6, 207, 1993.
43. Johnson, G. A., Defelice, M. S., and Helsel, Z. R., Cover crop management and weed control in corn (*Zea mays*), *Weed Technol.*, 7, 425, 1993.
44. Kumwenda, J. D. T., Radcliffe, D. E., Hargrove, W. L., and Bridges, D. C., Reseeding of crimson clover and corn grain yield in a living mulch system, *Soil Sci. Soc. Am. J.*, 57, 517, 1993.
45. Teasdale, J. R. and Daughtry, C. S. T., Weed suppression by live and desiccated hairy vetch *(Vicia villosa), Weed Sci.*, 41, 207, 1993.
46. Teasdale, J. R. and Mohler, C. L., Light transmittance, soil temperature, and soil moisture under residue of hairy vetch and rye, *Agron. J.*, 85, 673, 1993.
47. Kapusta, G. and Krausz, R. F., Weed control and yield are equal in conventional, reduced-, and no-tillage soybean *(Glycine max)* after 11 years, *Weed Technol.*, 7, 443, 1993.
48. Swanton, C. J., Clements, D. R., and Derksen, D. A., Weed succession under conservation tillage: a hierarchical framework for research and management, *Weed Technol.*, 7, 286, 1993.
49. Knowles, T. C., Hipp, B. W., Graff, P. S., and Marshall, D. S., Nitrogen nutrition of rainfed winter wheat in tilled and no-till sorghum and wheat residues, *Agron. J.*, 85, 886, 1993.
50. Francis, G. S. and Knight, T. L., Long-term effects of conventional and no-tillage on selected soil properties and crop yields in Canterbury, New Zealand, *Soil Tillage Res.*, 26, 193, 1993.
51. Grant, C. A. and Lafond, G. P., The effects of tillage systems and crop sequences on soil bulk density and penetration resistance on a clay soil in southern Saskatchewan, *Can. J. Soil Sci.*, 73, 223, 1993.

52. Hermawan, B. and Cameron, K. C., Structural changes in a silt loam under long-term conventional or minimum tillage, *Soil Tillage Res.*, 26, 139, 1993.
53. Pikul, J. L., Jr., Ramig, R. E., and Wilkins, D. E., Soil properties and crop yield among four tillage systems in a wheat-pea rotation, *Soil Tillage Res.*, 26, 151, 1993.
54. Vyn, T. J. and Raimbault, B. A., Long-term effect of five tillage systems on corn response and soil structure, *Agron. J.*, 85, 1074, 1993.
55. Mahboubi, A. A., Lal, R., and Faussey, N. R., Twenty-eight years of tillage effects on two soils in Ohio, *Soil Sci. Soc. Am. J.*, 57, 506, 1993.
56. Burton, G. W., Legume nitrogen versus fertilizer nitrogen for warm-season grasses, in *Biological N Fixation in Forage-Livestock Systems*, Hoveland, C. S., Ed., American Society of Agronomy, Madison, WI, 1976, 55.
57. Bauer, P. J., Camberato, J. J., and Roach, S. H., Cotton yield and fiber quality response to green manures and nitrogen, *Agron. J.,* 85, 1019, 1993.
58. Power, J. F., Follett, R. F., and Carlson, G. E., Legumes in conservation tillage systems: a research perspective, *J. Soil Water Conserv.*, 38, 217, 1983.
59. Millhollon, E. P. and Melville, D. R., *The Long-Term Effects of Winter Cover Crops on Cotton Production in Northwest Louisiana*, Louisiana Agric. Exp. Sta. Bulletin No. 830, Louisiana Agricultural Experiment Station, Baton Rouge, LA, 1991.
60. Biederbeck, V. O., Bouman, O. T., Looman, J., Slinkard, A. E., Bailey, L. D., Rice, W. A., and Janzen, H. H., Productivity of four annual legumes as green manure in dryland cropping systems, *Agron. J.,* 85, 1035, 1993.
61. Exner, D. N. and Cruse, R. M., Interseeded forage legume potential as winter ground cover, nitrogen source, and competitor, *J. Prod. Agric.*, 6, 226, 1993.
62. Bradow, J. M., Inhibitions of cotton seedling growth by volatile ketones emitted by cover crop residues, *J. Chem. Ecol.*, 19, 1085, 1993.
63. Hauser, S., Effect of *Acioa barteri, Cassia siamea, Flemingia macrophylla* and *Gmelina arborea* leaves on germination and early development of maize and cassava, *Agric. Ecosys. Environ.*, 45, 263, 1993.
64. Marten, G. C., Matches, A. G., Barnes, R. F., Brougham, R. W., Clements, R. J., and Sheath, G. W., Ed., *Persistence of Forage Legumes*, American Society of Agronomy, Madison, WI, 1989.
65. Ledgard, S. F. and Steele, K. W., Biological nitrogen fixation in mixed legume/grass pastures, *Plant Soil*, 141, 137, 1992.
66. Pitman, W. D., Portier, K. W., Chambliss, C. G., and Kretschmer, A. E., Jr., Performance of yearling steers grazing bahia grass pastures with summer annual legumes or nitrogen fertiliser in subtropical Florida, *Tropical Grasslands*, 26, 206, 1992.
67. McCalla, T. M., Peterson, J. R., and Lue-hing, C., Properties of agricultural and municipal wastes, in *Soils for Management of Organic Wastes and Waste Waters*, Elliott, L. F. and Stevenson, F. J., Ed., Soil Science Society of America, Madison, WI, 1977, 11.
68. Sharpley, A. N., Smith, S. J., and Bain, W. R., Nitrogen and phosphorus fate from long-term poultry litter applications to Oklahoma soils, *Soil Sci. Soc. Am. J.,* 57, 1131, 1993.
69. Kingery, W. L., Wood, C. W., Delaney, D. P., Williams, J. C., Mullins, G. L., and van Santen, E., Implications of long-term land application of poultry litter on tall fescue pastures, *J. Prod. Agric.*, 6, 390, 1993.
70. Schmitt, M. A., Sheaffer, C. C., and Randall, G. W., Preplant manure and commercial P and K fertilizer effects on alfalfa production, *J. Prod. Agric.*, 6, 385, 1993.

71. Vanderholm, D. H., Handling of manure from different livestock and management systems, *J. Anim. Sci.*, 48, 113, 1979.
72. Tunney, H., An overview of the fertilizer value of livestock waste, in *Livestock Waste: A Renewable Resource*, American Society of Agricultural Engineers, St. Joseph, MI, 1981, 181.
73. Edwards, D. R. and Daniel, T. C., Effects of poultry litter application rate and rainfall intensity on quality of runoff from fescuegrass plots, *J. Environ. Qual.*, 22, 361, 1993.
74. Motavalli, P. P., Kelling, K. A., Syverud, T. D., and Wolkowski, R. P., Interaction of manure and nitrogen or starter fertilizer in northern corn production, *J. Proc. Agric.*, 6, 191, 1993.
75. Paul, J. W. and Beauchamp, E. G., Nitrogen availability for corn in soils amended with urea, cattle slurry, and solid and composted manures, *Can. J. Soil Sci.*, 73, 253, 1993.
76. Freeze, B. S., Webber, C., Lindwall, C. W., and Dormaar, J. F., Risk simulation of the economics of manure application to restore eroded wheat cropland, *Can. J. Soil Sci.*, 73, 267, 1993.
77. Chang, C., Sommerfeldt, T. G., and Entz, T., Barley performance under heavy applications of cattle feedlot manure, *Agron. J.*, 85, 1013, 1993.
78. Hooda, P. S. and Alloway, B. J., Effects of time and temperature on the bioavailability of Cd and Pb from sludge-amended soils, *J. Soil Sci.*, 44, 97, 1993.
79. Chander, K. and Brookes, P. C., Residual effects of zinc, copper and nickel in sewage sludge on microbial biomass in a sandy loam, *Soil Biol. Biochem.*, 25, 1231, 1993.
80. Hauck, R. D., Slow-release and bioinhibitor-amended nitrogen fertilizers, in *Fertilizer Technology and Use,* 3rd ed., Engelstad, O.P., Ed., Soil Science Society of America, Madison, WI, 1985, 293.
81. Nelson, D. W., Huber, D. M., and Warren, H. L., Nitrification inhibitors— Powerful tools to conserve fertilizer nitrogen, in *Agrochemicals in Soils*, Banin, A. and Kafkafi, U., Eds., Pergamon Press, Oxford, 1980, 57.
82. Martin, H. W., Graetz, D. A., Locasio, S. J., and Hensel, D. R., Nitrification inhibitor influences on potato, *Agron. J.*, 85, 651, 1993.
83. Fox, R. H. and Piekielek, W. P., Management and urease inhibitor effects on nitrogen use efficiency in no-till corn, *J. Prod. Agric.*, 6, 195, 1993.
84. Sarkar, A. and Faroda, A. S., Modified form of urea for pearl millet-wheat sequence under subtropical conditions, *Tropical Agric.*, 70, 279, 1993.
85. Owens, L. B., Edwards, W. M., and Van Keuren, R. W., Nitrate levels in shallow groundwater under pastures receiving ammonium nitrate or slow-release nitrogen fertilizer, *J. Environ. Qual.*, 21, 607, 1992.
86. Randall, G. W., Wells, K. L., and Hanway, J. J., Modern techniques in fertilizer application, in *Fertilizer Technology and Use,* 3rd ed., Engelstad, O. P., Ed., Soil Science Society of America, Madison, WI, 1985, 521.
87. Stecker, J. A., Buchholz, D. D., Hanson, R. G., Wollenhaupt, N. C., and McVay, K. A., Application placement and timing of nitrogen solution for no-till corn, *Agron. J.*, 85, 645, 1993.
88. Zentner, R. P., Campbell, C. A., and Selles, F., Build-up in soil available P and yield response of spring wheat to seed-placed P in a 24-year study in the brown soil zone, *Can. J. Soil Sci.,* 73, 173, 1993.
89. Wibawa, W. D., Dludlu, D. L., Swenson, L. J., Hopkins, D. G., and Dahnke, W. C., Variable fertilizer application based on yield goal, soil fertility, and soil map unit, *J. Prod. Agric.*, 6, 255, 1993.

90. Cook, R. J., Challenges and rewards of sustainable agriculture research and education, in *Sustainable Agriculture Research and Education in the Field*, National Research Council, National Academy Press, Washington, D. C., 1991, 32.
91. Graves, J. B., Ottea, J. A., Leonard, B. R., Burris, E., and Macinski, S., Insecticide resistance management: an integral part of IPM in cotton, *La. Agric.*, 36, 3, 1993.
92. van Lenteren, J. C., Biological control in protected crops: where do we go?, *Pestic. Sci.,* 36, 321, 1992.
93. Deacon, J. W. and Berry, L. A., Biocontrol of soil-borne plant pathogens: concepts and their application, *Pesticide Sci.* 37, 417, 1993.
94. Feitelson, J. S., Payne, J., and Kim, L., *Bacillus thuringiensis*: insects and beyond, *Biotechnology*, 10, 271, 1992.
95. Powell, K., Is biological control the answer for sustainable agriculture?, *Chem. Ind.,* No. 5, 168, 1992.
96. Powell, K. A. and Jutsum, A. R., Technical and commercial aspects of biocontrol products, *Pestic. Sci.,* 37, 315, 1993.
97. Hedin, P. A., Use of natural products in pest control, in *Naturally Occurring Pest Bioregulators*, Hedin, P. A., Ed., American Chemical Society, Washington, D.C., 1991, 1.
98. Zuckerman, B. M., Dicklow, M. B., and Acosta, N., A strain of *Bacillus thuringiensis* for the control of plant-parasitic nematodes, *Biocontrol Sci. Technol.*, 3, 41, 1993.
99. Ehteshamul-Haque, S. and Ghaffar, A., Use of Rhizobia in the control of root rot diseases of sunflower, okra, soybean, and mungbean, *J. Phytopath.*, 138, 157, 1993.
100. Fuxa, J. R., Richter, A. R., and McLeod, P. J., Virus kills soybean looper years after its introduction into Louisiana, *La. Agric.*, 35 (No. 3), 20, 1992.
101. Bing, L. A. and Lewis, L. C., Occurrence of the entomopathogen *Beauveria bassiana* (Balsamo) Vuillemin in different tillage regimes and in *Zea mays* L. and virulence towards *Ostrinia nubilalis* (Hubner), *Agric. Ecosys. Environ.*, 45, 147, 1993.
102. Shaver, T. N. and Brown, H. E., Evaluation of pheromone to disrupt mating of *Eoreuma loftini* (Lepidoptera: Pyralidae) in sugarcane, *J. Econ. Entomol.*, 86, 377, 1993.
103. Hammond, A. M. and Aubrey, J. G., Using sex pheromones to trap insect pests, *Louisina Agric.,* 29 (No. 4), 6, 1986.
104. Chandler, L. D., Use of feeding stimulants to enhance insect growth regulator-induced mortality of fall armyworm (Lepidoptera: Noctuidae) larvae, *Fla. Entomol.,* 76, 316, 1993.
105. Croft, B. A., Messing, R. H., Dunley, J. E., and Strong, W. B., Effects of humidity on eggs and immatures of *Neoseiulus fallacis, Amblysieus andersoni, Metaseiulus occidentalis* and *Typhlodromus pyri* (Phytoseiidae): implications for biological control on apple, caneberry, strawberry and hop, *Exp. Appl. Acarol.*, 17, 451, 1993.
106. Jalali, S. K. and Singh, S. P., Susceptibility of various stages of *Trichogrammatoidea armigera* Nagaraja to some pesticides and effect of residues on survival and parasitizing ability, *Biocontrol Sci. Technol.,* 3, 21, 1993.
107. Sands, D. C. and Miller, R. V., Evolving strategies for biological control of weeds with plant pathogens, *Pestic. Sci.,* 37, 399, 1993.
108. Rattink, H., Targets for pathology research in protected crops, *Pestic. Sci.,* 36, 385, 1992.
109. Fokkema, N. J., Opportunities and problems of control of foliar pathogens with micro-organisms, *Pestic. Sci.,* 37, 411, 1993.

110. Wood, H. A. and Hughes, P. R., Biopesticides, *Science*, 261, 277, 1993.
111. Duke, S. O., Christy, A. L., Hess, F. D., and Holt, J. S., *Herbicide-resistant crops*, Council for Agricultural Science and Technology, Ames, IA, 1991.
112. Wilkinson, M., Harding, K., O'Brien, E., Dubbels, S., Charters, Y., and Lawson, H., Herbicides and transgenic rape, *Nature (London)*, 365, 114, 1993.
113. Wicks, G. A., Martin, A. R., and Mahnken, G. W., Control of triazine-resistant kochia *(Kochia scoparia)* in conservation tillage corn *(Zea mays)*, *Weed Sci.*, 41, 225, 1993.
114. Jordan, N., Prospects for weed control through crop interference, *Ecol. Appl.*, 3, 84, 1993.
115. Kremer, R. J., Management of weed seed banks with microorganisms, *Ecol. Appl.*, 3, 42, 1993.
116. Van Acker, R. C., Swanton, C. J., and Weise, S. F., The critical period of weed control in soybean [*Glycine max* (L.) Merr.], *Weed Sci.*, 41, 194, 1993.
117. Woolley, B. L., Michaels, T. E., Hall, M. R., and Swanton, C. J., The critical period of weed control in white bean *(Phaseolus vulgaris)*, *Weed Sci.*, 41, 180, 1993.
118. Pantone, D. J., Young, R. A., Buhler, D. D., Eberlein, C. V., Koskinen, W. C., and Forcella, F., Water quality impacts associated with pre- and postemergence applications of atrazine in maize, *J. Environ. Qual.*, 21, 567, 1992.
119. Schreiber, M. M., Hickman, M. V., and Gail, G. D., Starch-encapsulated atrazine: efficacy and transport, *J. Environ. Qual.*, 22, 443, 1993.
120. Kumar, B., Yaduraju, N. T., Ahuja, K. N., and Prasad, D., Effect of soil solarization on weeds and nematodes under tropical Indian conditions, *Weed Res.*, 33, 423, 1993.
121. Gullino, M. L., Integrated control of diseases in closed systems in the sub-tropics, *Pestic. Sci.*, 36, 335, 1992.
122. Micinski, S., Caldwell, W. D., Fitzpatrick, B. J., and Griffin, R. C., First Louisiana field trial of insect-resistant transgenic cotton, *La. Agric.*, 35 (No. 5), 8, 1992.
123. Pataky, J. K. and Eastburn, D. M., Comparing partial resistance to *Paccinia sorghi* and applications of fungicides for controlling common rust on sweet corn, *Phytopathology*, 83, 1046, 1993.
124. Moffat, A. S., Improving plant disease resistance, *Science*, 257, 482, 1992.
125. Bethlenfalvay, G. J. and Linderman, R. G., Eds., *Mycorrhizae in Sustainable Agriculture*, American Society of Agronomy, Madison, WI, 1992.
126. Gavito, M. E. and Varela, L., Seasonal dynamics of mycorrhizal associations in maize fields under low input agriculture, *Agric. Ecosys. Environ.*, 45, 275, 1993.
127. McGonigle, T. P. and Miller, M. H., Mycorrhizal development and phosphorus absorption in maize under conventional and reduced tillage, *Soil Sci. Soc. Am. J.*, 57, 1002, 1993.
128. Henzell, E. F., The role of biological nitrogen fixation research in solving problems in tropical agriculture, *Plant Soil*, 108, 15, 1988.
129. Fujita, K., Ofosu-Budu, K. G., and Ogata, S., Biological nitrogen fixation in mixed legume-cereal cropping systems, *Plant Soil*, 141, 155, 1992.
130. Chalk, P. M., The contribution of associative and symbiotic nitrogen fixation to the nitrogen nutrition of non-legumes, *Plant Soil*, 132, 29, 1991.
131. Moyer, D. D. and Niemann, B. J., Jr., Economic impacts of LIS technology upon sustainable natural-resource and agricultural management, *Surv. Land Inform. Syst.*, 51, 17, 1991.

132. McBride, R. A. and Bober, M. L., Quantified evaluation of agricultural soil capability at the local scale: a GIS-assisted case study from Ontario, Canada, *Soil Use Manage.*, 9, 58, 1993.
133. Ghosh, T. K., Environmental impacts analysis of desertification through remote sensing and land based information system, *J. Arid Environ.*, 25, 141, 1993.
134. Jones, C. A., Dyke, P. T., Williams, J. R., Kiniry, J. R., Benson, V. W., and Griggs, R. H., EPIC: an operational model for evaluation of agricultural sustainability, *Agric. Syst.*, 37, 341, 1991.
135. Ball, D. A. and Shaffer, M. J., Simulating resource competition in multispecies agricultural plant communities, *Weed Res.*, 33, 299, 1993.
136. Paterson, K. G. and Schnoor, J. L., Vegetative alteration of nitrate fate in unsaturated zone, *J. Environ. Eng.*, 119, 986, 1993.
137. CAST, *Alternative Agriculture, Scientists' Review*, Council for Agricultural Science and Technology, Ames, IA, 1990.
138. Papendick, R. I., Elliott, L. F., and Dahlgren, R. B., Environmental consequences of modern production agriculture: how can alternative agriculture address these issues and concerns?, *Am. J. Alternative Agric.*, 1, 3, 1986.
139. Mallawatantri, A. P. and Mulla, D. J., Herbicide adsorption and organic carbon contents on adjacent low-input versus conventional farms, *J. Environ. Qual.*, 21, 546, 1992.
140. Reganold, J. P., Palmer, A. S., Lockhart, J. C., and Macgregor, A. N., Soil quality and financial performance of biodynamic and conventional farms in New Zealand, *Science*, 260, 344, 1993.
141. Pimental, D., Economics and energetics of organic and conventional farming, *J. Agric. Environ. Ethics*, 6, 53, 1993.
142. Buttel, F. H., Gillespie, G. W., Jr., Janke, R., Caldwell, B., and Sarrantonio, M., Reduced-input agricultural systems: rationale and prospects, *Am. J. Alternative Agric.*, 1, 58, 1986.
143. Holmes, B., Can sustainable farming win the battle of the bottom line?, *Science*, 260, 1893, 1993.
144. Smith, R., Analyst warns U.S. agriculture it may miss 'greatest last opportunity', *Feedstuffs*, 65 (No. 41), 3, 1993.
145. Swaminathan, M. S., Environment and global food security, *Food Technol.*, (July), 89, 1992.
146. Ruttan, V. W., Concerns about resources and the environment, in *Sustainable Agriculture and the Environment, Perspective on Growth and Constraints*, Ruttan, V. W., Ed., Westview Press, Boulder, CO, 1992, 1.
147. Gilland, B., Cereals, nitrogen and population: an assessment of the global trends, *Endeavor*, 17, 84, 1993.
148. Hess, E. H., The U.S. Department of Agriculture commitment to sustainable agriculture, in *Sustainable Agriculture Research and Education in the Field*, National Research Council, National Academy Press, Washington, D.C., 1991, 13.
149. Abelson, P. H., Pathological growth of regulations, *Science*, 260, 1859, 1993.
150. Browner, C. M., Protecting the environment: EPA's role, *Science*, 261, 1373, 1993.
151. Carnegie Commission, *Environmental Research and Development, Strengthening the Federal Infrastructure*, Carnegie Commission of Science, Technology, and Government, New York, 1992.

152. Carnegie Commission, *Enabling the Future, Linking Science and Technology to Societal Goals*, Carnegie Commission of Science, Technology, and Government, New York, 1992.
153. Carnegie Commission, *Risk and the Environment, Improving Regulatory Decision Making*, Carnegie Commission of Science, Technology, and Government, New York, 1993.
154. Stone, R., Researchers score victory over pesticides — and pests — in Asia, *Science*, 256, 1273, 1992.
155. Graham, F., Jr., *Since Silent Spring*, Houghton Mifflin, Boston, 1970.

CHAPTER 8

Biotic Effects of Soil Microbial Amendments

David J. Glass

TABLE OF CONTENTS

0-87371-860-7/95/$0.00+$.50
© 1995 by CRC Press, Inc.

I. INTRODUCTION

This chapter will describe the impact of microbial soil amendments on biotic systems. A large number of naturally occurring microorganisms have been used for decades in a variety of agricultural and commercial applications. Inasmuch as they generally compete with traditional chemical inputs (in the case of agriculture) and established industrial practices (i.e., biological waste treatment methods), use of microbes usually represents a small segment of each affected market. Recent advances in biotechnology, fermentation technology, and related technologies have focused greater attention on microbial technologies and have enhanced the expectations for their performance in the marketplace.

In spite of their limited utility to date, the use of microorganisms in the environment offers great promise for a number of reasons. As will be described below, microbial products for agriculture are widely expected to lead to a gradual decrease in the amount of chemicals used to produce the world's food supply. Optimistic thinkers can envision a time, perhaps 50 years hence, when the chemistry of crop production is delivered largely through microorganisms: nitrogen-fixing microbes and other rhizobacteria to deliver nitrogen or to facilitate uptake of phosphorus and other nutrients; and soil- and leaf-dwelling bacteria and viruses to repel insect pests or kill invading weeds. Similarly, the use of indigenous and introduced microorganisms in the remediation and management of hazardous (and perhaps radioactive) wastes has become more respectable and has become an accepted tool in our anti-pollution arsenal.

The widespread use of microorganisms, natural and genetically engineered, may not come entirely without risk. The advent of genetic engineering in the 1970s and 1980s caused a number of questions to be raised about the potential environmental effects of genetically engineered microbes (GEMs) in the environment. The development of recombinant DNA techniques in the early and mid-1970s was met with an unprecedented examination of the implications of the technology for human health, environmental safety, and ethics. Although many of the earlier concerns later proved to be unfounded or overstated, an aura of risk persisted until the first environmental uses of engineered organisms were proposed in 1983.

Beginning at that time, a number of public interest activists began to examine such uses closely, raising questions over possible unintended perturbations in the microbial ecosystem that might arise from the introduction of GEMs. These activists have succeeded not only in creating a substantial public policy debate over GEMs in the environment, but also in framing that debate on their own terms. Since the mid-1980s, an extraordinary public discussion has been conducted under the shadow of a pervasive perception that genetic engineering is inherently risky, and that large-scale uses of engineered organisms in the environment posed unacceptable risks to society. In reality, many of these concerns should more appropriately be raised about any microbial introduction, most of which have historically never been regulated. As will be discussed below, the more reputable of these concerns deserve, and in many cases have received, serious scientific scrutiny. Although sound scientific risk assessment is gradually making headway, the tenor of the debate has created substantial problems of public perceptions that have hindered many branches of biotechnology.

This chapter, therefore, will examine the current and potential future uses of microorganisms in the environment, particularly as soil amendments in agriculture and waste remediation. The benefits of such uses will be weighed against the risks, real and imagined, with the hope of showing that the coming biological era in agriculture and waste management offers tangible environmental benefits to biotic systems.

II. HISTORICAL BACKGROUND

Microorganisms have a long history of use in the environment. Processes now known to be microbiological in nature have been used for years and were in many cases known to our ancestors. In recent decades, mankind has discovered and learned how to harness the responsible microbial agents.

For example, the role of bacteria in replenishing nutrients in agricultural soil was recognized as long ago as the time of the Romans, who discovered that cereal crops planted in a field used previously for legumes benefit from the residual nutrients. The microbiological basis for this, of course, was not discovered until the 19th century, when alfalfa farmers in Germany, and later in Wisconsin, found that crop growth was improved by inoculating a new field with soil from one in which alfalfa was previously grown.[1] The microorganisms responsible for this process are now known to be the rhizobia *(Rhizobium* and *Bradyrhizobium).* The discovery of these bacteria led to the formation in 1898 of the Nitragin Company, the first U.S. company to manufacture and sell bacterial cultures for use as soil amendments. Since that time, commercial products have been routinely available for farmers to supplement natural populations of rhizobia with additional bacteria added to the soil or seed at the time of planting (see below).

In the early decades of the twentieth century, other microbial species were investigated as crop additives, although most of these fell out of favor.[1] More recently, the insecticidal properties of naturally occurring microorganisms were discovered. In the 1940s, the U.S. Department of Agriculture registered the first bacterial insecticides: strains of *Bacillus popilliae* and *Bacillus lentimorbus* for the control of Japanese beetle larvae on turf.[2] The most common bacterium used in pest control today is *Bacillus thuringiensis*, which was first identified in the early part of the century as a deleterious agent against silkworms. Today, many distinct strains of *B. thuringiensis* are known, each of which produces a unique endotoxin protein that is toxic to a different order of insects (see Reference 3 and below). Over 20 different commercial products, comprising 4 different naturally occurring *B. thuringiensis* strains and several genetically modified versions, are registered in the U.S. and elsewhere as microbial pesticides.[4] Finally, more than a dozen species of naturally occurring insect viruses (baculoviruses) have been used in insecticidal preparations since the nineteenth century for the control of arthropod pests (see references in Chapter 7 of Reference 5). The viral nature of these preparations was first discovered in 1947.[6]

Microorganisms have long been used in aerobic treatment of waste waters, although historically this process has relied on naturally present microbial communities.[5] The development of freeze-drying techniques in the 1950s led to the availability of selected cultures for use in waste water treatment plants and other applications.[7] Additionally, beginning in the late 1960s microbial strains have been used to clean oily bilges in tankers and other ships.[8] More recently, many of these strains have begun to be used as soil amendments for the purpose of environmental remediation (Reference 7 and see below).

Similarly, microbial involvement in mining and metal leaching has been known for almost two and a half centuries. In 1752, it was discovered that the blue-green aqueous runoff from waste rock at the Rio Tinto copper mine in Spain actually contained 80 to 90% copper.[9,10] Although this phenomenon was used commercially

Table 1 Microbial Soil Amendments: Advantages And Disadvantages

Advantages	
	Specificity to target species (e.g., pest)
	Little to no toxicity to nontarget species
	Little to no environmental risks
	Low costs of application
Disadvantages	
	Range of action limited to specific target species
	Limited duration of action, requires multiple applications
	Low potency
	Limited shelf life
	Farmer unfamiliarity

Source: Glass & Lindemann, 1992 (ref. 4). Used by permission of Decision Reources, Inc.

as early as 1907 in a South African copper mine, it was not known until 1947 that a microorganism, *Thiobacillus ferridoxans,* was responsible.[9,10] Today, over 30% of the world's copper, as well as significant percentages of other minerals, are mined using biological methods,[9] although this generally involves only stimulation of indigenous populations, not addition of exogenous cultures.[10]

III. THE PROCESS OF COMMERCIALIZING MICROBIAL SOIL AMENDMENTS

A. THE NEED FOR STRAIN IMPROVEMENT

In spite of this history, microbial products have enjoyed only limited commercial usefulness, and it has only been since the early 1980s that increased attention has been focused on microbial soil amendments. This is because these products have real or perceived problems with effectiveness, range of activity, and duration and mode of action.[1,4] For example, since microbial products are composed of living organisms, they must be produced and formulated in ways that maintain their viability and biological activity. Many existing microbial products therefore have limited shelf lives. Another problem is that the living organisms must survive in the environment long enough to have the desired effect. When they don't, the product loses effectiveness, or must be applied several times during an agricultural growing season. Microbial products often have much more limited ranges of action, whereas chemical products may be applicable to a broader range of target species (see Table 1).

Another problem lies with the microbes themselves. Naturally occurring strains are rarely well suited for commercial use and often must be improved in the laboratory. Historically, this has been done by traditional mutation and selection. Mankind has been manipulating plants, animals, and microorganisms for centuries, using traditional techniques.[5,11] The advent of recombinant DNA genetic engineering created new opportunities to enhance microbial strain performance in specific, directed ways.

Table 2 Solving Problems Through Biotechnology

Deficiency	Solution
Limited shelf life	Improved preservation or coating technologies, improved scale-up through fermentation technology
Effectiveness	Improved toxin production through rDNA, improved scale-up through fermentation technology, improved delivery technology
Range of action	Enhanced host range through conjugation, rDNA
Duration of action	Enhanced environmental survival through rDNA, encapsulation

Source: Glass & Lindemann, 1992 (ref. 4). Used by permission of Decision Reources, Inc.

Once an optimal strain is obtained, methods of large-scale production must be developed in order to manufacture enough of the culture for commercial use, and ways must be found to preserve the cultures and their viability. This too can require significant laboratory experimentation. Advances in fermentation, preservation, and other delivery methods have occurred simultaneously with the development of new genetic technologies. These advances are generally believed to hold the potential to solve the problems that have historically hindered the adoption of microbial technologies (see Table 2). However, biotech-derived microbial products will still face many of the same obstacles as traditional microbials. In addition, they have to compete in the marketplace with dozens of more common chemical products, which are more familiar to the end user (e.g., the farmer).

The process of discovery, strain improvement, and commercialization of microbial products is discussed in detail elsewhere.[12] The following is a summary of several of the more relevant steps.

B. DISCOVERY AND DEVELOPMENT OF MICROBIAL SOIL AMENDMENTS

Development of a new soil microbial treatment begins with discovery of a useful naturally occurring organism. This is generally not serendipitous, but instead occurs as a result of extensive screening of natural soil isolates for strains having the desired traits. Discovery of a desired strain is expected to be a rare event, and it is not unusual to screen hundreds or thousands of bacterial colonies to find a single strain possessing a suitable combination of characteristics with commercial potential.

For pesticidal organisms, such screening can involve selection of healthy plants from areas otherwise infested with the targeted disease, followed by recovery of microbial colonies from the roots or rhizosphere on the assumption that isolates will include strains antagonistic to the pathogen. Isolates can then be screened for activity against the pest in a laboratory assay (e.g., growth inhibition on petri plates or plant assays). With nitrogen-fixing organisms, the objective is often to obtain novel strains that are superior at nodulating the host plant, particularly under competitive conditions. Biodegradative strains are commonly isolated from actual contaminated sites in the environment, or are often selected from strain collections that were isolated from unique environments, such as deep subsurface regions, hot springs, etc. (see, for example, Reference 13).

C. METHODS OF STRAIN IMPROVEMENT

Once soil isolates with desirable traits are obtained, they are usually improved through laboratory manipulation. Traditional mutation and selection involves subjecting microbial isolates to a mutagenizing agent, in order to induce a wider range of genetic variability than would arise spontaneously. Mutants with the desired properties are then selected by a suitable biochemical assay. Alternatively, cultures can be grown under some selective pressure whereby desired variants are better suited to survive. For example, by growing a biodegradative culture in a continuous fermenter with the target contaminant as the only food source, one can select for a population of organisms having enhanced ability to degrade the contaminant, because those organisms will grow and reproduce faster than other microbes.

Many of the same objectives can be addressed through genetic engineering. Genetic engineering, or recombinant DNA technology, allows the identification and isolation of genetic material of interest from anywhere in nature, whereupon the isolated gene can be transferred into a different organism to modify its behavior or biochemistry. Although often requiring extensive (and expensive) laboratory research, genetic engineering offers the potential for more powerful, more directed ways of strain improvement.

Although recombinant DNA techniques are well described elsewhere,[14] the following is a brief overview. The genetic material of any living organism resides in its DNA (deoxyribonucleic acid). DNA molecules are polymers where the monomeric units are four different carbohydrate-containing nucleotides — adenine, thymidine, guanine, and cytosine — joined at the sugars to create the backbone of the DNA molecule.

Each living cell includes one or more long molecules of DNA called chromosomes; on each chromosome are thousands of discrete units called genes that are the commonly recognized units of heredity. DNA is an information-carrying molecule, and each gene carries a distinct piece of information: the instructions that tell the cell how to construct a specific protein molecule. Proteins are ubiquitous molecules in all cells, which play a myriad of roles in the cell's growth and metabolism. Most proteins are enzymes: the biological catalysts responsible for all the cell's metabolic reactions, while other proteins play a structural role. In fact, biodegradative pathways are made up of sequential steps of enzyme-catalyzed reactions. All proteins are made up of linear chains of building blocks called amino acids, and each protein folds into a distinctive three-dimensional structure that gives the molecule its function.

The information of a gene is written in a four-letter code, in which the order of the four nucleotide bases tells the cell the order in which to insert the amino acid building blocks of the protein encoded by the gene. Each "triplet" of three successive nucleotides encodes a specific amino acid. In order to read this code, the cell first "transcribes" a copy of the DNA code from the chromosome into a complementary molecule of ribonucleic acid (RNA), preserving the order of the nucleotides. This RNA molecule, called a "messenger", is able to diffuse away from the chromosome to cellular structures called ribosomes, which are the site of protein synthesis. At the ribosomes, the cell reads the messenger RNA, in a process called "translation", and constructs a protein by inserting into the growing chain the amino acid that is specified by each triplet in the code. After protein synthesis is complete, the protein folds into its proper three-dimensional structure.

The key to genetic engineering is the fact that the genetic material of all species is chemically identical: the same genetic building blocks are used throughout nature, from single-celled bacteria to complex multicellular plant and animal life. Once the needed techniques and research tools were worked out, it became rather simple to isolate DNA from one organism and chemically splice it into the DNA of another. Because the genetic code is identical throughout nature, it is possible for the second organism to successfully read the genetic code of the DNA transplanted from the first organism.

As a practical matter, several key research tools are necessary to accomplish this, particularly the enzymes that cleave DNA in precise locations corresponding to specific sequences of nucleotides. Called "restriction enzymes", these are the key to creating DNA segments with matching ends that can be spliced one to the other. The other important elements of gene-splicing are small, naturally occurring DNA molecules called "plasmids". Because of their mobility and ease of use, these are commonly used as vectors to insert foreign DNA into an organism. Finally, it is often necessary to identify the genetic control switches (called "promoters") by which cells activate genes so that their message can be read. Generally, foreign DNA must be linked to a promoter that the host organism can recognize, in order for the new genetic material to be expressed in the host cell.

The commercial applicability of recombinant DNA techniques has been well documented.[15] The greatest commercial opportunities have relied on the identification and isolation of human genes involved in diseases or in maintaining the body's health, and then using genetically engineered bacteria to mass-produce the proteins that are the products of those genes. These proteins are then used as therapeutic products. The use of recombinant DNA in soil microbiology is somewhat different. Here the goal is to identify genes whose protein products (usually enzymes) would enhance the desired function of the soil microbe. For example, genes encoding a biodegradative enzyme might be inserted into a natural soil isolate, to enhance a naturally occurring biodegradative pathway or create an entirely new one.[16] Another example is nitrogen fixation in rhizobia, where recombinant DNA techniques have been used to introduce specific changes to key genes involved in the nitrogen fixation process.[17] In the future it may also be possible to modify or expand the host range of rhizobia by modifying genetic control of the nodulation process, and, perhaps, extend symbiotic nitrogen fixation beyond the legume family.

Another example would be to enhance naturally occurring pesticidal traits. As is discussed below, there are a number of bacterial species where insecticidal properties are conferred by a single protein endotoxin. As reviewed in Reference 18, many research groups are working to improve these microorganisms by transferring the protein toxins to other host microbes, to broaden the host range of the recipient, or to deliver the toxin to a part of the environment where it cannot currently be delivered (e.g., the root surface: see below).

The range of ways in which advanced biotechnologies are being used to improve soil microbial products is described in more detail elsewhere.[12,19,20] At this writing, genetically engineered soil microbial products are just beginning to move through the final stages of the commercialization process. Three pesticides comprising genetically engineered (killed) microbes have been approved by the U.S. EPA for commercial use, and others are in field testing. Although there is considerable academic and industrial research directed at improving microbial products through genetic engineering, a number of economic, regulatory, and public policy factors have hindered the entry of these products into the marketplace.[12]

Recombinant DNA is also being used to genetically engineer crop plants.[21,22] Most of these products affect traits that are outside the scope of this chapter (e.g., nutritional improvements). However, plants being created to be pest resistant will compete in the market with microbial pesticides, and those that are herbicide tolerant will affect agrichemical usage in other ways. Because of the potential environmental effects of these products, they will be briefly considered below.

D. FIELD TESTING AND OTHER STEPS TO COMMERCIALIZATION

Following strain selection, there are a number of other steps that are needed before a microbial strain can be commercialized.[12] Before a company can attempt to sell an agricultural product, it must be tested in situations that approximate actual farming conditions. Testing usually begins in greenhouses or growth chambers, which are contained, small-scale facilities that allow limited similarity to field conditions. Although this often provides a useful preliminary screen for selecting candidate microorganisms for further testing, it is very difficult to mimic true farming conditions in a greenhouse or growth chamber.[1] For this reason, well-designed field trials are necessary to prove the efficacy of any product. Field programs generally start with limited trials in one or a very few locations followed by expanded trials if promising results are obtained. Agricultural products often need to be tested in different climatic regions, to prove their efficacy for farmers under different growing conditions.

Biodegradative organisms are usually tested in the laboratory ("bench scale") and often in pilot-scale field use before they are used in large-scale commercial activities. However, this practice is not universal, since many owners of contaminated sites are unwilling to pay for lengthy research and development programs. There are government programs that provide financial or other incentives to field-test innovative waste treatment technologies (e.g., the SITE program administered by the U.S. Environmental Protection Agency).[23,24]

Field testing is often the first instance where significant government regulation is encountered. Although many agricultural products can be tested outdoors at small acreage, genetically engineered products are subjected in many countries to special regulatory scrutiny.[12,25] Microbial pesticides must be registered by the EPA (or equivalent foreign agencies) before commercial use, while use of microbial cultures for bioremediation often must meet cleanup criteria imposed by federal or state regulation or might actually require a permit under such laws.[26]

Manufacture of soil microbial products requires fermentation of the microorganism in commercial quantities.[27] Because the microbe is itself the product (as opposed to being the means to produce a downstream product), fermentation conditions must produce high cell titers while preserving high levels of cell viability.

Aqueous fermentation has historically been the preferred approach. However, several companies have recently developed other approaches, such as semisolid-state fermentation for commercial production of microbial products.[12]

E. PRODUCT FORMULATION AND DELIVERY TECHNOLOGIES

Microbial products must, of course, be sold in a form in which they can easily be used by farmers or other end users. In this regard, microbial products present unique problems of formulation and application.[27] Agricultural products, for example, must be supplied to farmers in a familiar format which does not require new equipment, new technology, or new application techniques. However, microbial products must also be formulated so as to preserve cell viability and maximum biological activity under prolonged conditions of shipment and storage. Many existing microbial products for agriculture have failed in this regard.

Microbial products on the market today are generally applied using methods developed for agricultural chemicals. Soil or root colonizing bacteria needed at the outset of plant growth are applied as seed coatings, planter box mixes, or in-furrow applications. Soil microbes needed later in the growth cycle (e.g., mycoherbicides or insecticides) are sprayed onto the soil or plant surfaces by conventional means. Because coatings are applied by the seed manufacturer prior to shipping and sale, formulations must provide a long shelf life to keep cells viable during six or more months of storage. It has been reported that viable numbers of rhizobia applied to legume seeds may drop one order of magnitude in the first hour alone, with continuing die-off in subsequent weeks.[28] Seed coatings must allow release of cells into the soil and to the growing root after germination.

Most biodegradative organisms are sold today as blended mixtures of distinct strains. For commercial sale, large-scale cultures are freeze-dried along with a mix of nutrients. Generally sold as dried powder, these products are prepared for field use by rehydration. At least one company, however, grows biodegradative cultures in individual fermentation broths tailored for the contaminant(s) to be degraded and ships the cultures in suspension culture in 55-gallon drums.[29]

IV. MICROBIAL PRODUCTS FOR AGRICULTURE

A. OVERVIEW

There has been considerable research and commercial activity in developing microbial products for agriculture, primarily for crop protection and growth promotion.

A wide range of bacteria, algae, fungi, and viruses have natural properties which make them potentially useful for these purposes, many of which have been successfully marketed over the past few decades. However, as noted above, these products have achieved only limited market penetration. In the belief that advanced biological technologies like rDNA will expand the usefulness of microbes in agriculture, the last 15 years have seen increased research and commercial activity in agricultural microbiology. This section will describe the current status of microbial soil products being sold or under development for agricultural use.

B. MICROBIAL PESTICIDES

1. Market Overview

Natural and genetically engineered microorganisms are used as biopesticides: insecticides, fungicides, herbicides, and nematicides. The 1991 worldwide market for all pest control agents was about $20 billion and was made up almost entirely of chemical products.[4] The U.S. market represented about $5 billion of this total. Microbial pesticides accounted for $100 to 150 million in worldwide sales and $40 to 60 million in the U.S. in 1990. The world biopesticide market might grow to $250 to 400 million by 1995 and $1 to 2 billion by 2000. The U.S. market may grow to $100 to 180 million by 1995, and perhaps as much as $0.5 to 1 billion by 2000.[4]

Many naturally occurring microorganisms and viruses possess biochemical traits which make them natural antagonists to organisms considered agricultural pests. As noted above, specific commercial products have been sold since the 1940s, and more recently, significant research and commercial activity has been directed at the development of better microbial pesticides. In the 1980s, a number of start-up biotechnology companies began to focus on microbial pesticide development, while established agrichemical companies have more recently become active.

2. Insecticides

Historically, this has been the most common use of microorganisms in pest control, since several types of naturally occurring microorganisms and viruses are known to have specific insecticidal properties. Chief among these are the various strains of *Bacillus thuringiensis* (B.t.), which produce specific proteins toxic to different insect orders (see Table 3). The earliest strains to be identified were *B. thuringiensis* var. *kurstaki*, which is toxic to Lepidoptera (caterpillars), and *B. thuringiensis* var. *israelensis*, toxic to Diptera (mosquitoes and flies).[3] More recently, two strains toxic to Coleoptera (beetles) have been found: the tenebrionis variety and the San Diego variety (which may, in fact, be the same).[3]

The existing market for B.t. products is dominated by Abbott Laboratories and Sandoz Agro (formerly Sandoz Crop Protection), both of which sell B.t. products for several different applications. More recently, other companies have entered the market, particularly Novo Nordisk, which has acquired several biopesticide businesses

Table 3 Major Microbial Pesticide Products Available in the U.S.

Microbe	Target Pest	Product Examples
Insecticides		
Bacillus popilliae	Japanese beetle larvae	Doom, Japidemic (Fairfax Biological Labs)
Bacillus thuringiensis var. aizawai	Wax moth larvae	Certan (Sandoz)
Bacillus thuringiensis var. israelensis	Mosquito, blackfly	Vectobac, Gnatrol (Abbott) Skeetal (Novo Nordisk) Teknar (Sandoz)
Bacillus thuringiensis var. kurstaki	Lepidopteran larvae	Dipel (Abbott) Biobit, Foray (Novo Nordisk) Javelin, Thuricide (Sandoz)
Bacillus thuringiensis var. San Diego	Coleopteran larvae (Colorado potato beetle)	M-One (Mycogen)
Bacillus thuringiensis var. tenebrionis	Coleopteran larvae (Colorado potato beetle)	Novodor (Novo Nordisk)
B. thuringiensis EG2348	Gypsy moth larvae	Condor (Ecogen)
B. thuringiensis EG2371	Lepidpoteran larvae	Cutlass (Ecogen)
B. thuringiensis EG2424	Lepidoptera and Coleoptera	Foil (Ecogen)
P. fluorescens w/ B. t. kurstaki toxin	Caterpillars	MVP (Mycogen)
P. fluorescens w/ B. t. San Diego toxin	Colorado potato beetle	M-Trak (Mycogen)
P. fluorescens w/ B.t. toxin	European corn borer	M-Peril (Mycogen)
Nosema locustae	Grasshoppers	NoLo Bait (Evans BioControl)
Metarrhizum anisopliae	Cockroaches	Bio-Path Chamber (EcoScience)
Pine sawfly nuclear polyhedrosis virus	Pine sawfly	Virox (Novo Nordisk) Noecheck (USDA Forest Service)
Heliothis nuclear polyhedrosis virus	Cotton bollworm Cotton budworm	Elcar (Sandoz)
Gypsy moth nuclear polyhedrosis virus	Gypsy Moth	Gypcheck (USDA Forest Service)
Tussock moth nuclear polyhedrosis virus	Douglas-fir tussock moth	Tm-BioControl-1 (USDA Forest Service)
Spodeptera exiqua virus	Beet army worm	Spod-X (Crop Genetics International)
Fungicides		
Pseudomonas fluorescens	Damping-off fungi	Dagger G (Ecogen)
Trichoderma harzianum/ T. polysporum	Wood rot microorganisms	BINAB T (BINAB USA, Inc.)
Trichoderma harzianum (protoplast fusion)	Damping-off fungi	F-Stop (Eastman Kodak)
Gliocladium virens	Damping-off fungi	unnamed (W.R. Grace)
Psueomonas cepacia	Fungi, nematodes	Blue Circle (Stine Seeds)
Herbicides		
Phytophthora palmivora	Strangler vine	DeVine (Abbott)
Colletotrichum gleosporiodes	Northern joint vetch	Collego (Ecogen)
Bactericides		
Agrobacterium radiobacter	Crown gall disease	Gall-trol A (AgBioChem)

Adapted from Glass and Lindemann, ref. 4, used by permission of Decision Resources, Inc.

since 1987, and has formed a subsidiary, Entotech, to conduct advanced research on biopesticides.[4]

In recent years, B.t. products have been introduced by two biotechnology start-up companies. Ecogen, a Langhorne, PA biopesticide company, is now marketing three *B. thuringiensis* strains improved through genetic conjugation: Cutlass® (for leafy vegetables), Condor® (for trees and soybeans), and Foil® (for potatoes). Foil has an enhanced host range, created by genetic conjugation to be toxic both to Coleoptera and Lepidoptera.[4,12] These are all used as foliar sprays, while a granular form of Condor received U.S. regulatory approval in 1993.

Mycogen, a biopesticide firm based in San Diego, has received EPA approval for four B.t. products. In addition to its Colorado potato beetle product, M-One®, based on wild type B.t. *San Diego*, Mycogen sells three genetically engineered products. These are recombinant *Pseudomonas fluorescens* strains expressing different versions of the B.t. endotoxin, in an encapsulated form which kills the cells while preserving toxin activity.[30] This strategy enables delivery of large amounts of toxin, while also avoiding the tougher regulation faced by live recombinant products. One product, MVP®, expresses the B.t. *kurstaki* endotoxin, and is intended to control caterpillars on vegetable crops, while another, M-Trak™, contains the B.t. *San Diego* toxin gene, and has replaced M-One for use against potato beetle.[4,12] More recently, Mycogen obtained EPA approval for a third recombinant product, trade-named M-Peril, in which a killed recombinant *Pseudomonas* expresses a B.t. toxin active against the European corn borer.[31]

Both Ecogen and Sandoz have begun field tests of B.t. strains modified through recombinant DNA,[18] and Crop Genetics International has conducted several years of field tests of recombinant endophytic bacteria *(Clavibacter xyli)* expressing the *B. thuringiensis* var. *kurstaki* endotoxin.[18,32] The company is developing this product, InCide™, for the control of European corn borer and intends to sell the products already incorporated into corn seeds so that the microbes can grow inside the stalks. Several companies, including Ecogen and Ciba-Geigy, are experimenting with transconjugant strains of B.t., which express multiple toxins to broaden the host range.[18]

Another strategy is to use insect viruses, which are of course naturally suited to combat insect pests. Among the nuclear polyhedrosis viruses registered by EPA are Sandoz's Elcar™ for cotton pests and several viral agents developed by the USDA Forest Service (see Table 3). The specificity of their host range makes such viruses potentially attractive targets for use as pesticides, but these products have generally not been commercially successful due to their low level of virulence, poor persistence in the environment and shelf life, and other factors.[6,18] As reviewed in References 6 and 18, several research groups are pursuing a variety of genetic engineering strategies to overcome these problems by creating recombinant viruses expressing, for example, the B.t. toxin or insect juvenile hormone.

The Boyce Thompson Institute at Cornell University became the first U.S. group to conduct a field test of a genetically engineered insect virus in 1989, with a second test begun in 1993. Similar tests were done by the National Environmental Research Council's Institute of Virology in the U.K. several years earlier.[18,33] All these tests involved only genetically "marked" or crippled viruses, for research purposes only, and there has not yet been a field test of a baculovirus engineered for enhanced activity.

Among commercial entities developing improved viral products are Crop Genetics International, American Cyanamid, and the French company Calliope,[18] as well as Sandoz Agro and Biosys, who are collaborating in a recently announced venture.[34] In 1991, Crop Genetics International acquired Espro, Inc., a small company that was a toll manufacturer for the USDA's viral insecticides, and has established a joint venture with DuPont to market these products. The joint venture's first new product, Spod-X, was approved in 1993 for the control of beet army worm.

Among other insecticides being sold is the protozoan *Nosema locustae*, used for control of grasshoppers in semi-arid rangeland, sold by Evans Biocontrol under the tradename NoLo Bait™.[4,12] Under development are two fungi, *Beauveria bassiana* for control of Colorado potato beetle and *Metarhizium anisopliae,* which has been used in Brazil and China and is reportedly toxic to 200 different insects.[35] In 1993, the EPA granted EcoScience approval to sell a strain of *M. anisopliae* for home pest control use, in a cockroach trap.

3. Fungicides

Many important plant diseases are caused by soil or root-inhabiting fungi. Many naturally occurring microorganisms have been isolated which appear to have fungicidal activity;[36] however, very few products based on such organisms are yet being marketed (Table 3). Ecogen received EPA registration in 1988 to sell a strain of *Pseudomonas fluorescens,* tradenamed Dagger G, for control of cotton damping-off caused by fungal attacks on cotton seedlings. The company sold this product for 1 year, but later suspended production and sales.[12]

Eastman Kodak and BINAB USA, Inc. have both obtained EPA registration for fungicidal uses of *Trichoderma* strains, including *T. harzianum*, to control diseases such as damping off and root rot in cucumber and peas.[12] Gustafson, Inc. (a subsidiary of Uniroyal) sells a product called Quantum-4000™, a preparation of *Bacillus subtilis* to reduce the levels of *Rhizoctonia* and *Fusarum* on the root systems of peanuts.[37] In 1992, Gustafson received EPA approval to sell a different strain of *B. subtilis* as a fungicide to protect cotton, peanut, and bean crops from seedling diseases, under the tradename Kodiak.[38] The Finnish company Kemira Oy is awaiting EPA approval for a microbial fungicide based on a *Streptomyces* isolate with activity against *Fusarum* diseases, for use on vegetable and agronomic crops.[4]

There is a great deal of academic and industrial research on the usefulness of microorganisms as fungicides. Monsanto Agricultural Company maintained the most active corporate program, conducting several field tests of strains of *Pseudomonas*, originally isolated at Washington State University, with suspected antifungal activity against wheat take-all.[18,33,39] However, Monsanto has since terminated its microbial programs. EcoScience Corporation has obtained the rights to antifungal agents discovered at the University of Massachusetts, for disease control on cranberries, strawberries, and turfgrasses.

4. Herbicides

Herbicides represent the greatest sector of the agrichemical business. Bioherbicides are based on soil microorganisms (generally fungi) capable of selectively killing weedy plants without harming crop plants. There are only two bioherbicide products currently registered in the U.S. (see Table 3). The first of these, trade-named DeVine™, is a product consisting of *Phytophthora palmivora*, and was registered in 1981 for the control of strangler (milkweed) vine in Florida citrus groves.[40] It is marketed by Abbott Laboratories. The second product, originally registered by Upjohn Company in 1982, and now sold by Ecogen, is a formulation of *Colletotrichum gloeosporiodes* sold as Collego®, an aerial spray for control of northern jointvetch in rice and soybeans.[40]

As has been reviewed elsewhere,[4,12,40-42] other mycoherbicides are under development. For example, Mycogen has been developing at least two such products, including Casst™, a strain of *Alternaria cassiae* for control of coffee senna and sicklepod in peanuts and soybeans, and a strain of *Xanthomonas* for control of annual bluegrass in managed turf. Sandoz has conducted field tests of mutated strains of *Sclerotina sclerotiorum*, a fungus believed to be effective against Canada thistle and spotted knapweed, originally developed at Montana State University. Philom Bios, a Canadian company, is developing a strain of *Colletotrichum gloeosporiodes* (distinct from Collego) for control of roundleaf mallow on small-grain crops in the U.S. and Canada. EcoScience Corporation is targeting aquatic weeds like Eurasian milfoil.

Mycoherbicides are potentially attractive alternatives to chemical pesticides. In this case, the specificity of biologicals is desirable: a bioherbicide is naturally targeted for the specific weedy plant and no other, while chemical herbicides tend to have broader toxicity than is desired. However, there are few mycoherbicides in the pipeline because the needed research can be extremely time consuming, due to the difficulty in working with these fungi, and because, as plant pathogens, additional regulatory hurdles may often be needed.

5. Nematicides

There has been limited interest in microbial nematicides. Mycogen has discovered novel strains of *B. thuringiensis* whose endotoxins target parasitic nematodes and is developing nematicides based on these strains. Stine Seed Company recently received EPA approval to market a strain of *Pseudomonas cepacia* for control of fungi and nematodes. Igene Biotechnology, Inc. sells a product known as ClandoSan®, which, while not itself a microorganism, contains the biological polymer chitin which stimulates the growth of soil microbes that can combat nematodes. In spite of the limited interest to date in biological nematicides, opportunities exist because of the lack of any existing chemical products offering protection from the $1-1.5 billion in damage nematodes cause to the major crops in the U.S.[4]

6. Bactericides

Soil microorganisms can also be used to combat bacterial diseases. *Agrobacterium radiobacter* was approved by the EPA in 1979 to control crown gall disease caused by *Agrobacterium tumefaciens* on a variety of plants. It is sold by AgBioChem as Galltrol-A™. An Australian company, Bio-Care Technology, received approval in 1989 to sell NoGall™, a genetically engineered strain of *A. tumefaciens* deleted for an antibiotic resistance gene.[43] This product, also aimed at crown gall disease, is the only live recombinant-derived soil microbial product approved for commercial agricultural use anywhere in the world.

C. NITROGEN FIXATION

The single largest class of microorganism used in commercial agriculture are bacteria that fix nitrogen symbiotically with leguminous plants.[44] Collectively known as "rhizobia", cultures of the genera *Rhizobium* and *Bradyrhizobium* have been used since the 1890s as seed treatments, or "inoculants", to provide nitrogen to legumes, by converting atmospheric nitrogen into ammonia.[1]

Nitrogen fixation is a multistep enzymatic reaction by which atmospheric nitrogen is converted into ammonia. This process occurs in a diverse group of prokaryotes which primarily inhabit soils and freshwater lakes. Included are such species as *Klebsiella pneumoniae*, *Azotobacter vinelandii*, *Clostridium pasteurianium*, *Frankia* spp., photosynthetic bacteria such as *Rhodospirillum rubrum*, and cyanobacteria such as *Anabaena* and *Nostoc*.[44-46] However, the rhizobia are the only nitrogen-fixing organisms of significant commercial importance due to their symbiotic association with agriculturally important leguminous plants.

Through a complex process of host recognition and root infection, rhizobia enter the growing roots of legume plants to form nodules. The bacteria then differentiate into the nitrogen-fixing bacteroid form within the root nodule,[47] where they are packaged within plant cells. Nodules are the site of nitrogen fixation, and the ammonia which results is readily assimilated by plant roots. The process is species specific: a given rhizobial species generally infects only a limited number of plants within its "cross-inoculation group". In fact, rhizobial nomenclature is based upon the legumes which act as hosts. Thus, soybeans are nodulated by *Bradyrhizobium japonicum* (and by certain species of *Rhizobium fredii*, although these are not yet used commercially); alfalfa by *R. meliloti*; and peas, beans, and clover by subspecies of *R. leguminosarum*.[48] Commercial products composed of these species have been routinely available for farmers to supplement natural populations of rhizobia with additional bacteria added to the soil or seed at the time of planting. Although beneficial in newly planted fields, these products often provide little benefit in major legume-growing regions, due to the presence of large indigenous rhizobial populations (see below).

The worldwide retail market for rhizobial inoculants in 1991 was about \$20 to 25 million, with the U.S. accounting for about \$15 million (down from about \$19 million in 1984).[4] In the U.S., most of this market ($10 million in 1991) is for soybeans, although today less than 15% of the planted acreage of this crop is treated with rhizobia each year. Reasons for the decline in the market are the trend for soybean farmers to eliminate marginal inputs to cut costs, and the fact that inoculating soybeans is not believed useful in fields where soybeans have previously been grown. In addition, many existing products have poor reputations.

The U.S. market for alfalfa inoculants has been more stable, although smaller than that for soybeans: 80 to 90% of each year's planted acreage is inoculated with rhizobia (since most alfalfa seed is sold precoated with rhizobial inoculant), representing a 1991 U.S. market of $3 to 5 million.[4] Smaller markets exist for other crops such as peanuts, peas, beans, and clovers. Rhizobia are also sold to home gardeners. The U.S. market, which once included dozens of suppliers, is now dominated by two companies, Nitragin (now owned by Lipha S.A. of France) and Research Seeds, Inc.

Several companies, including Agricultural Genetics Company (U.K.) and Helibioagri (Italy) supply the European market. AGC sells its own product, NPPL, for soybeans and beans in the U.K., with licensees such as Rustica Semences in France selling in other European countries, and MicroBio RhizoGen serving the Canadian market. Helibioagri sells a wide range of microbial inoculants, including some strains supplied by Agracetus. Today, the European market for all rhizobial products may be no more than $5 million.[4]

Improved rhizobial products created using genetic engineering may expand the market over the next decade. Genetic engineering work has focused on improving either nitrogen fixation or competitiveness: the ability of the bacteria to preferentially colonize the plant root. For most of the 1980s, BioTechnica International was the commercial leader in attempting to enhance nitrogen fixation using recombinant DNA to introduce specific changes to key genes involved in the nitrogen fixation process.[17,49] The company conducted seven field tests over 3 years of several strains designed either for improved nitrogen fixation, or to test its system of chromosomal genetic engineering.[33,50] Preliminary results from BioTechnica's laboratory, greenhouse, and field studies indicated that it may be feasible to increase legume yields by engineering rhizobia.[51] In 1991, BioTechnica sold its rhizobia program to Research Seeds, which continued field tests in 1992. Expanded field trials are scheduled for 1993, with initial product introduction to begin in 1994 or 1995.[52]

Although BioTechnica and other companies have worked on improving the competitiveness of rhizobia, most of the current research is carried out at universities. At the University of Wisconsin, Triplett and colleagues are investigating the possibility of using specific rhizobial toxins to enhance inoculant nodulation by eliminating competitive strains,[53] and have constructed and field tested engineered strains expressing these toxins.

There are tangible problems in applying recombinant biotechnologies to nitrogen fixation; the microbes are slow growers that are difficult to engineer, and the biochemical pathways involve at least 20 genes. Moreover, the real problem is not efficiency of nitrogen fixation so much as it is improving competition: a problem involving many more genes and still not well enough understood to be amenable to a molecular approach. Solving the competitiveness problem, through strain isolation, classical genetics, seed coating technology or biotechnology, will be critical in developing improved rhizobia products in the future.

D. MYCORRHIZAL FUNGI

A diverse group of fungal species are known to promote plant growth.[54,55] They form symbiotic associations, called mycorrhizae, with the roots of most plants. The mycorrhizae may be external to the roots of the plant (ectomycorrhizae) or may penetrate the cell walls of the root cortex (endomycorrhizae). The former are generally associated with the roots of woody species like oak and conifers, and the latter with most herbaceous plants. Because of their association with a large number of crop plants, it is generally the endomycorrhizae that have any commercial potential.[54,55]

Although the mechanisms of action are still unclear, mycorrhizal fungi promote the growth of inoculated plants relative to control plants, probably through increased uptake of phosphorus through the roots. Mycorrhizae may also impart greater resistance to environmental stress. Applications of mycorrhizae have been beneficial to newly transplanted nursery stock, bedding plants, and vegetable seedlings. Mycorrhizal inoculation can also benefit sterilized or fumigated soils, where indigenous fungal populations have been killed.[54,55]

Only a few mycorrhizal inoculants are available commercially. In general, these products are sold in formulations to be added to soil, potting media, etc. when plants are planted or transplanted. Native Plants, Inc. has sold a product called Nutri-Link®, composed of fungal spores, suitable for mixing with soil or soil-less growth media. The Canadian company Les Tourbieres Premier Ltée has introduced Mycori-mix, an endomycorrhizal product in a peat moss based carrier.[4]

The primary benefit conferred by mycorrhizal products is increased phosphate uptake. However, chemical phosphate fertilizers remain cheap, easy to use, and do not suffer from limited shelf lives, as microbial products would. The additional research that might improve mycorrhizal action is difficult to do, since the fungi are slow growers. Mycorrhizae are likely to remain only a minor product.

E. PLANT GROWTH PROMOTING RHIZOBACTERIA

A number of other bacteria and fungi have been investigated for use as growth promoters for different crop plants.[36,56] Although in some cases, lab and greenhouse experiments have shown significant growth improvements, mechanisms for many of these effects are not yet understood. It is suspected that mechanisms of action might involve limiting the growth of soil pathogens in the rhizosphere (the soil surrounding the roots) through the production of antibiotics or iron-sequestering compounds (siderophores). Most of the commercially available products are poorly characterized and have only small markets.

F. FROST-PREVENTING MICROORGANISMS

Genetically engineered bacteria capable of inhibiting ice nucleation on crop plants were the first altered microorganisms proposed for field testing, and gained a great deal of (unwanted) publicity because of the delays that resulted from activist lawsuits and other legal wrangling.

It has long been known that certain species of *Pseudomonas* express a protein that is responsible for forming the nucleus for the initiation of ice crystals.[57,58] The presence of these bacteria on plant leaves facilitates frost formation and is responsible for much of the crop damage caused by frost. It has been possible to create strains of bacteria that do not express the nucleation protein, both through classical mutation and recombinant DNA gene deletion.[59] Field tests have shown that both types of variant are capable of protecting crop plants from freezing.[60]

DNA Plant Technology (DNAP) is the only company pursuing this market, by virtue of its acquisition of Advanced Genetic Sciences (AGS). AGS conducted several highly publicized field trials of recombinant-derived "ice-minus" deletions in the 1980s,[61] and DNAP is now continuing development through its Frost Technology Corporation joint venture with 3A/Umbria Park. DNAP has obtained EPA approval for a product, known as Frostban™, that is based on naturally occurring versions of the microbes, and the firm apparently has no plans to market the recombinant version. Interestingly, Genencor International markets a technology complementary to Frostban: microbes that overexpress the ice nucleation protein that are sold under the tradename Snomax™, for artificial snow-making at ski resorts.

V. MICROBIAL PRODUCTS FOR ENVIRONMENTAL RESTORATION

A. BIOREMEDIATION

1. Overview

The major use of microorganisms in environmental restoration is in bioremediation: the degradation or detoxification of hazardous materials in the environment using biological processes. Biological methods have been used in municipal waste water treatment for decades,[62] while in-ground *("in situ")* bioremediation was first used to clean up hydrocarbon-contaminated aquifers more than 15 years ago.[63] Although bioremediation is considered an innovative waste treatment method, it has gained acceptance in recent years and is now generally regarded as a viable technology for certain defined uses.

Bioremediation relies on the ability of microorganisms to metabolize a diverse array of organic compounds. This ranges from simple hydrolytic reactions, catalyzed by enzymes such as proteases or cellulases, to more complex multi-step pathways capable of degrading fatty acids, petroleum hydrocarbons, and certain aromatic compounds. Field experience and extensive research at hundreds of laboratories around the world have shown that many classes of compounds can be biodegraded.[64,65] The simple hydrocarbons found in petroleum distillates,[66-69] as well as single ring aromatic compounds like benzene, toluene, and the xylenes,[70] are among the easiest chemicals for microbes to degrade. There have also been documented successes with more complex compounds, such as the multi-ring aromatic compounds found in coal tars and creosotes, and with pentachlorophenol.[71] Although highly chlorinated and high molecular weight compounds are more difficult for microbes to metabolize, research has elucidated numerous microbial pathways for the degradation of halogenated compounds, including haloaliphatics,[72,73] haloaromatics,[73,74] and even halogenated polycylics.[73,75]

Because microorganisms can evolve relatively quickly to develop a biochemical trait that confers a selective advantage, those microbes at a contaminated site that acquire the ability to use a contaminant as a food source will eventually outcompete other microflora. So, it is likely that new pathways would eventually evolve for those man-made ("xenobiotic") compounds for which a naturally occurring metabolic pathway does not exist. One example are the polychlorinated biphenyls (PCBs), a significant class of environmental pollutants which have long been thought not to be biodegradable. Biological pathways for PCB degradation have recently been discovered,[75-77] although a combination of aerobic and anaerobic processes are likely to be needed for complete degradation. Other examples are a number of synthetic, chlorinated pesticides, which are known to be susceptible to biodegradation.[75,78-80]

Natural biodegradation by indigenous microorganisms occurs slowly because there may be too few microorganisms having degradative ability, or because environmental conditions hinder degradation, particularly if one or more needed nutrients are found in lower than optimum conditions. To make bioremediation commercially viable, therefore, it is necessary to stimulate the natural biodegradation of hazardous compounds to yield practical remediation times. This is commonly done in the following ways.

2. Stimulation of Indigenous Populations by Nutrient Addition

The most common method of bioremediation of soils and aquifers involves the stimulation of naturally occurring bacteria residing at the site of contamination. By addition of the appropriate nutrients, principally oxygen, carbon, phosphorus, and nitrogen, and by maintaining optimum conditions of pH, moisture, and other factors, it is possible to trigger increased multiplication of the communities of indigenous organisms that together are capable of degrading the wastes. This method, called *in situ* bioremediation or "biostimulation", does not require addition of microorganisms to the site and has been proven effective in many instances, especially for contaminated aquifers.[81,82]

In situ bioremediation was first practiced on hydrocarbon-contaminated aquifers in the early to mid-1970s,[63] and it remains a useful technique for groundwater remediation. It can be more effective than "pump-and-treat" methods where groundwater is treated above-ground and returned to the aquifer, since microbial action can potentially degrade the often-significant contamination that is tightly adsorbed to the solid matrix of the aquifer formation. Because this contamination cannot be removed by pumping, it can become a long-term source of continued leaching into the groundwater.

Most aquifers exist under anaerobic (oxygen-poor) conditions; so, oxygen is usually the most critical additive to stimulate bioremediation of groundwater.[81,83] It is usually added in the form of hydrogen peroxide, which decomposes to create oxygen in water.[81] Oxygen can also be added in gaseous form. In one approach, called "bioventing", oxygen is injected into the unsaturated zone above a water table, in order to stimulate biodegradation by indigenous organisms.[84] When oxygen is injected into the saturated zone (i.e., below the water table), oxygen bubbles rise into the unsaturated zone, where natural biodegradation can be stimulated. This is called "biosparging".[81] In most cases, the limiting factor governing success is whether the hydrogeologic characteristics of the aquifer enable sufficient oxygen to be transported to the site of the contamination to promote biodegradation.[83]

Biostimulation can also be used for contaminated soils.[82] Chemical nutrients are usually added in aqueous solutions, simply by sprinkling into the soil, or through more sophisticated plumbing. Oxygen can be added by mechanical means, or, where the contamination is in the top soil layers, simply by tilling or turning the soil. In fact, a common variation, sometimes called land-farming, is where chemical nutrients are added to soil, often in an excavated pile, while adequate oxygenation is assured by frequent turning or disking of the soil.

Although *in situ* bioremediation is most often practiced under aerobic conditions, it is known that many hydrocarbons, particularly the alkylbenzenes, can be biodegraded under anaerobic conditions.[81,85] Anaerobic bioremediation requires an alternative electron acceptor to drive the enzymatic reactions. The most common electron acceptor to be demonstrated in the field is nitrate, which has been shown to be capable of promoting the degradation of hydrocarbons (Reference 86 and other references in References 81 and 85).

In biostimulation, the soil amendments are not microorganisms themselves, but are the nutrients needed to make indigenous microbes grow. These are hydrogen peroxide (or other peroxides) or gaseous oxygen; nutrients like carbon, nitrogen, and phosphorus (e.g., similar to agricultural fertilizers); surfactants to solubilize oily wastes; and, when used as an alternative electron acceptor, nitrates. The amounts of these materials used, and the potential market sizes, are hard to estimate. The ecological effects of these additives will be considered below.

3. Augmentation of Selected Microbial Cultures

The major alternative to biostimulation relies on the introduction of cultured microorganisms to the site of contamination, to augment indigenous populations and increase the rate of natural biodegradation. This can be done in two ways. Perhaps the best approach from a scientific standpoint is to select the best waste-degrading microorganisms from the site to be remediated, characterize them and increase their numbers in the laboratory, and return the enriched, scaled-up cultures to the site.[7]

While this may improve the effectiveness of biostimulation, this approach requires a greater amount of laboratory work, is generally more costly, and is practiced less frequently.

Another strategy, called "bioaugmentation", is to introduce non-native cultures previously selected from other sites for their ability to degrade specific wastes. In contrast to the pure cultures generally used in agricultural applications, prepackaged microbial biodegradation products are almost always blends of different species or strains, tailored for the types of compounds found in the target waste stream.[7,8] Initial products were used for municipal waste water treatment or for biotreatment of restaurant grease traps and sewer lines. More recently, several companies have begun selling microbial blends purported to be active against hazardous compounds, including use against industrial effluents and for *in situ* waste remediation.[7,8,87] Although precise formulations and often the names of species are kept as trade secrets, these products typically include select biodegradative strains of common soil bacteria such as *Pseudomonas putida, P. fluorescens, Bacillus subtilis,* and others.

Today, a few dozen manufacturers control the U.S. market for packaged microbial biodegradation products. By far the major participants are Polybac, Sybron, Solmar Corporation, and ERI/InterBio Group (which comprises the merged businesses of Microbe Masters and InterBio). The 1990 market for packaged microbial cultures was estimated at $30 to 50 million in the U.S.[88] There are several broad areas in which these products are or can be used.

Municipal waste water treatment was one of the first markets for prepackaged enzymes and microorganisms. Today, several manufacturers sell products rich in lipases, proteases, and cellulases for use in activated sludge treatment lagoons or on-line biological reactors for waste water treatment. Smaller, specialized markets exist for pretreatment of waste waters from specific industries such as meat packing and fruit and vegetable processing. Another early use for microbes was in the treatment of grease traps or sewer lines, often in restaurants or other establishments where fatty deposits clog plumbing. This continues to represent a major sector of the business.[7,8]

The most common products for *in situ* waste remediation are formulations for degradation of hydrocarbons and petroleum distillates. The earlier of these strains have been used to clean oily bilges in tankers and other ships since the 1960s[8] and have recently garnered attention for their possible usefulness against oil spills on land and sea.[89,90] Examples include BioChem ABR Petroleum Blend (Sybron Chemicals), MicroPro Marine D (Environmental Remediation), and Alpha BioSea (Alpha Environmental) (see Table 4). The National Environmental Technical Applications Corporation (NETAC) and its Bioremediation Product Evaluation Center have conducted evaluations of available microbial products for oil spills and have developed protocols to test their effectiveness.[91] However, the usefulness of such cultures for oil spills has not yet been proven, in spite of several well-publicized attempts.[90]

Several manufacturers claim that these and other products are active against aromatic compounds and other hazardous chemicals. Many of the oil-spill products are sold for *in situ* waste cleanup, and examples of other products are shown in Table 5. Virtually all of these products are blends of different microbial species and strains, and many are claimed to have very broad spectra of biodegradative activity.

Table 4 Oil Spill Products Selected for Evaluation by the National Environmental Technology Applications Corporation (NETAC).

June 1990. Two products chosen from among 10 for testing on beaches in Alaska	
BioChem ABR Petroleum Blend	Sybron Chemicals, Inc.
MicroPro Marine D	Environmental Remediation, Inc.
December 1991. Ten products chosen from among 33 for further protocol evaluation	
Alpha Biosea Process	Alpha Environmental, Inc.
Biolyte CX85	Interbio, Inc.
Bioversal	BioVersal USA, Inc.
DBC R5	Enviroflow, Inc.
Lockheed Product	Lockheed Missles & Space Co., Inc.
Medina Soil Activator	Medina Bioremediaton Products
Mycobac TX-20	Mycobac, Inc.
Oil Spill Eater II	OSEI Corporation
Petrobac, Hydrobac	Polybac Corporation
WST Bioblend H-JM	Waste Stream Technology, Inc.

From Glass, 1993, ref. 87. Used by permission of Decision Resources, Inc.

Table 5 Representative Microbial Products Sold in the U.S. for Bioremediation of Soils Contaminated with Hazardous Wastes

Product	Company
MicroPro "Super Cee"	Environmental Remediation, Inc.
MicroPro D	Environmental Remediation, Inc.
Advanced Bio Culture L-104	Solmar Corporation
ABR products	Sybron Chemicals, Inc.
MICROCAT XBS	Bioscience, Inc.
Phenobac	Polybac, Inc.
WST Bioblend M-4, M-5	Waste Stream Technology

From Glass, 1993, ref. 87. Used by permission of Decision Resources, Inc.

Certain pure biodegradative cultures have commercial utility in hazardous waste remediation. The two most notable examples are species of *Flavobacterium*, capable of degrading pentachlorophenol (PCP) and similar compounds,[71] and the white-rot fungus *Phanerochaete chrysosporium*, which can degrade lignins and many other aromatic compounds such as PCP and PAHs.[92] *Flavobacterium* is often used in bioreactors for PCP degradation, but is also used *in situ* as a soil additive. The white-rot fungus has been known for many years to have broad biodegradative activity, by virtue of its secretion of lignoperoxidases. More recently, related fungi like *P. sordida* and *Trametes hirsuta* have also been discovered to have similar activity.[93,94] Recent advances in fermentation technology have made large-scale commercial manufacture feasible, and there are at least two commercial ventures marketing such products: Groundwater Technology, Inc. (in partnership with Mycotech, the developer of the products) and L. F. Lambert Spawn Company, which is collaborating with the U.S. Department of Agriculture.

In addition, biodegradative pathways for a variety of other hazardous wastes have attracted considerable research attention, leading to the identification of other potentially useful microbial strains. Numerous microbial species capable of degrading chlorinated aliphatics have been discovered, although these microbes generally utilize unrelated pathways that fortuitously can metabolize the contaminants of interest.[72] Species that can biodegrade trichloroethylene (TCE; perhaps the most common pollutant of groundwater) include the methanotrophs, which degrade TCE under anaerobic conditions in the presence of methane.[95] More recently were discovered a different class of microbes, generally pseudomonads, that degrade TCE aerobically through pathways that have evolved for breakdown of toluene and other aromatic molecules.[96,97]

Although the advanced techniques of genetic engineering have been used in research on biodegradative pathways, there has never been a commercial application of genetically engineered microbes in waste treatment. Naturally occurring or classically selected microbes have generally been powerful enough for the tasks at hand, and there has been little need to resort to advanced biotechnology. There are, however, some situations where biotechnology offers potential advantages: for example, to engineer a naturally occurring degradative pathway so that it is continuously active in the bacteria, even in the absence of a molecule ordinarily needed to activate the pathway;[97] to artificially create hybrid degradative pathways for xenobiotics;[16] or to introduce into a microbe a variant enzymatic activity altered by *in vitro* protein engineering.[98] Lindow, et al.[19] and Ensley[20] review the possible strategies being contemplated for the use of advanced biotechnologies for biodegradation.

Although several companies practice bioaugmentation for hazardous waste remediation, biostimulation remains the preferred approach. There is controversy within the bioremediation field about the value of introducing selected microbial cultures into soil or groundwater.[81,99] Although manufacturers of prepackaged cultures often claim their effectiveness for *in situ* remediation of hazardous wastes, other observers question the efficacy of introducing laboratory-selected microbial cultures into the highly complex natural environments of contaminated soil or groundwater, and believe that such introduced organisms will have difficulty competing with indigenous microflora or handling the stresses present in natural environments.[100,101] For example, the high concentrations of toxic compounds at waste sites might hinder the growth of lab-bred organisms, but would not affect those natural bacteria acclimated to such stresses. It has often proven difficult to conduct well-controlled experiments to verify whether the addition of selected cultures results in accelerated or enhanced biodegradation relative to that shown by indigenous microbes.[90] This is one factor that has prevented the commercial use of some of the above-mentioned strains now undergoing laboratory research. In fact, many observers believe that the most effective use of preselected strains will be in bioreactors: contained vessels where relatively well-defined waste streams are passed.

B. OTHER ENVIRONMENTAL APPLICATIONS

In addition to the bioremediation applications discussed above, there are other, albeit minor uses, for microbes in the environment. Other bioremedial uses of microbes include the use of rhizosphere bacteria to enhance the uptake and transformation of heavy metals in the plants with which the bacteria associate.[7] At present, this is an experimental technology, although one that is beginning to attract some attention. More importantly, there are numerous microbial processes known whereby heavy metals can be removed from contaminated media. As reviewed in Reference 7, the two categories of such processes are biosorption of metals from solution into microbial biomass, or the mobilization, solubilization, or (alternatively) immobilization of heavy metals by redox reactions. Biosorption is already used commercially, with at least one company, Bio-Recovery Systems, marketing a bioreactor with algal biomass to concentrate heavy metals.[102]

The other approach finds its commercial usefulness in the mining and metallurgy industries. The iron-oxidizing bacterium *Thiobacillus ferrooxidans* and related species are known to be able to solubilize a variety of metal ions from ores, particularly including copper and gold, and this microbial process has been used for decades in mineral leaching.[9,10] It was estimated that over 30% of the copper mined in the U.S. in 1989 resulted from this process.[9] However, commercial use of bioleaching has almost exclusively made use of indigenous microbes, with only one reported attempt at inoculating a mine with non-native strains.[10] Although *T. ferrooxidans* is now amenable to genetic manipulation, it is a rock-dwelling bacterium that is difficult to culture in the laboratory and may therefore never see widespread use as a commercial inoculant or soil additive.

VI. EFFECTS ON BIOTIC SYSTEMS

A. OVERVIEW

Microorganisms, as natural products with a long history of safe use, are considered by many to represent safe, beneficial products when used correctly in the environment. In fact, microbial products like *Bacillus thuringiensis* and rhizobia have long been favored by the organic farming community and are often considered an important part of sustainable programs such as Integrated Pest Management.[103,104]

However, the use of microbes in the environment has been dogged by controversy since the early 1980s, when the first proposals were made to use genetically engineered microorganisms (GEMs) in agricultural field testing. These first proposals met substantial opposition by local communities and public interest groups, fueled in part by legitimate scientific dissent regarding the possible environmental impacts of introduced GEMs. Early opinion from among concerned scientists focused on the supposed parallels between use of GEMs in the environment and the ecological damage that sometimes resulted from introduction of exotic species.[105] While not a perfect model, this line of thinking led to an ongoing debate within the scientific community over the potential ecological impacts of engineered microorganisms introduced into the environment. The resulting scientific controversy lasted well into the late 1980s and has profoundly affected research and commercial development of genetically improved microbial soil amendments. Curiously, however, these issues are really more relevant to the general question of the impact of releasing any nonindigenous microorganism into a new environment; in practice, these concerns were leveled only at GEMs.

This section will describe some of the possible impacts, both positive and negative, of microbial products introduced into agricultural soils or elsewhere in the ecosystem. It draws substantially on a significant body of literature, including surveys and scientific reviews of the use of nonengineered microbes[5,106-108] as well as the growing track record in the peer-reviewed literature and elsewhere reporting the results of government-regulated field trials with GEMs.[18,33,109,110] This discussion is not intended to be restricted solely to impacts of environmental uses of GEMs; however, because government-sanctioned GEM field tests have almost all required environmental monitoring using marked strains (see below), many of these trials have yielded useful data on microbial behavior in the environment. Consideration will be given to those specific uses of microorganisms that are likely to have a beneficial impact on the environment.

B. HISTORICAL PERSPECTIVE: GEMs AND THE RISK ASSESSMENT DEBATE

1. *Historical Perspective: Biotechnology and the Environment*

For the last ten or more years, consideration of the impacts of microorganisms introduced into the environment has been shaped by the debate over the ecological safety of GEMs, with many critics believing that introduction of novel organisms into the environment poses potential hazards requiring special assessment. Microorganisms (as well as plants and animals) have been genetically manipulated for centuries using classical techniques.[5,11] The modern era of biotechnology was ushered in in 1973, with the discovery of the means to manipulate the genetic make-up of living organisms at the molecular level, through the techniques of recombinant DNA (rDNA).[14] Public and media reaction to this discovery was considerable and led to an unprecedented public debate over the safety and ethical implications of recombining genes from different sources in nature.[111] In response to this controversy, leading scientists in the field convened an international symposium at the Asilomar conference center in California, to review the available scientific evidence about the possible risks of recombinant DNA technology. As described in more detail elsewhere,[111] this ultimately led in 1976 to the promulgation by the National Institutes of Health (NIH) of guidelines for the safe conduct of rDNA experiments.[112] These guidelines attempted to classify all possible experiments using rDNA and to assign physical and biological controls to each category of experiment, in order to ensure as much as possible that recombinant organisms would be contained within laboratories. These guidelines originally prohibited the deliberate outdoor release of any recombinant organism. Although the guidelines were binding only on institutions receiving federal funds, they have served as a safety standard for industry and academic researchers alike.

At first, the major safety concern involved potential public health hazards that might occur should altered organisms accidentally leave the laboratory.[111] Later risk assessment experiments and an accumulated record of safety lessened the level of concern over public health implications.[111] As the guidelines were progressively relaxed, deliberate releases of engineered organisms into the environment became possible on a case-by-case basis with the express permission of the NIH. In 1983, the NIH approved the first field tests of recombinant microorganisms (the "ice-minus" bacteria),[59] and this action attracted public controversy and litigation.[113] The resulting legal and regulatory uncertainty eventually led to the adoption of a comprehensive federal regulatory framework to ensure that all engineered organisms proposed for outdoor use receive some prerelease review (see Reference 25 and below).

Under this framework, almost all proposed GEM field tests have been subjected to a thorough risk assessment, focusing on the potential environmental effects of the introduced microorganism. Such assessments examined generally applicable questions such as the toxicity and/or infectivity of the GEM and its effects on other organisms in the biota, as well as traits that affect the ecological behavior of introduced strains, for example, whether the introduced microbes persist in the environment or outcompete natural populations.

The possibility that introduced GEMs could out-compete indigenous populations struck a receptive chord for many critics; examples abound of plants, animals, and, in some cases, microbes or viruses, which flourished upon introduction into a new habitat lacking its natural predators.[105] Many dispute the applicability of this model to GEMs, which in most cases would not be "exotic" but would be modifications to common soil microbes (see, for example, Reference 114, but note that this view has been vigorously disputed by proponents of the exotic species paradigm).[115]

However, it is this potential threat that captured the fancy of the public. Scare scenarios about microbes multiplying out of control were commonly voiced by public interest groups opposing field testing plans. The potential for microbes in the environment to multiply uncontrollably was cited by the EPA as one basis to justify its decision to subject small-scale field testing to government oversight even at acreages far below the usual regulatory cutoff levels.[116] As will be discussed below, the available evidence now contradicts this hypothesis.

Other environmental considerations arise for more specific uses of microbes. Environmental release of certain GEMs might entail the possibility of unwanted "horizontal" transfer of the introduced genetic material into other soil organisms. The concern is that a gene that is harmless in one host may persist in the environment after the original host is dead and have unexpected adverse effects when transferred to and expressed in another host. Use of biodegradative organisms might result in accumulation of toxic intermediates or by-products. For the special case of insecticidal microorganisms, there is now evidence that overuse of such agents could accelerate the creation of resistance in the target insect populations. All these potential risks are discussed in more detail below, in the context of the expected benefits of each proposed application.

Table 6 Factors Determining Environmental Impact of Microorganisms Used in the Environment

Release	Will the microbe be released?
Survival	Will the microbe survive in the environment?
Multiplication	Will the microbe proliferate?
Dissemination	Will the microbe be dispersed to distant sites?
Transfer	Will introduced genetic information be transferred?
Harm	Will the microbe be harmful?

Adapted from Alexander 1985, ref. 117.

2. *Risk Assessment Frameworks*

Several paradigms have been suggested over the years for the analysis or assessment of these risks. Alexander[117] suggested a six-part test to determine the probability that a GEM could create a deleterious effect (see Table 6). Alexander's first criterion is whether the microbe is accidentally released from a containment system or, as in the case of soil amendments, deliberately released into the environment. The next two criteria are the related questions of whether the GEM can survive and multiply in the environment, necessary factors for there to be a large-scale deleterious effect. Alexander then considered dissemination to points distant from the test site and the possibility that introduced genetic material could transfer to other species. Finally comes the consideration of "harm": will the organism be harmful (i.e., toxic, infective, etc.)?

Regulatory risk assessments have made good use of this paradigm, since such assessments calculate risk as the product of "exposure" and "hazard". Alexander's first five criteria address exposure questions: whether the organism or the introduced gene can persist long enough or be transported to a location suitable for it to cause a negative effect. Only the final criterion addresses the innate hazard of the GEM. However, in focusing on questions like survival and persistence, the paradigm may be misleading, since these traits in and of themselves are not necessarily deleterious. In fact, persistence is a required trait for soil amendments to have their desired effect. Many would argue that the final test, the harm of the organism, is the most relevant.[118]

Beginning in 1987, more in-depth studies were undertaken to survey the available literature regarding the ecological effects of introduced microorganisms (and plants) in the environment. A 1987 report from the National Academy of Sciences[107] presented a brief summary of some of the available literature and concluded that the potential risks associated with engineered organisms were no different than those of unmodified organisms or organisms modified using traditional techniques, and that assessment of such risks should be based on the make-up of the organism itself rather than the process of its construction. In light of the contemporaneous public policy debates, this document can best be considered to have served a political purpose in framing the scientific issues for further study.

A more detailed study was published in 1988 by the Congressional Office of Technology Assessment (OTA).[106] This report presented a fairly thorough review of available knowledge to date about the potential impacts of GEMs, although it too is a political document (OTA reports are aimed at members of Congress and present policy options along with lay explanations of science). Among the conclusions of this report were that although uncertainties existed about the potential impacts of GEMs in the environment, suitable knowledge and technologies were available to perform adequate risk assessments on a case-by-case basis. In addition, the report asserted that most small-scale field tests were unlikely to result in uncontrollable environmental problems, and, in fact, small-scale testing is probably the only way to truly assess the potential risks of larger scale activities.[106]

Later publications provided more detailed, peer-reviewed scientific assessments, beginning with a report published by Tiedje et al. under the auspices of the Ecological Society of America.[108] Tiedje et al. surveyed a fair amount of available information on the potential ecological effects of releasing transgenic organisms into the environment and offered a balanced, cautious approach to risk assessment. The monograph focused on many of the same phenomena first articulated by Alexander[117] (e.g., persistence), while also speculating to a greater degree on ecosystem effects. In proposing a scheme for regulatory risk assessments, Tiedje et al. offered the concept that the attributes of the environment (i.e., the ecosystem) into which an organism is to be released are just as important as the attributes of the GEM and its parents. Tiedje et al. supported the use of advanced biotechnology to develop environmentally sound products, but advocated a careful, case-by-case risk assessment, based on appropriate scientific methods and criteria.

Later in 1989, the National Academy of Sciences (NAS) also published a detailed study, entitled *Field Testing of Genetically Modified Organisms* (Reference 5). This monograph reiterated some of the conclusions of Tiedje et al.,[108] particularly the importance of considering the attributes of the environment in conducting a risk assessment. In surveying the available literature regarding field experience, the NAS Report[5] noted that although there had been greater experience in the introduction of novel plants than for novel microorganisms, there is information available for certain well-studied microbes (e.g., rhizobia, mycorrhizae) that supports their safety in the environment, and, furthermore, that any ecological uncertainties associated with a proposed release could be addressed by appropriate risk assessment methods.

This study proposed its own regulatory and risk assessment framework, in which the concept of "familiarity" was introduced. Certain proposed uses of GEMs could be considered familiar enough to past introductions with a safe history so as not to require special oversight. Other releases could be conducted after consideration of the potential for environmental effects and the ability to control the release.[5]

3. *Field Tests of GEMs: The Available Record*

The release of these reports coincided to some degree with the cooling-off of the debate over the appropriate regulatory and risk assessment framework for environmental release of GEMs in the U.S. In 1986, the U.S. government finalized its framework for the regulation of biotechnology products.[119] Regulation of environmental uses of engineered organisms fell primarily to the EPA and the U.S. Department of Agriculture (USDA) under existing statutes such as the pesticide law and the Plant Pest Act. Later in the 1980s, several states enacted their own specific biotechnology laws,[25] although state reviews have largely conformed to those of federal agencies.

The procedures for obtaining approval for field tests under these laws are described in more detail elsewhere.[12,25,50] Government approvals usually require detailed case-by-case risk assessments, focusing on the exposure and hazard issues discussed above, and weighing potential risks against the possible benefits of the test. Although early approvals were controversial and required considerable regulatory review time, and were sometimes delayed by litigation,[61,113] by the late 1980s, the process became more routine as the agencies had gained experience with the scientific issues involved.

The first authorized field test of a transgenic plant took place in 1986, and the first three field tests of live recombinant microorganisms occurred a year later.[18,33,61,113] Numerous subsequent field tests began to be approved in the U.S.; between 1987 and 1991, there were about 100 field tests of genetically modified microorganisms, 40 of which involved GEMs. Another ten GEM field tests took place in 1992. This record is complemented by the number of microbial field tests that have taken place elsewhere in the world (see, for example, Reference 120 for an early summary, and several papers in References 109 and 110 for reports of tests in specific countries).

Drahos[33] and Kostka and Drahos[18] provide recent summaries of some of the earliest field tests of GEMs in the U.S. and elsewhere. The first engineered microbes to be field tested were *Pseudomonas syringae*, deleted using recombinant techniques for the gene responsible for ice-nucleation.[60] These "ice-minus" tests were conducted by research groups at the University of California at Berkeley and a private company, Advanced Genetic Sciences. A variety of other *Pseudomonas* species have been engineered to contain the *lac*ZY gene cassette for identifying and quantitating microbes from soil samples,[121] and these strains have been field tested in several locations by Monsanto Company and several academic collaborators.[18,33,39] Several field trials have also taken place of recombinant *Rhizobia* and *Bradyrhizobia*, engineered by BioTechnica International to contain marker genes or for enhanced nitrogen fixation,[17,33,50] as well as *Bacillus thuringiensis* strains with the insecticidal endotoxin modified using various genetic techniques, including rDNA.[18] Among other insecticides tested have been modified Baculoviruses[122] and the endophytic bacterium *Clavibacter xyli,* engineered to express the *B. thuringiensis* endotoxin.[32] Among the field trials conducted overseas have been tests of nitrogen-fixing bacteria in China,[123] *Pseudomonas*, *Rhizobium,* and *Agrobacterium radiobacter* in Australia,[43] and *Pseudomonas* in the Netherlands.[124]

The government agencies approving field tests usually required that environmental monitoring accompany each trial. These monitoring plans involved soil, air, or water sampling, coupled with an appropriate microbiological assay to enumerate populations of the introduced strain. This generally required the use of some marker system to distinguish introduced microorganisms from indigenous microflora. Commonly used systems included antibiotic resistance genes[125] or the *lac*ZY system.[121]

The results of these field trials are reported to the EPA and its international counterparts, and they have provided a new set of data regarding the environmental effects of microorganisms. More recently, these results have been presented at scientific meetings[60,109,110,122,126] and in the peer-reviewed scientific literature.[127] To the best of anyone's knowledge, all these tests have been conducted safely, and none have resulted in any untoward environmental effects. As more tests began to be successfully conducted, the public policy debates and the controversy diminished in the U.S.,[128] although public concerns have not necessarily subsided in some European countries due to the influence of the Green Party.[129]

Although it is difficult to make generalizations about such a varied collection of tests, it is possible to conclude the following about the environmental behavior of the GEMs in these tests. First, the engineered microorganisms behaved in the environment extremely similarly to their parental counterparts and similarly to expectations. Second, enhanced long-term persistence of the GEM was not seen (although some tests have shown seasonal or transient population increases). Third, although limited dispersal from the test site was sometimes seen, it was generally possible to confine the vast majority of the released bacteria to the test site. Fourth, significant populations of released microbes were found only on the habitats intended, i.e., plant surfaces or the rhizosphere, and were not found on nontarget plants or off-site. Finally, in no case were unexpected adverse environmental effects reported. These findings are discussed in more detail below.

It should be noted that virtually all the field tests of GEMs to date have been at small scale (i.e., generally less than 10 acres). Although many believe that small-scale testing is the best way to collect the data to assess the environmental effects of large-scale field releases, others are concerned that larger scale field uses pose risks that cannot be predicted.[106] Although the data on GEMs to be discussed below indeed focuses on a relatively few small-scale tests, the historical record of the use of microbes in commercial agriculture gives some comfort that the potential environmental impacts of large-scale use of GEMs can be predicted.

C. CONSIDERATION OF THE POTENTIAL ADVERSE EFFECTS OF INTRODUCED MICROORGANISMS

This section will present a discussion of the potential adverse effects that may result from the introduction of microorganisms into the environment, focusing on those risks alleged to result from the use of GEMs in the environment. This discussion is based not only on the available record from 6 years of field testing of genetically engineered microbes, but also the accumulated record of relevant introductions of naturally occurring or traditionally modified microorganisms. Although an attempt has been made to discuss the issues completely and fairly, a comprehensive review of what has now become a substantial body of literature is clearly outside the scope of this section.

1. Toxicity, Pathogenicity of the Introduced Microorganism

Toxicity or pathogenicity is largely a characteristic of the host organism. As such, this will only present a problem when the chosen host has some toxic or infective character. This will naturally be judged on a case-by-case basis. This will rarely be a problem, since most microbial soil amendments will be derived from well-characterized nonpathogenic soil isolates. For example, *Bacillus thuringiensis* strains that have been used for decades as insecticides (particularly var. *kurstuki*) are known to have virtually no toxicity to vertebrates or to insects outside their normal host range.[130,131] Rhizobial and bradyrhizobial inoculants have been used in commercial agriculture for almost a century, with no known toxic effects on humans or animals (although some rhizobial strains produce molecules toxic to other rhizobia,[53] while other, specific strains have shown a chlorotic effect on soybeans).[132] However, some soil isolates are related to primary or opportunistic pathogens (e.g., *Klebsiella*, *Pseudomonas*), and other microbial inoculants may be related to plant pathogens, so that toxicity may, in rare cases, be a legitimate concern.

Concern has often been expressed that recombinant manipulations might enhance or create pathogenic function in the recipient microorganism. Microbial pathogenicity is controlled by a large portion of the prokaryotic genome, and it is believed to be highly unlikely that single gene changes can impart pathogenicity to an otherwise nonpathogenic host.[5,19,133]

2. Persistence and Establishment in the Environment

As noted above, persistence itself is rarely a detrimental effect; however, it is usually a prerequisite for an introduced organism causing environmental harm. In spite of the widespread layman's perception that microorganisms are capable of growing out of control in the open environment, the available record suggests the opposite. There is a sizable amount of evidence showing that although nonindigenous microbial populations may become established and persist in the environment, these populations do not exhibit net growth above initial inoculum sizes, and in many cases, populations decline rapidly.[5]

For example, many years' experimentation and field experience with rhizobia bear this out (as reviewed in References 134, 135, and 136). In general, rhizobia grow poorly in fallow soil (i.e., in the absence of growing plants),[134] and there are numerous reports that rhizobia inoculated into unamended nonsterile soil steadily decreased in numbers, and generally reached a steady-state level 1 to 2 orders of magnitude below the initial inoculum size (reviewed in Reference 135). One such study, however, showed a transient increase in population size.[137] In numerous studies, there has never been any reported long-term net growth (i.e., persistence at levels above the initial inoculum size) of an introduced rhizobial strain (see references in References 135 and 136).

Where longer-term persistence patterns have been studied, populations generally fluctuated over the course of a growing season, rising during active plant growth due to the rhizosphere effect, and declining later in the season.[138-140] Rhizobial strains introduced into normal agricultural soils at the time of planting often do not persist through the winter in high enough numbers to provide any benefit to the subsequent year's crop.[141]

However, there are reports that rhizobia introduced into soils where legumes had never previously been grown are capable of long-term establishment (reviewed in References 134, 139, and 142). Indeed, the prevalence of *B. japonicum* 123 in the soils of the U.S. midwest was the result of such an introduction; widespread use by farmers in the middle part of the century, promoted by the USDA, resulted in the establishment of this strain in midwestern soils, which has hindered the ability of other strains with more efficient nitrogen fixing ability to nodulate soybeans when used as inoculants.[137] This implies that intraspecific competition may determine whether an introduced strain can become established. In fact, there are numerous reports of higher levels of rhizobial persistence in sterile soils compared with nonsterile soils (e.g., Reference 143 and other references in Reference 134), implying that competition and possibly predation by other soil microflora are important factors limiting survival.

Persistence of rhizobia has also been studied in groundwater and sewage, where survival has been poor.[144,145] Other research has identified numerous specific factors as potentially limiting, including temperature, desiccation, salinity, pH, and the presence of herbicides or pesticides (reviewed in References 134 to 136).

Field and microcosm studies of other microorganisms provide findings similar to those for rhizobia. Van Elsas et al.[146] summarizes a number of studies of persistence of various *Pseudomonas* species and strains in soil, all of which showed gradual drop-offs in population sizes over time, with decay rates ranging from 0.2 to 1.1 log decline in a 10-day period. Other references cited by van Elsas et al. show the same behavior for other species, such as *Salmonella typhimurium* and *Klebsiella pneumoniae,*[145] and *Flavobacterium* and *Alcaligenes.*[146] Van Elsas also reports that although the fluorescent pseudomonads also show the rhizosphere effect during early stages of plant growth, these populations also decline to low numbers, often below detection limits.[146] Stotzky[147] reports that plasmid-containing *E. coli* strains introduced into nonsterile soil decreased by 2 to 3 logs during a 4-week time course, although the strains continued to be detectable. *Klebsiella* and *Pseudomonas* species showed somewhat better survival in similar experiments, and, in all cases, survival in sterilized soil was several orders of magnitude higher than in nonsterile soils and did not drop off from these high levels.[147]

Fuxa[148] has reviewed the literature relating to the persistence of introduced insecticidal bacteria and viruses, and has summarized the results of dozens of studies. Long-term persistence has been seen for *Bacillus popilliae* and *B. thuringiensis* because their spores remain viable in soil for many years, but these spores lose activity in hours or days on foliage.[148] *B. thuringiensis* is known not to be able to multiply vegetatively on plant leaves.[5] *B. thuringiensis* can grow in soil in the laboratory, but generally does not grow saprophytically in the environment (see references in Reference 148). Baculoviruses show similar patterns: long term survival in soil, but limited persistence on foliage.[148]

The available record to date from field releases of GEMs affords some additional evidence. In virtually all cases to date, the introduced GEMs showed nearly identical persistence patterns to their wild-type parent, and in no case was long-term persistence or permanent net growth seen of the released GEM. For example, in BioTechnica International's 1989 field tests of engineered *R. meliloti*, sharp population increases of both engineered and wild-type strains in the rhizosphere were seen in the first month after inoculation, after which levels dropped slightly to a steady-state level that persisted through to the subsequent winter.[17,51] Bacterial numbers were about an order of magnitude higher in the "inner rhizosphere" closest to the root surface than the "outer rhizosphere", and significant populations were not found in soils outside the rhizosphere.[17,51] In Monsanto's 1987 to 1988 trial of *lac*ZY-marked *Pseudomonas* on winter wheat, initial population densities rose to approximately 10^6 CFU per g root at the time of late summer planting, gradually dropped 2 orders of magnitude through the fall and winter, and continued to decline the following spring.[33] Only low levels of the marked strains were found on the roots of crops planted in succeeding years. Similar results have been reported for a different pseudomonad, also marked with *lac*ZY.[39] Other field tests reported in References 18 and 33 report similar, limited persistence patterns that resembled the wild-type strain.

3. Competition and/or Displacement of Natural Microflora

A related concern is whether the introduced GEM will compete with or displace natural microflora. As summarized in Reference 5, there are documented instances where modified microbes are disadvantaged and poorly competitive relative to their wild counterpart, but also where genetic modifications have resulted in increased fitness. However, the available record indicates that it is extremely difficult for introduced microorganisms to displace or outcompete indigenous species in the absence of large-scale introduction, particularly in soil environments.[100,101]

As a practical matter, inter- and intraspecific competition is only one factor affecting persistence and survival of an introduced strain. It is therefore difficult to design an experiment to specifically address the effect of competition on persistence. Experiments comparing sterile soil with nonsterile soil begin to address these questions and do indicate that introduced strains face tough competition from indigenous strains (see above).

Competition can be specifically studied for rhizobia, since inoculated strains must compete with indigenous strains for nodule formation, and the numbers of marked inoculant populations in nodules can be directly measured. The available literature indicates that rhizobial strains that are or have become indigenous in soil consistently outcompete introduced strains for nodule occupancy (e.g., references in Reference 149). Bromfield et al.[149] report that an introduced strain of *R. meliloti* occupied only 10 to 38% of alfalfa nodules in the first 2 years after inoculation, and two other studies report that introduced *B. japonicum* strains occupied less than 17% of nodules in the first year after inoculation.[150,151] However, this bias towards indigenous strains can be overcome by massive introductions of the nonindigenous strain over several years,[150] or by providing the inoculant strain in at least 1000-fold excess over the indigenous organisms.[152]

In BioTechnica's field tests of engineered rhizobia, introduced strains were subject to competition for nodulation from indigenous strains, to varying degrees depending on the depth of the nodules.[17,51] No significant differences in nodulation ability were seen, however, between the GEMs and their wild-type parent.[17,51]

Competition has also been studied in the "ice-minus" field tests, since the introduced GEM was designed to displace its wild-type counterpart to prevent frost damage. In the U.C. Berkeley test, the ice-nucleation deficient *P. syringae* mutant successfully displaced the wild-type Ice-plus bacteria at the site of the release, accounting for as much as 90% of the total bacteria on treated plants up to 4 weeks after the application.[60,153] However, released Ice-minus microbes that were dispersed off-site (up to 30 m from the point of application) did not displace Ice-plus populations on plant leaves, presumably due to the high indigenous populations of the Ice-plus strains on these off-site plants.[60,153]

4. Dissemination in the Environment

The potential for environmental dissemination of microbes cannot be discounted. The presence of solids in soils limits microbial motility,[146] and microorganisms have generally been shown to have limited horizontal and vertical motility.[154,155] However, microorganisms can be disseminated long distances by wind or other aerosols and are also transported through soil by rain and water runoff, burrowing animals, etc. (see literature reviewed in Reference 154). Like persistence, however, dissemination of a microorganism away from the site of release is an "exposure" issue and presents environmental risks only if the organism has an unpredicted or uncontrollable negative impact at the distal site.

Although some laboratory studies have been done addressing movement of microbes through the soil, the most appropriate studies must take place in the field. Field tests conducted to date have shown extremely limited dissemination of introduced strains. BioTechnica's *Rhizobium* field tests showed no significant inoculant populations anywhere except the rhizosphere, and only extremely limited vertical movement seen to depths of 2 and 10 in.[51] No differences were seen between GEMs and wild type. No significant aerial dispersal of the microbes was seen, even though the *R. meliloti* experiments involved spray-application of inoculants.

In Monsanto's 1987 to 1988 *Pseudomonas* field test, very low levels (less than 10^3 CFU per g root) of introduced bacteria were seen in border rows 18 and 36 cm from the nearest treated row, and the engineered microbes only migrated vertically in connection with the root system.[18] Here, too, no differences were seen between the GEM and wild type.

Aerial dispersal was specifically studied in the "ice-minus" experiments. In Advanced Genetic Sciences' test, less than 0.001% of the recombinant *P. syringae* entered the aerosol cloud above the plants, and, of these, 92% settled back onto the plants and soil of the test plot.[33] The numbers of bacteria dispersed from the site were estimated at 1.2×10^6, however. In the University of California test, only minimal levels of introduced bacteria were detected at distances greater than 20 m from the point of application.[60]

5. Horizontal Transfer of Introduced Genes

A concern that is unique to use of GEMs in the environment is the potential for transfer of introduced genes from the original host organism into other soil microorganisms (sometimes called "horizontal gene transfer"). Should the transferred gene have an unexpected or negative effect in its new host, an otherwise benign introduction might have deleterious effects.

Gene transfer between organisms of the same or different species is known to be common in nature.[156] There are four major mechanisms that have been identified: conjugation (direct "sexual" transfer from one organism to another); transduction (DNA transfer mediated by bacteriophages); transformation (the transfer of naked, unpackaged DNA molecules); and protoplast fusion.[157] Although these mechanisms have been extensively studied in the laboratory, the extent to which they occur in nature is not well understood because of a lack of reliable data.[147] It is generally believed that gene transfer occurs more frequently between closely related microbial species than it does between more distantly related species,[5] and it is well accepted that "promiscuous" genetic exchange is often seen in related families, such as the Gram-negative genera.[156]

Early studies involving GEMs utilized organisms engineered through the use of recombinant plasmids, which were specifically created for use as cloning vectors. Because of the need for them to mobilize into, and replicate in, their new hosts, these vectors were often based on naturally occurring plasmids known to be widely transmissible across species lines. A common series of cloning vectors for Gram-negative soil microorganisms (e.g., *Rhizobium*, *Pseudomonas*, etc.) was constructed from R-factor plasmids originally isolated from opportunistic pathogens.[158,159] Certain of these vectors were constructed to lack the trans-acting functions necessary for self-transmissibility, and were thus believed to be nonself-mobilizable. In at least one field test using *R. meliloti* engineered with such vectors, no evidence was seen for transfer of an antibiotic resistance marker residing on the vector into indigenous soil microbes.[51]

More recently, most engineering of soil microbes is accomplished using chromosomal integration by homologous recombination.[49,121] There is one school of thought that holds that chromosomally located genes are less likely to be transferred to other microbes than plasmid-borne genes (although some of these chromosomal constructs contain transposon sequences that potentially pose a risk of horizontal transfer). However, recent studies by Stotzky and his colleagues appear to contradict this thinking.[147] Transfer of chromosomal genes by conjugation was found to occur with frequencies between 10^{-5} to 10^{-4}, while conjugal transfer of plasmid-borne genes was between 10^{-7} and 10^{-5}.[147] It was hypothesized that this may be because there is not enough time in the wild for an entire plasmid to conjugate into a new cell. Stotzky also found that the gene transfer frequencies for transduction ranged from 10^{-5} to 10^{0}, and frequencies for transformation from 10^{-7} to 10^{-5}.[147] This group has also found evidence that naked DNA can persist in the environment, e.g., by binding clay minerals, and is capable of transforming other microbial cells in the environment.[147]

The GEM field trials to date have attempted to detect and quantitate gene transfer in the soil. In the Monsanto 1987 to 1988 field test, over 10,000 presumptive recipients of the *lac*ZY gene cassette (i.e., soil isolates having a lactose-positive phenotype) were obtained from rhizosphere soil and were probed with a radioactive copy of the *lac*ZY cassette. Of the 557 isolates showing hybridization to the probe, all were found to be the original introduced GEM, so that no transfer of the *lac*ZY genes could be detected to other soil microbes.[18] In recombinant *C. xyli* experiments, no gene transfer was reported to have been detected, in spite of the instability of a chromosomally introduced construct.[18]

The work of Stotzky and others makes it clear that one cannot discount the possibility of unexpected horizontal gene transfer in the environment. The fact that it is expected to be a rare event makes it difficult to detect or monitor in the field. However, as an "exposure" issue, the persistence or spread of introduced genetic information is not necessarily itself an adverse effect, and the transferred gene must behave differently in its new host in order for an adverse effect to be manifested.

6. *Host Range Modifications*

One of the advantages of microbial agents is their specificity; *Bacillus thuringiensis*-based insecticides, rhizobial inoculants, and mycoherbicides generally all have defined host ranges and are usually incapable of infecting plants or animals other than their usual target. Accidental changes to host range might result in unexpected, perhaps negative, environmental effects.

There are relatively few microbes that have such selective relationships with plants, and the genetics of the symbiotic or the pathogenic process are well understood.[19,133] The best example is the situation with rhizobia, where a combination of several specific genes control the nodulation process and determine host specificity.[19,133] Although host range remains an important issue to consider in microbial releases, it is unlikely for there to be a change in host range unless these genes are altered in the genetic manipulation.

D. ENVIRONMENTAL RISKS AND BENEFITS FOR SPECIFIC BIOTECHNOLOGY APPLICATIONS

1. *Overview*

Specific uses of microbes and genetically engineered plants in the environment offer the potential for tangible environmental benefits. The increased usage of the agricultural microorganisms now on the market or under development is expected to reduce or replace synthetic chemicals now used in commercial agriculture. Microbial pest control agents would replace specific chemical pesticides; nitrogen-fixing bacteria might reduce the usage of synthetic nitrogenous fertilizers; and other microbial inoculants like mycorrhizae could conceivably be used to supply other key nutrients, like phosphorus, in place of chemical products. Transgenic plant products may have similar beneficial effects on agrichemical usage. Bioremediation (using either microbial inoculants or nutrient addition) is expected to lead to cost-effective methods of destruction of hazardous environmental contaminants.

These expected benefits need to be weighed against the potential adverse environmental impacts discussed above. In some cases, other risks more specific to the application may need to be considered. This section will consider the risk/benefit considerations for certain specific uses of biotechnology in the environment.

2. Biological Pesticides: Microbes and Transgenic Plants

The greatest potential to replace chemical inputs lies in biopesticides. Commercial agriculture today has come to rely on a wide range of synthetic chemical products for pest and weed control, and the impressive crop yields possible in the U.S. and other industrialized countries are at least partially the result of the use of these products.[160] It has been estimated that over 900 million lb of chemical pesticides were applied to U.S. forests and fields in 1987,[161] and the health risks, both real and potential, associated with many of these chemicals is becoming clear. Often, their broad toxicity creates risks to pesticide applicators, who must take specific precautions when handling the more toxic agricultural chemicals, or to nontarget organisms within agricultural fields (e.g., beneficial insects). This is compounded by the long half-lives of many agrichemicals in soils. In addition, agricultural chemicals migrate away from the point of application and are found in groundwater, lakes, and streams in many farming communities, while pesticide residues in foods have become a major public health concern.[160] While it is true that not all commercially available pesticides contribute to environmental or health hazards, it is clear that overuse of certain agrichemicals has had long-lasting deleterious effects.

In contrast, biological pesticides are believed to have low inherent toxicity beyond their target species,[130,131,148] generally have narrow host ranges, affording specificity for the target pest,[3,6] and often persist poorly in the environment.[148] There are no known general occupational hazards to field workers (although there is a single documented incident of a corneal ulcer developing after a commercial *B. thuringiensis* product was splashed in the eye of a farm worker).[131] The major demonstrated environmental impacts both of viral and bacterial pesticides are consistent with the purpose of the application: an increase in entomopathogen numbers and a decrease of insect pest populations.[148] On the surface, biopesticides would seem to be an attractive substitute for synthetic chemicals.

Today, biological pesticides make up perhaps 2% of the $20 billion worldwide pesticide market,[4] and even with optimistic growth scenarios, it is hard to see biopesticides taking more than 5 to 10% share by the end of the decade, because of the historical problems such as poor shelf life and low effectiveness in the field.[1,4] Still, even the 2% current market share may represent the equivalent of replacing 18 million lb of pesticides each year in the U.S. alone. One cannot assume, however, that all microbial pesticides will be used to directly replace chemical inputs. Most biopesticides are being developed to meet specific market needs, in some cases because chemical products have been withdrawn from the market or have become obsolete through development of resistance in target pest populations,[160] or in other cases to address specific pests that are unable to be controlled by any existing chemical products.[12] Only a subset of biopesticides will directly replace (or even compete in the marketplace with) synthetic chemical products, perhaps limiting the potential for reduction of chemical usage.

Other biotechnology strategies are being pursued to replace chemical pesticides. Using recombinant DNA and other techniques, genes encoding pesticidal molecules are being inserted into crop plants.[21,22] Most common are the *B. thuringiensis* insecticidal toxins, viral coat proteins as anti-viral agents, and antifungal molecules like the chitinases (reviewed in References 4, 21, and 22). Field introduction of transgenic plants has not engendered quite the same controversy as has deliberate release of GEMs, but a number of potential environmental impacts have been raised, including the potential for engineered plants to become weeds, or for introduced genes to spread to weedy relatives of crop plants through spread of pollen.[5,162,176] A recent report has shown that genomic recombination is possible between transgenic plants expressing a portion of a viral coat protein and a defective plant virus inoculated on those plants, creating concern over whether widespread use of such transgenic plants could lead to the unintentional creation of new plant viruses.[177] In general, these risks are believed to be minimal and/or manageable,[5,178] with substantial field experience with transgenic plants buttressing this contention.[109,110] Public interest groups have also expressed concerns over the food use of engineered plants,[163] with particular concern over the presence of B. t. endotoxins in food crops, since the available toxicology data for B. t. does not address the truncated form of the toxins that is commonly engineered into plants.[164] Risk assessment schemes for food use of engineered plants have been proposed by several groups around the world, most of which are similar to Reference 165, and conclude that any risks can be effectively monitored and managed.

When "pesticidal" transgenic plant products begin to reach the market in the mid to late 1990s, some will compete with synthetic agrichemicals in the market, and may lead to additional reduction in the use of chemicals. However, transgenic plants expressing B. t. or other insecticidal proteins will compete with microbial pesticides for the same market, thus perhaps limiting the impact of microbials.[4] Transgenic plants expressing antiviral proteins, however, represent a new market, since plant viruses cannot be effectively controlled with synthetic chemicals.

An important potential limitation to market penetration for microbial pesticides and transgenic plants is the prospect that target pests will develop resistance to these biological pest control agents. Through natural selection, insects have frequently acquired resistance to chemical pesticides, thereby limiting the usefulness of certain agrichemicals.[166] In contrast, B. t. products and naturally occurring baculoviruses have been used commercially for decades with no reports of resistance in field populations,[148] leading some to speculate that resistance would not occur.[167] However, recent years have seen numerous reports of resistance arising in field and laboratory populations of several insect species that were exposed to B. t. toxins (as reviewed in References 160 and 168). More troubling are studies showing that insects exposed to one B. t. subspecies have acquired resistance to other subspecies expressing different endotoxins.[168]

Concern over the very real possibility of B. t. resistance stems from the expected widespread use of transgenic plants expressing these endotoxins. Because early gene constructs have the toxins expressed constitutively (i.e., continuously) in all tissues of the plant, insects would be exposed to a greater selective pressure to acquire resistance. The limited environmental half-lives of microbial B. t. products are not expected to create the same potential for resistance and are probably one reason why resistance has been so slow to develop until now.

Several strategies have been proposed to limit the onset of insect B. t. resistance, including creating transgenic plants expressing different toxins (to target different active sites in the pest), ensuring that farmers plant insect-sensitive crops mixed in with insect-resistant plants (to limit selective pressure), and engineering the endotoxin to be expressed only at the time when needed and in the appropriate plant tissues.[160,169] As reviewed in Reference 160, the efficacy of certain of these strategies has been placed in doubt by recent data; so, it is not yet clear if B. t. resistance can effectively be avoided or managed.

Taken together, the market introduction of microbial pesticides and pesticide-expressing transgenic plants will undoubtedly lead to some level of reduction in chemical pesticides, although this reduction will certainly occur only slowly. Care must be taken to prevent insect resistance from limiting the usefulness of this promising pest control approach.

3. *Herbicide-Tolerant Plants*

Another class of biotechnology products having the potential to cause significant effects on the environment are herbicide-tolerant plants, engineered to detoxify or afford resistance to specific chemical herbicides.[21,22,170] Since many herbicides have broad specificity, toxicity to a given crop plant may prevent use of that herbicide against weeds prevalent with that crop. Engineering herbicide tolerance into such crops could therefore broaden the uses of specific herbicides.

There is substantial debate over the environmental impact of the widespread use of herbicide-tolerant plants. Numerous public interest groups and academic researchers maintain that commercial use of herbicide-tolerant plants will lead to increased use of chemical herbicides, with concomitant negative environmental effects.[171] These groups generally favor increased emphasis on nonchemical weed control and sustainable agriculture, and see herbicide-tolerant plants as incompatible with such practices.

Other observers, including those from industry, counter that herbicide tolerance is likely to increase usage only of those herbicides known to be less hazardous to the environment (e.g., with lower toxicity, short half-life in the soil, compatibility with no-till practices, or a less-frequent application rate).[160,170] This is especially true because herbicides are applied to such a vast majority of crops already that the adoption of herbicide tolerance is likely only to shift usage from one category to another, rather than increase use.[160] Most of the reasonably well-advanced commercial development programs are focusing on herbicide categories that are generally believed to be more environmentally benign (i.e., glyphosate, bromoxynil, glufosinate, and the sulfonylureas; but see References 160 and 171 regarding possible health risks of bromoxynil), and allegations to the contrary by environmental groups are largely based on outdated information (e.g., references in Reference 171 to long-terminated research programs on atrazine resistance). The environmentalists counter that, without government intervention, there is no assurance that companies will continue to develop plants resistant only to those more benign herbicides.[171]

Taken together, References 160, 170, and 171 provide a much fuller summary of this debate than is possible here. It is likely that the truth will lie somewhere between these competing positions. Because the use of herbicide-tolerant plants will be under considerable scrutiny (at least in the short term), it seems likely that the earliest crops on the market will not be ones having negative environmental effects, and that the overall impact of herbicide tolerance will not be deleterious.

4. Nitrogen-Fixing Microorganisms

Rhizobia and *Bradyrhizobia* and other nitrogen-fixing microorganisms provide a source of nitrogen nutrient to legume plants, so that it is possible to consider whether broader use of these microbial inoculants could lead to decreases in nitrogenous fertilizer use. It has been estimated that the rhizobia and bradyrhizobia collectively fix the equivalent of 40 million tons of nitrogen fertilizer per year worldwide, with the equivalent of perhaps 10 to 20 million tons in the U.S. alone.[172] Use of nitrogenous fertilizer (principally in the form of nitrate) is associated with known environmental and health risks, such as run-off of nitrates into groundwater, health risks to livestock and people, and occupational exposure to workers involved in fertilizer production. Even a modest increase in the usage of rhizobial inoculants could theoretically lead to a reduction of hundreds of thousands of tons of nitrogen fertilizer a year.

When legumes are grown in fields with sufficient rhizobial populations, or where sufficient inoculant is added, the plants generally receive enough nitrogen for growth and should not require chemical nitrogen fertilizer. However, it is likely that many legume farmers add as much as 20 lb per acre of chemical nitrogen anyway,[52] although there is some evidence that this practice may in fact inhibit rhizobial nodulation.[173]

In any event, most observers do not expect increased use of rhizobial inoculants to lead to significant reductions in nitrogen fertilizer on legume crops. Rather, the potential benefit is expected to come in decreased fertilizer use on the cereal crops that are grown in rotation with legumes. Because some of the nitrogen fixed by legumes leaks into the soil where it is available for the subsequent year's crop, it is common for farmers to reduce the amount of preplant nitrogen fertilizer for the rotation crop to take advantage of this "nitrogen credit". Broader adoption of nitrogen-fixing inoculants, as well as the development of more efficient strains, could eventually lead to significant reductions in fertilizer usage in cereal crops, particularly corn, on which more nitrate fertilizer is used than on any other crop. Because current rhizobial products have fallen out of favor, and because progress on strain improvement has been slow, these benefits will certainly be slow to develop.

Greater reduction could come if nitrogen-fixing inoculants could be developed for nonlegume crops. This might be done by altering the host range of rhizobia, so that they can infect and nodulate cereal plants, or by introducing the nitrogen fixation genes into microbes indigenous to the rhizosphere of cereal plants. Technical problems like the complexity of the nitrogen fixation gene family will likely prevent such developments in the near future, but it is conceivable that such products could be created in a 10 to 20-year time frame.

5. Remediation of Hazardous Environmental Contaminants

As discussed above, indigenous and introduced microorganisms are being used to remediate a variety of hazardous materials that are presently contaminating soils, sediments, and groundwater all over the world. The magnitude of the hazardous waste problem worldwide is staggering. The site remediation market in the U.S. alone is $5 to 12 billion a year,[87] and some estimates of total cleanup costs run in the hundreds of billions of dollars over the next 10 years.

To the extent biological methods can assist in these cleanup efforts, there will clearly be a net environmental benefit. Bioremediation can be used to clean up a number of compounds of known human toxicity or carcinogenicity (e.g., trichloroethylene),[72] and can also be used to remediate contamination by select pesticides.[75,78-80] Bioremediation is becoming an accepted alternative treatment technology that is proven to be successful for certain well-defined uses. Bioremediation is a natural process that, when practiced correctly, can lead to complete destruction of the contaminant, to the innocuous end products carbon dioxide and water.[81,82] More importantly for its ultimate commercial acceptance, it is estimated to be far less costly to implement than incineration or other technologies,[174] is sometimes quicker than traditional methods (particularly "pump and treat" methods for groundwater treatment),[81] and *in situ* bioremediation can be implemented without removing or damaging existing buildings on site.

However, biotreatment has its limitations. As noted above, biological methods are only effective against certain organic chemicals, and generally have only limited utility against toxic heavy metals. In addition, the rate of microbial biodegradation slows down as the concentration of the contaminant decreases, so that bioremediation is not always capable of reducing contaminant levels more than 2 or 3 logs. In spite of having greater than 99% efficiency, this is often not sufficient to meet minimum concentrations specified in government regulations. Bioremediation is also temperature sensitive, so that it is more likely to succeed in warmer climates. More generally, not every site will have the appropriate geological, chemical, and biological characteristics to make it amenable to bioremediation, and preliminary lab analysis is usually needed to make this determination, thereby adding to the costs of implementation. Although bioremediation may be applicable only to a limited percentage of hazardous waste sites, it should nevertheless play an important role in environmental remediation in coming years.

A potential drawback may arise in those cases where biodegradation is incomplete, perhaps because of a lack of one or more needed members of a microbial consortium, leading to accumulation of one or more intermediates in the pathway. This would only be of concern if a given intermediate were toxic; for example, one pathway for the biodegradation of trichloroethylene and tetrachloroethylene includes the problematic compound vinyl chloride as an intermediate.[94] In the case of bioremediation of oil spills, potentially troublesome breakdown products include the quinones and the naphthalenes, but it is expected that such compounds would eventually be themselves broken down in the environment.[90] Since the progress of bioremediation is often monitored using laboratory assays that measure total toxicity of the sample, problems of this nature can often be detected and controlled.

As discussed above, bioremediation is most often practiced through the addition of nutrients or chemical additives to soils or groundwater to stimulate indigenous microbial populations, rather than by adding exogenous microorganisms. While this can be accomplished similarly to the application of agricultural chemicals, certain additives might give cause for caution or might raise additional environmental concerns. For example, addition of nitrate to soils or groundwater may be restricted by regulation, because of its known toxic effects. Nitrogen is a required nutrient for *in situ* bioremediation, but it is usually supplied in the form of ammonia. However, in some forms of anaerobic bioremediation, nitrate itself is used in place of oxygen as the electron acceptor,[86] and this would need to be done in a manner consistent with applicable regulations.

A more general concern is whether the nutrients or fertilizers added as stimulants might themselves have negative impacts, such as toxicity to nontarget organisms or the promotion of eutrophication, leading to algal blooms and oxygen depletion.[90] The possibility of such algal blooms was studied during the 1989 to 1990 biostimulation experiments on the Exxon Valdez oil spill in Alaska (see references in Reference 90). These investigations showed no significant difference between treated and control plots, with no evidence of algal blooms resulting from use of the fertilizer. These experiments also evaluated the potential toxicity of the fertilizers used, and aside from an apparently transient mild toxicity to oyster larvae, no toxic effects were noted on any of the marine species in the vicinity of the test.[90] It was noted, however, that one component, butoxyethanol, had to be handled with care by applicators and required wildlife deterrent devices, but that it generally evaporated from beach surfaces within 24 h.

Most observers believe that, when practiced correctly at an appropriate site, bioremediation is an effective method for hazardous waste remediation that results in a net benefit to the environment. Because of some of the concerns discussed above, as well as a general discomfort with the prospects of genetic engineering, there are some critics who are skeptical of bioremediation's potential and its environmental impacts.[175]

VII. CONCLUSIONS

Although progress has been slower than expected, microbial soil amendment products and other biological processes are beginning to assume an important role in agriculture and environmental management, as new technologies promise to overcome historical problems. As a result of the debate over the use of genetically engineered organisms in the environment, many future uses of microbes in environmental applications will be subjected to greater scrutiny than in the past, and the ecological impact of such products may be called into question. Based on the evidence available to date, many of the earliest fears about microbial field releases have proven unfounded, but it remains to be seen if this will still be true as more microbial products move from small-scale testing to expanded commercial use. In most cases, any potential risks will be negligible or manageable, and will be far outweighed by the expected benefits to the environment and biotic systems these products will bring.

REFERENCES

1. Brill, W. J., The use of microorganisms for crop agriculture, in *Agricultural Biotechnology at the Crossroads*, Fessenden MacDonald, J., Ed., National Agricultural Biotechnology Council, Ithaca, NY, 1991, page 91.
2. Betz, R., Levin, M., and Rogul, M., Safety aspects of genetically-engineered microbial pesticides, paper presented at American Chemical Society Annual Meeting, Washington, D.C., September 1, 1983.
3. Hofte, H. and Whitely, H. R., Insecticidal crystal proteins of Bacillus thuringiensis, *Microbiol. Rev.,* 53, 242, 1989.
4. Glass, D. J. and Lindemann, J., *Biotechnology in Agriculture: The Next Decade*, Decision Resources, Inc., Burlington, MA, 1992.
5. National Academy of Sciences, *Field Testing Genetically Modified Organisms: Framework for Decisions*, National Academy Press, Washington, D.C., 1989.
6. Wood, H. A. and Granados, R. R., Genetically engineered baculoviruses as agents for pest control, *Annu. Rev. Microbiol.*, 45, 69, 1991.
7. Skladany, G. J. and Metting, F. B., Bioremediation of contaminated soil, in *Soil Microbial Ecology*, Metting, F. B., Ed., Marcel Dekker, New York, 1993, chap. 17.
8. Grubbs, R. B., Enhanced biodegradation of aliphatic and aromatic hydrocarbons through bioaugmentation, paper presented at 4th Annual Hazardous Materials Management Conference, Atlantic City, NJ, June 2–4, 1986.
9. Debus, K. H., Mining with microbes, *Technol. Rev.,* 93 (6), 50, 1990.
10. Brierley, J. A., Use of microorganisms for mining metals, in *Engineered Organisms in the Environment: Scientific Issues*, Halvorson, H. O., Pramer, D., and Rogul, M., Eds., American Society for Microbiology, Washington, D.C., 1985, 141.
11. Hardy, R. W. F. and Glass, D. J., Genetic enginering: our investment: what is at stake?, *Issues Sci. Technol.*, 1 (3), 69, 1985.
12. Glass, D. J., Commercialization of soil microbial technologies, in *Soil Microbial Ecology*, Metting, F. B., Ed., Marcel Dekker, New York, 1993, chap. 22.

13. Brockman, F. J., Denovan, B. A., Hicks, R. J., and Frederickson, J. K., Isolation and characterization of quinoline-degrading bacteria from subsurface sediments, *Appl. Environ. Microbiol.*, 55, 1029, 1989.
14. Sylvester, E. J. and Klotz, L. C., *The Gene Age*, Charles Scribner's Sons, New York, 1983.
15. U.S. Congress Office of Technology Assessment, *Biotechnology in a Global Economy*, U.S. Government Printing Office, Washington, D.C., 1991.
16. Rojo, F., Pieper, D. H., Engesser, K. H., Knackmuss, H. J., and Timmis, K. N., Assemblage of ortho-cleavage route for simultaneous degradation of chloro- and methylaromatics, *Science*, 238, 1395, 1987.
17. Ronson, C. W., Bosworth, A., Genova, M., Gudbrandsen, S., Hankinson, T., Kwiatkowski, R., Ratcliffe, H., Robie, C., Sweeney, P., Szeto, W., Williams, M., and Zablotowicz, R., Nitrogen fixation: achievements and objectives, in *Proceedings of the 8th International Conference on Nitrogen Fixation*, Gressfhoff, P. M., Roth, L.E., Stacey, G., and Newton, W. E., Eds., Chapman & Hall, New York, 1990, 397.
18. Kostka, S. J. and Drahos, D. J., Genetic engineering of microorganisms for pest control: survival and potential effectiveness of such microorganisms in crop systems, in *Pesticide Interactions in Crop Production: Beneficial and Deleterious Effects*, Altman, J., Ed., CRC Press, Boca Raton, FL, 1993, chap. 21.
19. Lindow, S. E., Panopoulos, N. J., and McFarland, B. L., Genetic engineering of bacteria from managed and natural habitats, *Science*, 244, 1300, 1989.
20. Ensley, B., Genetic strategies for strain improvement, in press.
21. Gasser, C. S. and Fraley, R. T., Genetically engineering plants for crop improvement, *Science*, 244, 1293, 1989.
22. Gasser, C. S. and Fraley, R. T., Transgenic crops, *Sci. Am.,* 266 (6), 62, June 1992.
23. De Young, H. G., EPA and industry push for new Superfund solutions, *Chem. Week,* February 24, 1988, page 28.
24. Dean, M. and Kremer, F., Advancing research for bioremediation, *Environ. Protection*, 3 (7), 19, 1992.
25. Glass, D. J., Impact of government regulation on commercial biotechnology, in *The Business of Biotechnology: From the Bench to the Street*, Ono, R. D., Ed., Butterworth-Heinemann, Boston, 1991, chap. 10.
26. Bakst, J. S., Impact of present and future regulations on bioremediation, *J. Ind. Microbiol.,* 8, 13, 1991.
27. Walter, J. F. and Paau, A. S., Microbial inoculant production and formulation, in *Soil Microbial Ecology*, Metting, F. B., Ed., Marcel Dekker, New York, 1993, chap. 21.
28. Burton, J. C., Nodulation and symbiotic nitrogen fixation, in *Alfalfa Science and Technology*, Hanson, C. H., Ed., American Society of Agronomy, Madison, WI, 1972, 229.
29. Roy, K.A., Petroleum company heals itself — and others, *Hazmat World*, 5 (5), 75, 1992.
30. Gaertner, F. and Kim, L., Current applied recombinant DNA projects, *Trends Biotechnol.*, 6, 54, 1988.
31. *AgBiotechnology News*, 10 (3), 4, 1993.
32. Kostka, S. J., The design and execution of successive field releases of genetically-engineered microorganisms, in *The Biosafety Results of Field Tests of Genetically Modified Plants and Microorganisms*, MacKenzie, D. R. and Henry, S. C., Eds., Agricultural Research Institute, Bethesda, MD, 1991, 167.

33. Drahos, D. J., Field testing of genetically engineered microorganisms, *Biotechnol. Adv.*, 9, 157, 1991.
34. *Ag Biotechnology News*, 10 (1), 10, 1993.
35. Bayer's first biopesticide, *Agrow*, October 6, 1989, 28.
36. O'Sullivan, D. J. and O'Gara, F., Traits of fluorescent Pseudomonas spp. involved in suppression of plant root pathogens, *Microbiol. Rev.*, 56 (4), 662, 1992.
37. Backman, P.A., Turner, J.T., Crawford, M.A., and Clay, R.P., New biological seed treatment fungicide increases peanut yields, *Highlights Agric. Res.,* 31, 4, 1984.
38. Gustafson's Kodiak Receives EPA Registration, *AgBiotechnology News*, 9 (6), 11, 1992.
39. Cook, R.J., Weller, D. M., Kovacevich, P., Drahos, D., Hemming, B., Barnes, G., and Pierson, E., Establishment, monitoring, and termination of field tests with genetically altered bacteria applied to wheat for biological control of take-all, in *The Biosafety Results of Field Tests of Genetically Modified Plants and Microorganisms*, MacKenzie, D. R. and Henry, S. C., Eds., Agricultural Research Institute, Bethesda, MD, 1991, 177.
40. TeBreest, D.O. and Templeton, G.E., Mycoherbicides: progress in the biological control of weeds, *Plant Disease*, 69, 6, 1985.
41. Charudattan, R. and Walker, H.L., *Biological Control of Weeds with Plant Pathogens*, John Wiley & Sons, New York, 1982.
42. Strobel, G. A., Biological control of weeds, *Sci. Am.,* July 1991, 72.
43. Millis, N. F., Australian experience in the release of live modified organisms, in *The Biosafety Results of Field Tests of Genetically Modified Plants and Microorganisms*, Casper, R. and Landsmann, J., Eds., Biologische Bundesanstalt fur Land- und Forstwirtschaft, Braunschweig, Germany, 1992, 81.
44. McCardell, A., Sadowsky, M. J., and Cregan, P. B., Genetics and improvement of biological nitrogen fixaton, in *Soil Microbial Ecology*, Metting, F. B., Ed., Marcel Dekker, New York, 1993, chap. 6.
45. Postgate, J.R., Evolution within nitrogen-fixing systems, *Symp. Soc. Gen. Microbiol*, 24, 263, 1974.
46. Sprent, J.I., *The Biology of Nitrogen-Fixing Organisms* , McGraw-Hill, London, 1979.
47. Vincent, J.M., Factors controlling the legume Rhizobium symbiosis, in *Nitrogen Fixation*, vol. 2, Newman, W. E. and Orme-Johnson, W. H., Eds., University Park Press, Baltimore, MD, 1980, 103.
48. Keyser, H. H., Somasegaran, P., and Bohlool, B. B., Rhizobial ecology and technology, in *Soil Microbial Ecology*, Metting, F. B., Ed., Marcel Dekker, New York, 1993, chap. 8.
49. Williams, M.K., Cannon, F., McLean, P., and Beynon, J., Vector for the integration of genes into a defined site in the Rhizobium meliloti genome, in *Molecular Genetics of Plant-Microbe Interactions*, Palacios, R. and Verma, D., Eds., American Phytopathological Society Press, St. Paul, MN, 1988, 198.
50. Glass, D. J., Regulating biotech: a case study, *Forum for Appl. Res. Public Policy*, 4 (3),92, 1989.
51. Hankinson, T., unpublished results, 1992.
52. Wacek, T., personal communication, 1993.
53. Triplett, E. W., Construction of a symbiotically effective strain of Rhizobium leguminosarum bv. trifolii with increased nodulation competitiveness, *Appl. Environ. Microbiol.*, 56, 98, 1990.

54. Jarstfer, A. G. and Sylvia, D. M., Inoculum production and inoculation strategies for vesicular-arbuscular mycorrhizal fungi, in *Soil Microbial Ecology*, Metting, F. B., Ed., Marcel Dekker, New York, 1993, chap. 13.
55. O'Dell, T. E., Castellano, M. A., and Trappe, J. M., Biology and applications of ectomycorrhizal fungi, in *Soil Microbial Ecology*, Metting, F. B., Ed., Marcel Dekker, New York, 1993, chap. 14.
56. Kloepper, J. W., Plant growth-promoting rhizobacteria as biological control agents, in *Soil Microbial Ecology*, Metting, F. B., Ed., Marcel Dekker, New York, 1993, chap. 10.
57. Maki, R. L., Galyon, E. L., Chang-Chien, M., and Caldwell, D. R., Ice nucleation produced by Pseudomonas syringae, *Appl. Environ. Microbiol.*, 28, 456, 1974.
58. Lindow, S. E., Arny, D. C., and Upper, C. D., Distribution of ice nucleation active bacteria on plants in nature, *Appl. Environ. Microbiol.*, 36, 831,1978.
59. Lindow, S. E., Ecology of Pseudomonas syringae relevant to the field use of Ice-deletion mutants constructed in vitro for plant frost control, in *Engineered Organisms in the Environment: Scientific Issues*, Halvorson, H. O., Pramer, D., and Rogul, M. G., Eds., American Society for Microbiology, Washington, D.C., 1985, 23.
60. Lindow, S. E. and Panopoulos, N. J., Field tests of recombinant ice – Pseudomonas syringae for biological frost control in potato, in *The Release of Genetically Engineered Microorganisms,* Sussman, M., Collins, C. H., Skinner, F. A., and Stewart-Tull, D. E., Eds., Academic Press, New York, 1988, 246.
61. Piller, C., *The Fail-Safe Society: Community Defiance and the End of American Technological Optimism*, Basic Books, 1991, chap. 4.
62. Winkler, M. A., *Biological Treatment of Waste-Water*, Ellis Horwood Ltd., Chichester, U.K., 1981.
63. Raymond, R. L., Jamison, V. H., and Hudson, J. O., *Beneficial Stimulation of Bacterial Activity in Ground Waters Containing Petroleum Products*, API Publication Number 4427, American Petroleum Institute, Washington, D.C., 1975.
64. Alexander, M., Biodegradation of chemicals of environmental concern, *Science,* 211, 132, 1981.
65. Gibson, D. T., *Microbial Degradation of Organic Compounds*, Marcel Dekker, New York, 1984.
66. Atlas, R. M., Microbial degradation of petroleum hydrocarbons: an environmental perspective, *Microbiol. Rev.,* 45, 180, 1981.
67. Atlas, R. M., Stimulated petroleum biodegradation, *CRC Crit. Rev. Microbiol.,* 5, 371, 1977.
68. Colwell, R. R. and Walker, J. D., Ecological aspects of microbial degradation of petroleum in the marine environment, *CRC Crit. Rev. Microbiol.,* 5, 423, 1977.
69. Leahy, J. G. and Colwell, R. R., Microbial degradation of hydrocarbons in the environment, *Microbiol. Rev.*, 54 (3), 305, 1990.
70. Evans, W. C. and Fuchs, G., Anaerobic degradation of aromatic compounds, *Annu. Rev. Microbiol.,* 42, 289, 1988.
71. Crawford, R. L. and Mohn, W. W., Microbiological removal of pentachlorophenol from soil using a Flavobacterium, *Enzyme Microbiol. Technol.,* 7, 617, 1985.
72. Ensley, B. D., Biochemical diversity of trichloroethylene metabolism, *Annu. Rev. Microbiol.*, 45, 283, 1991.
73. Chaudhry, G. R. and Chapalamadugu, S., Biodegradation of halogenated organic compounds, *Microbiol. Rev.*, 55 (1), 59, 1991.

74. Reineke, W. and Knackmuss, H. J., Microbial degradation of haloaromatics, *Annu. Rev. Microbiol.*, 42, 263, 1988.
75. Mohn, W. W. and Tiedje, J. M., Microbial reductive dehalogenation, *Microbiol. Rev.*, 56 (3), 482, 1992.
76. Bedard, D. L., Haberl, M. L., May, R. J., and Brennan, M. J., Evidence for novel mechanisms of polychlorinated biphenyl metabolism in Alcaligenes eutrophus H850, *Appl. Environ. Microbiol.,* 53, 1103, 1987.
77. Quensen, J. F., Tiedje, J. M., and Boyd, S. A., Reductive dechlorination of polychlorinated biphenyls by anaerobic microorganisms from sediments, *Science,* 242, 752, 1988.
78. Stevens, T. O., Crawford, R. L., and Crawford, D. L., Selection and isolation of bacteria capable of degrading dinoseb, *Biodegradation*, 2, 1, 1991.
79. Topp, E., Xun, L., and Orser, C. S., Biodegradation of the herbicide bromoxynil by purified pentachlorophenol hydroxylase and whole cells of Flavobacterium sp. strain ATCC 39723 is accompanied by cyanogenesis, *Appl. Environ. Microbiol.*, 58, 502, 1992.
80. Borow, H.S. and Kinsella, J. V., Bioremediation of pesticides and chlorinated phenolic herbicides — above ground and *in situ* — case studies, in *Proceedings of Superfund '89,* Hazardous Materials Control Research Institute, Silver Spring, MD, 1989, 325.
81. Lee, M. D., Thomas, J. M., Borden, R. C., Bedient, P. B., Ward, C. H., and Wilson, J. T., Biorestoration of aquifers contaminated with organic compounds, *CRC Crit. Rev. Environ. Control,* 18, 29, 1988.
82. Sims, J. L., Sims, R. C., and Matthews, J. E., *Bioremediation of Contaminated Surface Soils*, Publication Number EPA-600/9-89/073, U.S. Environmental Protection Agency, Ada, OK, 1989.
83. Brown, R. A. and Crosbie, J. R., Oxygen sources for *in situ* bioremediation, in *Proceedings of Superfund '89*, Hazardous Materials Control Research Institute, Silver Spring, MD, 1989, 338.
84. Hinchee, R. E., Downey, D. C., and Miller, R. N., Enhancing biodegradation of vadose zone JP-4 through soil venting, in *Proceedings of Hazardous Wastes and Hazardous Materials*, Hazardous Materials Control Research Institute, Silver Spring, MD, 1990, 387.
85. Thomas, J. M. and Ward, C. H., *In situ* biorestoration of organic contaminants in the subsurface, *Environ. Sci. Technol.*, 23 (7), 760, 1989.
86. Batterman, G., A large scale experiment on *in situ* biodegradation of hydrocarbon in the subsurface, in *Ground Water in Water Resources Planning. Volume II. Proc. Int. Symp.,* IASA Publication 142, International Association of Hydrological Sciences, London, 1983, 983.
87. Glass, D. J., Hazardous waste bioremediation. I. U.S. market, in *Spectrum Environmental Management Industry*, Decision Resources, Waltham, MA, March 29, 1993.
88. Glass, D. J., The promising hazardous waste bioremediation market in the United States, in *Spectrum Environmental Management Industry*, Decision Resources, Burlington, MA, August 19, 1991.
89. Snyder, J. D., Bioremediation: respectable at last?, *Environ. Today*, November/December 1990, 17.
90. U.S. Congress Office of Technology Assessment, *Bioremediation for Marine Oil Spills — Background Paper*, U.S. Government Printing Office, Washington, D.C., May 1991.
91. Ten oil spill products chosen for protocol validation, *BioTreatment News*, 2 (2), 3, 1992.

92. Bumpus, J. A., Tien, M., Wright, D., and Aust, S. D., Oxidation of persistent environmental pollutants by white rot fungus, *Science*, 228, 1434, 1985.
93. Davis, M. W., Glaser, J.A., Evans, J. W., and Lamar, R. T., Field evaluation of lignin degrading fungi to treat creosote contaminated soil, *Environ. Sci. Technol.*, in press.
94. Lamar, R. T., Evans, J. W., and Glaser, J. A., Solid-phase treatment of a pentachlorophenol-contaminated soil using lignin-degrading fungi, *Environ. Sci. Technol.*, in press.
95. Vogel, T. M. and McCarty, P. L., Biotransformation of tetrachloroethylene to trichloroethylene, dichloroethylene, vinyl chloride, and carbon dioxide under methanogenic conditions, *Appl. Environ. Microbiol.*, 49, 1080, 1985.
96. Nelson, M. J. K., Montgomery, S. O., O'Neill, E. J., and Pritchard, P. H., Biodegradation of trichloroethlyene by a bacterial isolate, *Appl. Environ. Microbiol.*, 53, 949, 1987.
97. Winter, R. B., Yen, K.-M., and Ensley, B., Efficient degradation of trichloroethylene by a recombinant Escherichia coli, *Bio/Technology*, 7, 282, 1989.
98. Ornstein, R. L., Rational redesign of biodegradative enzymes for enhanced bioremediation: overview and status report for cytochrome P450, in *Proceedings of National Research and Development Conference on the Control of Hazardous Materials*, Hazardous Materials Control Research Institute, Greenbelt, MD, 1991, 314.
99. Leavitt, M. and Brown, K., Biostimulation vs. bioaugmentation: let the incumbents reign, paper presented at *In Situ* and On-Site Bioreclamation, The Second International Conference, San Diego, April 5–8, 1993.
100. Goldstein, R. M., Mallory, L. M., and Alexander, M., Reasons for possible failure of inoculation to enhance biodegradation, *Appl. Environ. Microbiol.*, 50, 977, 1985.
101. Ramadan, M. A., El-Tayeb, O. M., and Alexander, M., Inoculum size as a factor limiting success of inoculation for biodegradation, *Appl. Environ. Microbiol.*, 56, 1392, 1990.
102. U.S. Environmental Protection Agency, *Emerging Technologies: Bio-Recovery Systems Removal and Recovery of Metal Ions from Groundwater*, Publication EPA/540/5-90/005a, Washington, D.C., 1990.
103. Hileman, B., Alternative agriculture, *Chem. Eng. News*, March 5, 1990, page 26.
104. Massachusetts Department of Food and Agriculture, *The Massachusetts Farm-And-Food System: A Five-Year Policy Framework*, Boston, MA, 1988.
105. Sharples, F. E., Spread of organisms with novel genotypes: thoughts from an ecological perspective, *Recomb. DNA Tech. Bull.*, 6, 43, 1983.
106. U.S. Congress Office of Technology Assessment, *New Developments in Biotechnology — Field-Testing Engineered Organisms: Genetic and Ecological Issues*, U.S. Government Printing Office, Washington, D.C., May 1988.
107. Kelman, A., Anderson, W., Falkow, S., Federoff, N.V., and Levin, S., *Introduction of Recombinant DNA-Engineered Organisms into the Environment: Key Issues*, National Academy Press, Washington, D.C., 1987.
108. Tiedje, J.M., Colwell, R.K., Grossman, Y.L., Hodson, R.E., Lenski, R.E., Mack, R.N., and Regal, P.J., The planned introduction of genetically engineered organisms: ecological consideration and recommendations, *Ecology*, 70, 297, 1989.
109. MacKenzie, D. R. and Henry, S. C., Eds., *The Biosafety Results of Field Tests of Genetically Modified Plants and Microorganisms*, Agricultural Research Institute, Bethesda, MD, 1991.
110. Casper, R. and Landsmann, J., Eds., *The Biosafety Results of Field Tests of Genetically Modified Plants and Microorganisms*, Biologische Bundesanstalt fur Land- und Forstwirtschaft, Braunschweig, Germany, 1992.

111. Krimsky, S., *Genetic Alchemy: The Social History of the Recombinant DNA Controversy*, MIT Press, Cambridge, MA, 1982.
112. U.S. Department of Health, Education and Welfare, Guidelines for research involving recombinant DNA molecules, *Fed. Regist.*, 41, 27911, 1976.
113. Krimsky, S. and Plough, A. , The release of genetically engineered organisms into the environment: the case of ice-minus, in *Environmental Hazards: Communicating Risks as a Social Process*, Auburn House Publishing, Dover, MA, 1988, chap. 3.
114. Brill, W. J., Safety concerns and genetic engineering in agriculture, *Science*, 235, 1329, 1987.
115. Colwell, R. K., Norse, E. A., Pimental, D., Sharples, F. E., and Simberloff, D., Genetic engineering in agriculture, *Science*, 229, 111, 1985.
116. U.S. Environmental Protection Agency, Statement of policy: microbial products subject to the Federal Insecticide, Fungicide and Rodenticide Act and the Toxic Substances Control Act, *Fed. Regist.*, 51, 23313, 1986.
117. Alexander, M., Genetic engineering: ecological consequences: reducing the uncertainties, *Issues Sci. Technol.*, 1(3), 57, 1985
118. Davis, B. D., Bacterial domestication: underlying assumptions, *Science*, 235, 1329, 1987.
119. U.S. Office of Science and Technology Policy, Coordinated framework for regulation of biotechnology: announcement of policy and notice for public comment, *Fed. Regist.*, 51, 23302, 1986.
120. Gesselschaft für Biotechnologische Forschung mbH, *GENTEC Update*, Braunschweig, Germany, 1990.
121. Drahos, D.J., Hemming, B.C., and McPherson, S., Tracking recombinant organisms in the environment: beta-galactosidase as a selectable non-antibiotic marker for fluorescent pseudomonads, *Bio/Technology* , 4, 439, 1986.
122. Bishop, D. H. L., Entwistle, P. F., Cameron, I. R., Allen, C. J., and Possee, R. D., Field trials of genetically-engineered baculovirus insecticides, in *The Release of Genetically Engineered Microorganisms*, Sussman, M., Collins, C. H., Skinner, F. A., and Stewart-Tull, D. E., Eds., Academic Press, New York, 1988, 143.
123. Chen, Z. L., Field releases of recombinant bacteria and transgenic plants in China, in *The Biosafety Results of Field Tests of Genetically Modified Plants and Microorganisms*, Casper, R. and Landsmann, J., Eds., Biologische Bundesanstalt fur Land- und Forstwirtschaft, Braunschweig, Germany, 1992, 53.
124. Bakker, P. A. H. M., Schippers, B., Hoekstra, W. P. M., and Salentijn, E., Survival and stability of a Tn5 transposon derivative of Pseudomonas fluorescens WCS374 in the field, in *The Biosafety Results of Field Tests of Genetically Modified Plants and Microorganisms*, MacKenzie, D. R. and Henry, S. C., Eds., Agricultural Research Institute, Bethesda, MD, 1991, 201.
125. Prentki, P. and Krisch, H. M., *In vitro* insertional mutagenesis with a selectable DNA fragment, *Gene*, 29, 303, 1984.
126. Baum, R., Field tests of recombinant organisms shown safe, *Chem. Eng. News*, 67 (17), 30, 1989.
127. Kluepfel, D. A., Kline, E. L., Skipper, H. D., Hughes, T. A., Gooden, D. T., Drahos, D. J., Barry, G. F., Hemming, B. C., and Brandt, E. J., The release and tracking of genetically engineered bacteria in the environment, *Phytopathology*, 81, 348, 1991.
128. Sun, M., Preparing the ground for biotech tests, *Science*, 242, 503, 1988.
129. Dixon, B., Who's who in European antibiotech, *Bio/Technology*, 11, 44, 1993.

130. Hadley, W. M., Burchiel, S. W., McDowell, T. D., Thilsted, J. P., Hibbs, C. M., Whorton, J. A., Day, P. W., Friedman, M. B., and Stoll, R. E., Five-month oral (diet) toxicity/infectivity study of Bacillus thuringiensis insecticides in sheep, *Fundam. Appl. Toxicol.*, 8, 236, 1987.
131. Green, M., Heumann, M., Sokolow, R., Foster, L. R., Bryant, R., and Skeels, M., Public health implications of the microbial pesticide Bacillus thuringiensis: an epidemiological study, Oregon, 1985–86, *Am. J. Publ. Health*, 80, 848, 1990.
132. Vest, G., Weber, D. F., and Sloger, C., Nodulation and nitrogen fixation, in *Soybeans: Improvement, Production and Use*, Caldwell, B. E., Ed., 1972, 353.
133. Keen, N. T. and Staskawicz, B., Host range determinants in plant pathogens and symbionts, *Annu. Rev. Microbiol.*, 42, 421, 1988.
134. Stacey, G., The Rhizobium experience, in *Engineered Organisms in the Environment: Scientific Issues*, Halvorson, H. O., Pramer, D., and Rogul, M., Eds., American Society for Microbiology, Washington, D.C., 1985, 109.
135. BioTechnica International, Inc., *Proposal for Field Test of Genetically Engineered Rhizobium Meliloti*, PMN Numbers P87-568, 569, 570, available from Environmental Protection Agency, Washington, D.C., 1987.
136. BioTechnica Agriculture, Inc., *Proposal to Field Test Genetically-Engineered Strains of Bradyrhizobium Japonicum*, PMN Numbers P89-340/341, available from Environmental Protection Agency, Washington, D.C., 1989.
137. Ellis, W. R., Ham, G. E., and Schmidt, E. L., Persistence and recovery of Rhizobium japonicum inoculum in a field soil, *Agron. J.*, 76, 573, 1984.
138. Reyes, V. G. and Schmidt, E. L., Population densites of Rhizobium japonicum strain 123 estimated directly in soil and rhizospheres, *Appl. Environ. Microbiol.*, 37, 854, 1979.
139. Hiltbold, A. E., Patterson, R. M., and Reed, R. B., *Soil Sci. Am. J.*, 49, 343, 1985.
140. Mahler, R. L. and Wollum, A. G., Seasonal variation of Rhizobium meliloti in alfalfa hay and cultivated field in North Carolina, *Agron. J.*, 74, 428, 1982.
141. Osa-Afiana, L. O. and Alexander, M., *Soil Sci. Am. J.*, 43, 925, 1979.
142. Kluepfel, D. A., The behavior of nonengineered bacteria in the environment: what can we learn from them?, in *The Biosafety Results of Field Tests of Genetically Modified Plants and Microorganisms*, Casper, R. and Landsmann, J., Eds., Biologische Bundesanstalt fur Land- und Forstwirtschaft, Braunschweig, Germany, 1992, 37.
143. Danso, S. K. A. and Alexander, M., Regulation of predation by prey density: the protozoan-Rhizobium relationship, *Appl. Environ. Mirobiol.*, 29, 515, 1975.
144. Sinclair, J. L. and Alexander, M., Role of resistance to starvation in bacterial survival in sewage and lake water, *Appl. Environ. Microbiol.*, 48, 410, 1984.
145. Liang, L. N., Sinclair, J. L., Mallory, L. M., and Alexander, M., Fate in model ecosystems of microbial species of potential use in genetic engineering, *Appl. Environ. Microbiol.*, 44, 708, 1982.
146. van Elsas, J. D., Heijnen, C. E., and van Veen, J. A., The fate of introduced genetically engineered microorganisms (GEMs) in soil, in microcosm, and the field: impact of soil textural aspects, in *The Biosafety Results of Field Tests of Genetically Modified Plants and Microorganisms*, MacKenzie, D. R. and Henry, S. C., Eds., Agricultural Research Institute, Bethesda, MD, 1991, 67.
147. Stotzky, G., Gene transfer among and ecological effects of genetically modified bacteria in the soil, in *The Biosafety Results of Field Tests of Genetically Modified Plants and Microorganisms*, Casper, R. and Landsmann, J., Eds., Biologische Bundesanstalt fur Land- und Forstwirtschaft, Braunschweig, Germany, 1992, 122.

148. Fuxa, J. R., Fate of released entomopathogens with reference to risk assessment of genetically engineered microorganisms, *Bull. Entomol. Soc. Am.*, Winter 1989, 12.
149. Bromfield, E. S. P., Sinha, I. B., and Wolynetz, M. S., *Appl. Environ. Microbiol.*, 51, 1077, 1986.
150. Dunigan, E. P., Bollich, P. K., et al., *Agron. J.*, 76, 463, 1984.
151. Ham, G. E., Cardwell, V. B., and Johnson, H. W., *Agron. J.*, 63, 301, 1971.
152. Weaver, R. W. and Frederick, L. R., *Agron. J.*, 66, 229, 1984.
153. Lindow, S. E., Environmental release of pseudomonads: potential benefits and risks, in *Pseudomonas: Molecular Biology and Biotechnology*, Galli, E., Silver, S., and Witholt, B., Eds., American Society for Microbiology, Washington, D.C., 1992, 399.
154. Madsen, E. L. and Alexander, M., Transport of Rhizobium and Pseudomonas through soil, *Soil Sci. Soc. Am. J.*, 46, 557, 1982.
155. Ames, P. and Bergman, K., Competitive advantage provided by bacterial motility in the formation of nodules by Rhizobium meliloti, *J. Bacteriol.*, 148, 728, 1981.
156. Stotzky, G. and Babich, H., Fate of genetically-engineered microbes in natural environments, *Recomb. DNA Tech. Bull.*, 7, 163, 1984.
157. Slater, J. H., Gene transfer in microbial communities, in *Engineered Organisms in the Environment: Scientific Issues*, Halvorson, H. O., Pramer, D., and Rogul, M., Eds., American Society for Microbiology, Washington, D.C., 1985, 89.
158. Ditta, G., Stanfield, S., Corbin, D., and Helinski, D. R., *Proc. Natl. Acad. Sci. U.S.A.*, 77, 7347, 1980.
159. Priefer, U. B., Simon, R., and Puhler, A., *J. Bacteriol.*, 163, 324, 1985.
160. Krimsky, S. and Wrubel, R., *Agricultural Biotechnology: An Environmental Outlook*, Department of Urban and Environmental Policy, Tufts University, Medford, MA, 1993.
161. Schneider, K., Fear of chemicals is turning farmers to biological pesticides, *New York Times*, June 11, 1989.
162. Hoffman, C. A., Ecological risks of genetic engineering of crop plants, *BioScience*, 40 (6), 434, 1990.
163. Hopkins, D. D., Goldburg, R. J., and Hirsch, S.A., *A Mutable Feast: Assuring Food Safety in the Era of Genetic Engineering*, The Environmental Defense Fund, New York, 1991.
164. Goldburg, R. J. and Tjaden, G., Are B.t.k. plants really safe to eat?, *Bio/Technology*, 8 (11), 1011, 1990.
165. International Food Biotechnology Council, Biotechnologies and food: assuring the safety of foods produced by genetic modification, *Regul. Toxicol. Pharmacol.*, 12 (3), S1, 1990.
166. Lambert, B. and Peferoen, M., Insecticidal promise of Bacillus thuringiensis, *Bioscience*, 42 (2), 112, 1992.
167. McGaughey, W. H., Insect resistance to the biological insecticide Bacillus thuringiensis, *Science*, 229, 193, 1985.
168. Tabashnik, B. E., Finson, N., Johnson, M. W., and Moar, W. J., Resistance to toxins from Bacillus thuringiensis subsp. kurstaki causes minimal cross-resistance to B. thuringiensis subsp. aizawai in the diamondback moth (Lepidoptera: Plutellidae), *Appl. Environ. Microbiol.*, 59, 1332, 1993.
169. McGaughey, W.H. and Whalon, M. E., Managing insect resistance to Bacillus thruingiensis toxins, *Science*, 258, 1451, 1992.

170. Duke, S. O., Christy, A. L., Hess, F. D., and Holt, J. S., *Herbicide-Resistant Crops*, Council for Agricultural Science and Technology, Ames, IA, 1991.
171. Goldburg, R., Rissler, J., Shand, H., and Hassebrook, C., *Biotechnology's Bitter Harvest: Herbicide Tolerant Crops and the Threat to Sustainable Agriculture*, The Biotechnology Working Group, Washington, D.C., 1990.
172. Hardy, R. W. F., personal communication, 1987.
173. Eardly, B. D., Hannaway, D. B., and Bottomley, P. J., *Agron. J.*, 77, 57, 1985
174. Flathman, P. E. and Githens, G. D., *In situ* biological treatment of isopropanol, acetone, and tetrahydrofuran in the soil/groundwater environment, in *Groundwater Treatment Technology* , Nyer, E. K., Ed., Van Nostrand Reinhold, New York, 1985.
175. The California Biotechnology Action Council, The Overselling of Bioremediation: A Primer for Policy Makers and Activists, Sacramento, CA, 1992.
176. Rissler, J. and Mellon, M., *Perils Amidst the Promise: Ecological Risks of Transgenic Crops in a Global Market,* Union of Concerned Scientists, Cambridge, MA, 1993.
177. Greene, A. E. and Allison, R. F., Recombination between viral RNA and transgenic plant transcripts, *Science,* 263, 1423, 1994.
178. Falk, B. W. and Bruening, G., Will transgenic crops generate new viruses and new diseases?, *Science,* 263, 1395, 1994.

Index

A

D

E

F

G

M

N

O

P

Q

R

S

T

X

Z